1992

HAZARDOUS AND TOXIC WASTES: Technology, Management and Health Effects

The Pennsylvania Academy of Science Publications
Proceedings and Books

Editor: Shyamal K. Majumdar
Professor of Biology
Lafayette College
Easton, Pennsylvania 18042

1. *Proceedings* of the Pennsylvania Academy of Science. Two issues per year; current volume (1984) is 58. ISSN: 0096-9222. Editor: Shyamal K. Majumdar.

2. *Energy, Environment, and the Economy,* 1981. ISBN: 0-9606670-0-8. Editor: Shyamal K. Majumdar.

3. *Pennsylvania Coal: Resources, Technology, and Utilization,* 1983. ISBN: 0-9606670-1-6. Editors: Shyamal K. Majumdar and E. Willard Miller.

4. *Hazardous and Toxic Wastes: Technology, Management and Health Effects,* 1984. ISBN: 0-9606670-2-4. Editors: Shyamal K. Majumdar and E. Willard Miller.

Volumes on *Solid and Liquid Wastes* and *Radioactive Materials and Wastes* are in preparation.

HAZARDOUS AND TOXIC WASTES: Technology, Management and Health Effects

EDITED BY
SHYAMAL K. MAJUMDAR, Ph.D.
Professor of Biology
Lafayette College
Easton, Pennsylvania 18042
and
E. WILLARD MILLER, Ph.D.
Professor of Geography and
  Associate Dean for Resident
  Instruction (Emeritus)
The Pennsylvania State University
University Park, Pennsylvania 16802

Founded on April 18, 1924

A Publication of
The Pennsylvania Academy of Science

Library of Congress Cataloging in Publication Data

Hazardous and Toxic Wastes: Technology, Management and Health
   Effects

   Bibliography
   Appendix
   Includes Index

   I. Majumdar, Shyamal K., 1938-   , ed.
   II. Miller, E. Willard, 1915-   , co-ed.

Library of Congress Catalog Card No.: 83-83117

ISBN 0-9606670-2-4
   Copyright © 1984 By The Pennsylvania Academy of Science

   Printed in the United States of America by

   Typehouse of Easton
   Phillipsburg, New Jersey 08865

**COMMONWEALTH OF PENNSYLVANIA
DEPARTMENT OF ENVIRONMENTAL RESOURCES**
P.O. Box 2063
Harrisburg, PA 17120

The Secretary

Nicholas DeBenedictis
Secretary, Department of Environmental Resources

FOREWORD

Implementation of comprehensive federal and state laws to regulate solid, hazardous and radioactive wastes has made business, government and the scientific community examine our current waste disposal methods and evaluate alternative techniques for the future. The preservation of our environment, the health of our economy and the maintenance of our quality of life are dependent upon the safe and proper disposal of the waste produced by all members of our society.

Efforts to establish and develop proper disposal sites have been hampered by public opposition, much the result of problems associated with old, abandoned dumps. To gain the public's confidence, government, industry and community leaders must be willing to lead a constructive, open dialogue to site and manage new and innovative waste facilities.

Our immediate task is to identify and cleanup those dumps that pose a threat to the public and the environment and to prevent a repeat of past mistakes. Only then can the confidence of the public be gained to develop new technologies and facilities.

With limited space remaining in many landfills across the Commonwealth and opposition to opening new sites, attention is being focused on resource recovery and waste-to-energy projects.

142, 522

Approximately 300 permitted landfills are in operation in Pennsylvania; over 800 new inadequate sites have been closed by the state since 1968. The inability to locate new landfills demonstrates the need for more energy-efficient solid waste disposal.

In 1983, Pennsylvania awarded a record $873,000 to 23 communities for solid waste planning grants to study long-range disposal alternatives and evaluate recycling possibilities. This year, more communities than ever before in the program's 14-year history applied for grants, which serve as pilot studies for other municipalities as well as guiding immediate help for solid waste problems.

The safe disposal of hazardous wastes will present a difficult challenge in the future, as public concerns over "midnight dumping" and groundwater contamination from illegal and unregulated disposal flourishes. Much of the scientific community is skeptical of the long-term safety of total dependence on landfills for hazardous waste disposal, creating another impetus for alternatives to land disposal.

The nation's industries could reduce the volume of hazardous wastes through reprocessing of used materials and the evaluation of manufacturing methods, but the development of new treatment and disposal methods cannot be overemphasized. Research into high-temperature incineration, biological treatment and detoxification of hazardous wastes may provide the best disposal alternatives available for the future.

The scientific problems presented by the disposal of radioactive wastes are only part of the social and political considerations surrounding the issue that must be faced in the coming years. For too long, band-aid solutions to the disposal of radioactive wastes have been accepted.

The nation, including Pennsylvania, is required by law to establish low-level radioactive waste disposal sites, either alone or in cooperation with other states, by January 1986. With the assistance of community leaders, scientists and elected officials, Pennsylvania can properly locate a suitable site for low-level radioactive wastes.

The Pennsylvania Academy of Science is to be commended for its efforts in the study of waste management, through the expertise of government, industry and the academic communities. Development of new techniques in waste disposal is dependent on new ideas, such as are included in these volumes. The academy will play an important role in the development of solutions essential to a clean environment and a sound economy.

Sincerely,

Nicholas DeBenedictis
Secretary
Pennsylvania Department
of Environmental
Resources

PREFACE

This is one of three books published by the Pennsylvania Academy of Science on Hazardous and Toxic Wastes, Solid and Liquid Wastes, and Radioactive Materials and Wastes. These volumes are comprehensive, authoritative source books describing and analyzing the technology and disposal of waste products, one of the most critical problems of our times. They consist of an assemblage of quality papers contributed by a group of leading experts from five countries and over fifteen states.

The problem of the disposal of the various types of wastes created by an industrial society is not a new one, but one that has reached a critical dimension in our times. The modern industrial wastes can not only devastate the natural environment, but have the potential to affect the health of millions of human beings. The seriousness of the problem becomes evident when waste materials have not been disposed of for many years and entire neighborhoods become contaminated, or when scores of abandoned hazardous waste sites are discovered in densely populated regions and the total environment is subject to contamination.

Each of the three volumes in this series considers a different type of waste material. This volume on Hazardous and Toxic Wastes is divided into five parts. Part one considers waste types, treatment and disposal methods and covers such aspects as industrial waste incineration, destruction of toxic chemical wastes and the management of hazardous wastes. The sites of hazardous and toxic wastes as to their distribution, selection and geological considerations are discussed in part two. Part three covers transportation, emergency response and preparations needed in a toxic spill emergency. Part four includes chapters on management, regulations and economic considerations, and the last part is devoted to environmental and health effects of hazardous wastes.

The book on Solid and Liquid Wastes classifies the waste types, discusses management possibilities, describes pretreatment and treatment methods, evaluates environmental and health impacts, examines recycling and energy recovery potential from wastes, and considers laws, regulations and socioeconomic problems.

The third book of the series is devoted to the treatment of Radioactive Materials and Wastes. The book contains chapters on the types of radioactive wastes and detection methodology, management, treatment and transportation of nuclear wastes, handling, storage and disposal problems, socio-political aspects, the preparations and planning considerations, and environmental and health effects of radioactive materials and wastes.

These books will be of value to a wide audience. Such individuals as engineers, scientists, medical doctors, social scientists, and environmentalists will find them useful. They provide a wide perspective so that a specialist in one field, can be informed about developments and trends in another branch of the subject. The volumes will also be of value to individuals who want to be informed of some of the most critical problems of the day.

We express our deep appreciation for the excellent cooperation and dedication of the contributors, who recognize the importance of solving the critical problems of waste disposal. For a task of this magnitude many individuals in addition to the authors made contributions, and we are most pleased to acknowledge them. The advice and guidance of the members of the editorial and advisory boards are gratefully acknowledged. Gratitude is extended to Dr. Robert S. Chase, Head, Department of Biology, Lafayette College, and to Dr. Gregory Knight, Head, Department of Geography, The Pennsylvania State University for providing facilities for editorial work to the editors of the three volumes. Thanks are due to Caryn Golden of Lafayette College, and Nina McNeal and Joan Summers of The Pennsylvania State University for competent secretarial assistance. S. K. Majumdar and E. W. Miller extend heartful thanks to their wives Jhorna and Ruby, respectively, who graciously shared weekends and evenings with the preparation of the series and provided help and encouragement.

Shyamal K. Majumdar, Ph.D.
Lafayette College
and
E. Willard Miller, Ph.D.
The Pennsylvania State University
Editors
March 1984

HAZARDOUS AND TOXIC WASTES:
Technology, Management and Health Effects

Table of Contents

CONTRIBUTORS

D.G. Ackerman Jr., (Chapter 4). TRW Inc., California

R.F. Anderson, (Chapter 11). Boston University, Massachusetts

R.L. Bittle, (Chapter 15). Pennsylvania Department of Environmental Resources, Pennsylvania.

K. Caputo, (Chapter 15). Pennsylvania Department of Environmental Resource, Pennsylvania

R. Carnes, (Chapter 2). U.S. Environmental Protection Agency, Arkansas

J.J. Cudahy, (Chapter 3). I.T. Enviroscience, Tennessee

R.G. Dinsmore, (Chapter 19). Western Electric Company, Pennsylvania

A. Fisher, (Chapter 21). U.S. Environmental Protection Agency, Washington, D.C.

C.L. Fraust, (Chapter 19). Western Electric Company, Pennsylvania

G.R. Galida, (Chapter 16). Pennsylvania Department of Environmental Resources, Pennsylvania

T.M. Grabowski, (Chapter 12). Sun Refining and Marketing Company, Pennsylvania

M.R. Greenberg, (Chapter 11). Rutgers, The State University, New Jersey

C.W. Heath, (Chapter 26). Emory University, Georgia

Teh-Wei-Hu, (Chapter 22). The Pennsylvania State University, Pennsylvania

R.A. Hunter, (Chapter 23). RAH Consulting, Pennsylvania

D.L. Kaltrieder, (Chapter 22). The Pennsylvania State University, Pennsylvania

P.M. King, (Chapter 20). P.P.G. Industries, Inc., Pennsylvania

D. Lesak, (Chapter 14). Hazard Management Associates, Pennsylvania

T. Longosky, (Chapter 5). Envirosafe Service, Inc., Pennsylvania

G. Lumb, (Chapter 27). Hahnemann Medical University, Pennsylvania

J.E. McClure, (Chapter 18). GRW Engineers, Inc., Kentucky

C.H. McPherson, Jr. (Chapter 13). U.S. Environmental Protection Agency, Georgia

M.S. Monmonier, (Chapter 8). Syracuse University, New York

E.T. Oppelt, (Chapter 2). U.S. Environmental Protection Agency, Ohio

J.W. Osheka, (Chapter 20). P.P.G. Industries, Inc., Pennsylvania

R.R. Parizek, (Chapter 10). The Pennsylvania State University, Pennsylvania

D. Pitts, (Chapter 3). I.T. Enviroscience, Tennessee

A.J. Raymond, (Chapter 12). Sun Refining and Marketing Company, Pennsylvania

R.W. Regan, (Chapter 1). The Pennsylvania State University, Pennsylvania

H.G. Rückel, (Chapter 6). Zweckverband Sondermullplatze Mittlefranken, West Germany

B.M. Russell, (Chapter 19). Western Electric Company, Pennsylvania

A.J. Smith, (Chapter 13). U.S. Environmental Protection Agency, Georgia

C.L. Smith, (Chapter 5). Envirosafe Service, Inc., Pennsylvania

A. Scharsach, (Chapter 7). Von Roll Ltd., Switzerland

R.F. Schmalz, (Chapter 9). The Pennsylvania State University, Pennsylvania

G.A. Schnell, (Chapter 8). State University of New York, New York

G.A. Tokuhata, (Chapter 24). The University of Pittsburgh, Pennsylvania

G. VanderVelde, (Chapter 4). Chemical Waste Management, Illinois

R.R. Van Stockum, Jr., (Chapter 17). Louisville, Kentucky

J.F. Villaume, (Chapter 25). Pennsylvania Power and Light Company, Pennsylvania

J.A. Weaver, (Chapter 22). The Pennsylvania State University, Pennsylvania

E. Webber, (Chapter 23). V.F.L. Technology Corporation, Pennsylvania

Hazardous and Toxic Wastes: Technology, Management and Health Effects

Editors: Dr. Shyamal K. Majumdar,
Professor of Biology, Lafayette College
Easton, Pennsylvania 18042

Dr. E. Willard Miller,
Professor of Geography and Associate Dean for Resident
Instruction (Emeritus), The Pennsylvania State University,
University Park, Pennsylvania 16802

EDITORIAL BOARD

Dr. Dale E. Baker, Department of Agronomy, 221 Tyson Bldg., The Pennsylvania State University, University Park, PA 16802

Dane C. Bickley, Coal Specialist, Pennsylvania Governor's Energy Council, 1625 North Front St., P.O. Box 8010, Harrisburg, PA 17105

Dr. C.P. Blahous, Attorney, Vice-President of Environment, Health and Safety, PPG Industries, Inc., 1 Gateway Center, Pittsburgh, PA 15222

Dr. Ricahrd J. Bord, Department of Sociology, 413 Lbieral Arts Tower, The Pennsylvania State University, University Park, PA 16802

Dr. Robert S. Chase, Head, PAS Past-President, Department of Biology, Lafayette College, Easton, PA 18042

Dr. Donald J. Epp, Assistant Director, Institute for Research on Land and Water Resources, 8E Weaver Bldg., The Pennsylvania State University, University Park, PA 16802

Gary Galida, Chief, Division of Hazardous Wastes, Pennsylvania Department of Environmental Resources, P.O. Box 2063, Harrisburg, PA 17120

Dr. Tah-Wei-Hu, Professor of Economics, Department of Economics, The Pennsylvania State University, University Park, PA 16802

Dr. Al P. Dufour, Chief Bacteriology Section, Microbiology Branch, HERL, Environmental Protection Agency, 26 West St., Clair St., Cincinnati, Ohio 45268

Sister M. Gabrielle, PAS Past-President, Grove and McRobert Road, Pittsburgh, PA 15234

Donald A. Lazarchick, P.E. Director, Bureau of Solid Waste Management, Pennsylvania Department of Environmental Resources, P.O. Box 2063, Harrisburg, PA 17120

Mark M. McClellan, Executive Director, Citizen Advisory Council, PA. DER, 8th Floor, Executive House Apartment, 2nd and Chestnut Streets, Harrisburg, PA 17120

Dr. Bruce D. Martin, PAS Past-President, Associate Vice President for Academic Affairs, Duquesne University, Pittsburgh, PA 15219

Dr. Heinz G. Pfeiffer, Manager, Technology & Energy Assessment, Pennsylvania Power and Light Co., Two North Ninth St., Allentown, PA 18101

Dr. Arun K. Sharma, Professor, Calcutta University, President, Indian National Science Academy, New Delhi, India.

Daniel E. Wiley, Manager, R/D, PPG Industries, Inc., Industrial Chemical Division, Pittsburgh, PA 15222

Dr. J.B. Yasinsky, General Manager, Advanced Power Systems Divisions, Nuclear Center, Westinghouse Electric Corporation, P.O. Box 355, Pittsburgh, PA 14230

Prof. Stanley J. Zagorski, PAS President, Associate Dean, College of Science and Engineering, Gannon University, Perry Square, Erie, PA 16541

SYMPOSIA

I *Hazardous and Radioactive Wastes.* Holiday Inn, Grantville, Pennsylvania, October 29, 1982.
Chairman: Donald Zappa, President, Vector Corporation, Pittsburgh, PA

II *Solid and Hazardous Wastes.* Host Corral, Lancaster, Pennsylvania, April 10, 1983.
Chairman: Justice John P. Flaherty, Supreme Court of Pennsylvania. Chairperson of The Pennsylvania Academy of Sciences' Advisory Council

III *Radioactive Materials and Wastes.* Marriott Hotel, Monroeville, Pennsylvania, October 27, 1983.
Chairman: Dr. George C. Shoffstall, PAS President Elect. Director of Education and Organizational Development, Western Pennsylvania Hospital, Pittsburgh, PA

ACKNOWLEDGMENTS

The Pennsylvania Academy of Science published this book in association with the Pennsylvania Department of Environmental Resources (DER). Any opinions, findings, conclusions, or recommendations expressed are those of the author(s) and do not necessarily reflect the views of the DER, or The Pennsylvania Academy of Science.

The publication of this book was aided by contributions from The Pennsylvania Power and Light Company, Allentown, Pennsylvania, U.S. Ecology, Louisville, Kentucky and other companies.

OFFICERS OF THE
PENNSYLVANIA ACADEMY OF SCIENCE

Message from

JOHN P. FLAHERTY
Justice
Supreme Court of Pennsylvania
Chairman, Advisory Board
Pennsylvania Academy of Science

A society advances, indeed survives, measured by its control and disposition of societal waste. As is quite evident by even a cursory look at history, nothing is more destructive to life than uncontrolled human, industrial and, a fortiori, radioactive waste. During the German blitz of London in 1941-42, for example, Winston Churchill's greatest fear was a breakdown of the sewer system! No amount of explosive then known could have caused the human destruction of which such an event was capable. If we are, thus, to accommodate an increasing population on our now highly urbanized and industrialized planet, it is essential that the scientific community devote itself *with priority* to neutralizing the inundating waste which, unabated, will cause catastrophe to mankind, unparalleled in history.

The Pennsylvania Academy of Science, recognizing the importance of the subject, has endeavored by this publication to stimulate the scientific reader to further innovation, as well as to provide an anthology of present methodology.

Highly industrialized, with a large urban population, Pennsylvania is particularly an appropriate situs for this work, as within its borders occurred an event of stark terror, presaging potential future disaster — *Three Mile Island!*

Message from the President of the Pennsylvania Academy of Science

Professor Stanley J. Zagorski
Associate Dean, College of Science and Engineering, Gannon University, Erie, Pennsylvania

The Pennsylvania Academy of Science (PAS) is proud to present this book as one of three volumes on Solid, Hazardous and Radioactive Wastes. The Academy feels that these three books are the first of their kind ever published, and is happy to make them available to the citizens of the Commonwealth and to the nation.

The PAS, representing all aspects of science in Pennsylvania, has a duty and an obligation to present to its citizens information on topics of current environmental concern and interest. Since the impact of Three Mile Island and its aftermath of publicity, the President and the Executive Board of the Academy have decided that a topic of tremendous importance to the people of Pennsylvania, the nation and the world would be solid, hazardous and radioactive wastes, and their handling, transportation and environmental significance.

To begin acquiring authoritative scientific information on these subjects, the first in a series of three symposia was organzied by Donald Zappa, President of Vector Corporation and the Academy's Industrial Relations Chairman. This first symposium was held at the Holiday Inn in Grantville, Pennsylvania, on October 29, 1982. Three prominent speakers were engaged. Dr. John B. Yasinsky, General Manager, Advanced Power Systems Division, Westinghouse Electric Corporation, spoke on "Radioactive Waste Management—A Manageable Task." "Toward a Reasonable Solution to Our Waste Utilization and Disposal Problems," was the topic of John A. Bartone, President of Organic

Processing Systems, Inc., of Erie, Pennsylvania. Patrick F. Mutch, Assistant Executive Director at The Western Pennsylvania Hospital in Pittsburgh, Pennsylvania, addressed the meeting on "Solid Waste Handling in Hospitals."

Shortly after this symposium, Dr. Shyamal K. Majumdar and Dr. E. Willard Miller, Editors of previous Academy publications, began to solicit manuscripts from authorities in the fields of solid, hazardous and radioactive wastes. The response was overwhelming, and many more authoritative papers than could possible be published as a single volume were received.

The Academy needed help in this area and turned to the Pennsylvania Department of Environmental Resources (DER). The Executive Board then met with two representatives of the DER, Mr. Donald Lazarchik, Director of the Bureau of Solid Waste Management, and Mr. Gary Galida, Chief of the Division of Hazardous Waste Management. From this initial meeting grew an outline that was so diverse and covered so many aspects of solid, hazardous and radioactive wastes, that the Academy decided to publish three books of about 500 pages each, instead of the original single volume. The Department of Environmental Resources representatives agreed to join in this venture and assist the Academy as co-sponsors of the publications.

A second symposium on the subject areas was then organized by Mr. Justice J. P. Flaherty, Chairperson of the Academy's Advisory Council. The symposium was held at the annual meeting of the Academy at the Host Corral in Lancaster, Pennsylvania, on April 10, 1983. The speakers included Dr. J. B. Bundock, Scientific Advisor, Ministry of the Environment, Quebec, Canada, who spoke on "Fluorides, Water Fluoridation and Environmental Quality: A Retrospective Study in the Environment." "Pennsylvania's Regulation of Hazardous Waste Management" was the topic presented by Gary R. Galida, Chief of the Hazardous Waste Management Division, Department of Environmental Resources of the Commonwealth of Pennsylvania. Jack Dingman, President of Organic Bio-Conversion, Inc., of Madisonville, Kentucky, addressed the topic of "Solid Waste and Sewage Sludge Management and Recovery." The symposium was a tremendous success, and again stimulated the submission of more authoritative papers on the three subject areas.

A third symposium was held at the joint fall meeting of the Pennsylvania Science Teachers Association and the PAS Executive Council. The location was the Marriott Hotel in Monroeville, Pennsylvania, and the date was October 27, 1983. Organizing the symposium and presiding was Dr. George C. Shoffstall, President-Elect of the PAS. Dr. George K. Tokuhata, Director of the Division of Epidemiological Research, Pennsylvania Department of Health, addressed the group on "Epidemiology Methods of Investigating Health Hazards of Toxic Wastes." "Emergency Preparation and Response to a Nuclear Accident" was the topic of E. Preston Rahe, Jr., Manager, Nuclear Safety Department, Nuclear Technology Division, Westinghouse Electric Corporation. Andy Sabo, Director of Licensing, Safeguards and Licensing, Water Reactor Division of

Westinghouse Electric Corporation spoke on "Radiation Protection Standards and Radiation Risks." "Pennsylvania Low Level Waste Disposal" was the topic addressed by Dr. Ralph DiSibio of Hittman Nuclear Development Corporation, Columbia, Maryland. "Management of Radioactive Materials Spillage" was presented by Stephen Marchetti, Manager of Environmental and Occupational Safety, West Valley Nuclear Service Company, New York.

At a time when the quality of the environment is in constant jeopardy of being destroyed, and when the quality of life on earth is under constant influence of toxic and hazardous wastes, these three books should provide the information to the nation that we must take measures now to prevent any tragedy that would change the quality of our lives or our future as human beings on this planet called Earth.

As President of the Academy, I feel that these volumes are of national importance because they provide the information on wastes that was lacking in the past, and at the same time, can provide the scientific information that is necessary to help make corrections of damage that may have been done to the environment by these wastes.

The Academy has again provided the impetus in these three volumes for legislation on both the state and national levels to stand up and be counted so that hazardous and toxic wastes in the environment can be handled and disposed of properly, with minimal effect on state and national populations. This, I believe, is one of the responsibilities of the Pennsylvania Academy of Science: to present to the citizens of Pennsylvania and the nation the problems of solid, hazardous and radioactive wastes and the current research in progress to help solve these problems, so that the citizens of the Commonwealth can make educated judgments on the solutions to such problems, which requires a well-informed citizenry.

PART 1

Waste Types, Treatment and Disposal Technology

The disposal of hazardous and toxic wastes is now recognized as one of the nation's major environmental problems. Part One provides the foundation chapters for the discussion of hazardous wastes. The initial chapter presents a classification and properties of hazardous wastes based on the Environmental Protection Agency and the Pennsylvania Department of Environmental Resources hazardous waste listing. The 363 hazardous substances are classified into 35 reactivity groups, chemical classes and chemical reactivities. This is followed by a discussion of the hazardous waste treatment and disposal options based on the chemical structure of the reactivity groups.

Because of the seriousness of the hazardous and toxic waste problems there have been many technological developments in recent years. Chapters two through seven consider some of these processes. Traditionally hazardous wastes were deposited in landfills or disposal sites rather than by incineration. As a consequence many poor disposal practices evolved creating environmentally hazardous waste conditions. It is now recognized that a viable alternative to the landfill is incineration of the hazardous materials.

A number of incineration systems have been developed in recent years. Of these, the rotary kiln and the liquid injection system have proved most satisfactory. In the development of waste incineration, technology considerations include such aspects as burner, nozzle, and combustion chamber design.

Besides the incineration of hazardous wastes on land, techniques have been developed for the incineration of liquid organic wastes at sea. This is a recent development practiced first in Europe, but is now carried out in other parts of the world, including the United States. As the practice has grown it is now controlled by national and some international regulations. As a result of numerous tests EPA has concluded that oceanic incineration has not resulted in any measureable environmental impact.

A number of unique methods are being developed to control the spread of hazardous wastes through the ground at landfills. The pozzolanic cementation process, in commercial operation since 1980, is one of these efforts. In this technique hazardous wastes are blended into a pozzolanic cement and placed in a lined landfill. The pozzolanic cement develops a clay-like permeability and effectively seals off the waste from weathering and leaching. Since its initial development, the pozzolanic process has been used in the disposal of about a half million tons of hazardous materials. Results are most promising for its future utilization.

Part One concludes with two chapters on the disposal of hazardous wastes in the German Federal Republic. Such applied technologies are discussed as chemico-physical treatment of special waste, emulsion-splitting-effluent purification, cyanide de-toxication, heavy metal precipitation and thermodestruction of wastes.

Hazardous and Toxic Wastes: Technology, Management and Health Effects. Edited by S.K. Majumdar and E. Willard Miller. © 1984, The Pennsylvania Academy of Science.

Chapter One

Classification and Properties of Hazardous and Toxic Wastes

Raymond W. Regan, Ph.D., P.E.

Associate Professor of Civil Engineering
The Pennsylvania State University
University Park, PA 16802

The disposal of municipal and industrial solid wastes in sanitary landfills (SLF) is a widely used waste management practice. During past years, it had become evident that there had been possible adverse human health effects and serious environmental damage associated with improper disposal of hazardous waste (HW).

Congress, by the passage of the Resource Conservation and Recovery Act (RCRA) of 1976 (PL94-580), directed the Environmental Protection Agency (EPA) to promulgate regulations to protect human health and the environment from improper solid waste (including HW) management practices. EPA published listings of industrial HW sources, chemical names and properties on May 19, 1980(1). The Pennsylvania Department of Environmental Resources (DER) published similar listings (2).

The objective of this chapter was to summarize recent information concerning the classification and properties of hazardous and toxic wastes, based on the EPA and DER HW listings. The specific purpose include, (a) summarizing EPA criteria for HW identification, (b) categorizing HW into reactivity groups according to molecular functional groups, chemical classes or chemical reactivities and (c) assessing the application of the reactivity groups based on chemical structure for various HW treatment and disposal options.

The primary sources used for this chapter include Rotz (3), Bonner et al. (4), and Hatayama et al. (5).

CLASSIFICATION AND PROPERTIES

The overall HW listings include approximately 363 chemical (organic and inorganic) substances from a wide range of industrial sources. This section outlines the specific oriteria for HW identification as listed by the regulatory authorities, and summarizes the use of reactivity groups for classifying HW.

Regulatory authorities

As defined by the DER (6), hazardous waste is "a solid waste, or combination of solid wastes, which because of its quantity, concentration, or physical, chemical or infectious characteristics may (a) cause, or significantly contribute to an increase in mortality or an icrease in serious irreversible, or incapacitating reversible illness; or (b) pose a substantial present or potential hazard to human health or the environment when improperly treated, stored, transported, or disposed of, or otherwise managed."

A waste may be designated as a HW if it meets one of the following conditions (1,2):

1. Acute HW; a waste which has been found to be fatal to humans in low doses or, in the absence of data on humans, has been found to have this effect on laboratory animals. The criteria used to make this judgement include, (a) an oral LD50 of less than 50 mg/kg, (b) an inhalation LD50 of less than 2 mg/L, or (c) a dermal LD50 of less than 200 mg/kg.
2. A waste was hazardous if it contains any of the toxic constituents listed in the regulations (363 compounds listed).
3. Specific criteria for HW identification were listed under several categories including ignitability, corrosivity, reactivity and toxicity.
A. Ignitability
   A waste is ignitable if it is,
   1. a liquid solution which contained less than 24 percent alcohol by volume and has a flash point less than 60 °C,
   2. not a liquid solution but under standard temperature and pressure may cause a fire through friction, etc.,
   3. an ignitable compressed gas, an oxidizer, such as peroxide.
B. Corrosivity
   A waste is corrosive if it is,
   1. aqueous and has a pH less than or equal to 2, or greater than or equal to pH 12.5,
   2. or if it was a liquid which corroded steel at a specified rate within a 200-h test period,
C. Reactivity
   A waste is reactive if it is,
   1. normally unstable and readily undergoes violent change without detonating,

TABLE 1

Hazardous Waste Definition Threshold Levels[a]

Contaminant	Maximum concentration ($\mu g/L$)
Arsenic	5,000
Barium	100,000
Cadmium	1,000
Chromium	5,000
Lead	5,000
Mercury	200
Selenium	1,000
Silver	5,000
Endrin (1,2,3,4,10,10-hexachloro-1 7-epoxy-1,4,4a,5,6,7,8 8a-octahydro-1,4-endo, endo-5, 8-dimethano naphthalene)	20
Lindane (1,2,3,4,5,6-hexachlorocyclohexane, gamma isomer)	400
Methoxychlor (1,1,1-trichloro-2,2-bis p-methoxyphenyl ethane	10,000
Toxaphene ($C \star \, ''H_{10}C_{18}$, technical chlorinated camphene, 67-69% chlorine)	500
2,4-D, (2,4-Dichlorophenoxyacetic acid)	10,000
2,4,5-TP; Silvex (2,4,5-Trichlorophenoxypropionic acid)	1,000

[a]Concentrations listed were 100 times the NIPDWS values.

2. reacted violently with water,
3. formed explosive mixtures with water or formed toxic gases or fumes when mixed with water,
4. cyanide or sulfide waste (fumes),
5. capable of detonation,
6. or if it contained forbidden explosives as defined by law.

D. Extraction Procedure (EP) Toxicity Test

Toxicity of wastes was determined by an Extraction Procedure (EP), involving a 24-h agitation of a 20:1 mixture of waste in aqueous media, adjusted if possible to a pH of 5.0 with 0.5 N acetic acid (400 mL maximum mL maximum addition) (6). The EP was intended to provide an accelerated laboratory simulation of the contamination potential of toxic substances leached from industrial hazardous solid waste flowing into the groundwater below a nonsecure municipal landfill. The EP extract was analyzed for the eight metals (As, Be, Cd, Cr VI, Pb, Hg, Se, and Ag), four pesticides (Endrin, Lindane, Methoxychlor, and Toxaphene), and two herbicides (2,3-D and 2,4,5-TP(Silvex)) for which National Interim Primary Drinking Water Standards (NIPDWS) have been established (7). Hazardous waste definition threshold levels of 100 times the NIPDWS have been established for each of the species to account for attenuative processes expected to occur during the movement of leachate through the underlying strata and groundwater

TABLE 2

Reactivity Groups Based on Molecular Structure (After Hatayama et al. (5))

Reactivity Group Number	Group Name
1	Acids, Mineral, Non-Oxidizing
2	Acids, Mineral, Oxidizing
3	Acid, Organic
4	Alcohols and Glycols
5	Aldehydes
6	Amides
7	Amines, Aliphatic and Aromatic
8	Azo Compounds, Diazo Compounds and Hydrazines
9	Carbamates
10	Caustics
11	Cyanides
12	Dithiocarbamates
13	Esters
14	Ethers
15	Fluorides, Inorganic
16	Hydrocarbons, Aromatic
17	Halogenated Organics
18	Isocyanates
19	Ketones
20	Mercaptans and Other Organic Sulfides
21	Metals, Alkali and Alkaline Earth, Elemental Alloys
22	Metals, Other Elemental and Alloys in the Form of Powders, Vapors or Sponges
23	Metals, Other Elemental, and Alloy, as Sheets, Rods, Moldings, Drops, etc.
24	Metals and Metal Compounds, Toxic
25	Nitrides
26	Nitriles
27	Nitro Compounds
28	Hydrocarbon, Aliphatic, Unsaturated
29	Hydrocarbon, Aliphatic, Saturated
30	Peroxides and Hydroperoxides, Organic
31	Phenols and Cresols
32	Organophosphates, Phosphothioates and Phosphodithioates
33	Sulfides, Inorganic
34	Epoxides
35	Miscellaneous (after Rotz, (3))

aquifer (Table 1).

At a later time other criteria such as carcinogenicity, mutagenicity (change in hereditary trait), teratogenicity (deformities), aquatic toxicity, phytotoxicity and mammalia toxicity are to be included in the EPA regulations as test protocols become established.

TABLE 3

Reactivity Groups Based on Chemical Classes or Reactivities (after Hutayama et al. (5))

Reactivity Group Number	Group Name
101	Combustible and Flammable Materials, Miscellaneous
102	Explosives
103	Polymerizable Compounds
104	Strong Oxidizing Agents
105	Strong Reducing Agents
106	Water and Mixture Containing Water
107	Water Reactive Substances

Reactivity Groups

Hatayama et al. (5) categorized various HW into reactivity group number (RGN) according to molecular functional groups, chemical classes or reactivities. Groups RGN 1 to 34 were based on the molecular structure of the compound (Table 2). Rotz (3) created a RGN 35 for HW on the DER list that were not defined by the previous author. HW substances represented by chemical classes and chemical reactivities (5) were classified as RGN 101 through 107 (Table 3). Generic compounds included in each RGN were reported (3,5).

Chemical compounds bearing similar molecular functional groups may react with other compounds in similar ways. Treatment and disposal methods for HW processing depend on chemical reactions for waste destruction. In such cases, chemical structure (RGN 1 through 34) may be a major factor for determining whether a substance could be detoxified successfully or destroyed by a particular process. Therefore, the assessment process for suitable treatment options for 34 RGN was determined on the basis of similar molecular structure for 340 organic HW.

DISCUSSION

In the following discussion HW detoxification and treatment processes were summarized and the technical feasibility for their use with candidate organic HW substances were evaluated based on RGN 1 through 34. Finally, the overall impact of thermal processing for these RGN were estimated.

Processes Developed for HW Applications

Incineration technology appears to be a fairly advanced means for providing thermal detoxification of HW. In recent years, applying incineration technologies to HW has emerged as the preferred alternative to other treatment and disposal methods. Incineration has been defined as an engineered process using thermal oxidation of a waste material to produce a less bulky, non-toxic

TABLE 4

Classification of Thermal Processing Technologies by Stage of Development
(Adapted from Berkowitz et al. (9))

Process is developed for hazardous waste treatment	Liquid injection incineration Rotary kiln incineration Fluidized bed incineration Multiple earth incineration
Process is developed, but not used for hazardous wastes	Calcination Cement kilns Lime kilns Wet air oxidation
Process appears useful for hazardous wastes, but requires development work	Pyrolysis Molten salt destruction
Process might work in 5 to 10 years but needs further research	Infrared furnace incineration Microwave discharge

TABLE 5

Operating Parameters and Applicable Physical Forms for Thermal Processing Methods
(Kiang and Metry, (10))

	Temperature Range, F	Residence Time Range	Applicable Physical Forms
Liquid injection incineration	1800-3000	0.1-2 sec.	gas liquid slurry
Rotary kiln incineration	1500-3000	liquids and gases: seconds solids: hours	gas liquid slurry sludge solid
Fluidized bed incineration	1400-1800	seconds- hours	gas liquid slurry sludge solid

residual (Bonner et al. (4)). The principal products of combustion were carbon dioxide, water vapor and ash. HW containing sulfur, nitrogen, halogens and heavy metals may appear as solid, liquid or gaseous by-products of the process, thereby necessitating additional treatment prior to ultimate disposal or discharge. Incineration should be applicable to most organic wastes, as well as to some inorganics.

Numerous commercial and industrial incinerators have been available for HW processing (Table 4). For this section only three of the developed processes for

HW treatment were addressed, including liquid injection, rotary kiln, and fluid-ized bed incinerators. Operating parameters and applicable physical forms for these thermal processing methods were listed (Table 5). More complete evalua-tions of the available technology have been reported by Kiang and Metry (8) and Bonner et al. (4).

Assessment of Organic RGN's for Applicable Incineration Technologies

Bonner et al. (4) presented information identifying whether a given HW was good (G), potential (P), or poor (N) candidates for incineration. Based on tech-nical considerations such as applicable physical form, the list identifies appropri-ate incineration technologies, including liquid injection (L), rotary kiln (R) and fluidized bed (F), respectively. Background documents for specific wastes, trial burn data and engineering judgment based on chemical formulas of compounds present in the waste were considered in the preparation of Bonner's report. The following criteria concerning chemical composition were employed:

Waste Containing:	*Incineration Category:*
-Carbon, hydrogen and/or oxygen	Good
-Carbon, hydrogen, ≤ 30 percent by weight chlorine and/or oxygen	Good
-Carbon, hydrogen and/or oxygen, >30 percent by weight chlorine, phosphorus, sulfur, bromine, iodine or nitrogen	Potential
-Unknown percent of chlorine	Potential
-Inorganic compounds	Poor
-Compounds containing metals	Poor

Bonner, et al. (4) stressed that this evaluation method for incinerator tech-nology should be used for indicative guidance rather than conclusive decisions.

The reactivity groups consisting of organic substances appropriate for inciner-ation include RGN's 3 to 9, 12 to 14, 16 to 20, 26 to 32 and 34. Although primari-ly organic in nature, a large number of compounds in RGN 11 (cyanides) and all of those in RGN 24 (metal compounds) were identified as poor incineration can-didates. These two RGN have been excluded from the discussion which follows.

The information reported by Rotz (3) indicated that approximately 98 percent of the overall 340 commercial chemical products or manufacturing chemical in-termediates listed totalling 396 RGN either were G or P candidates for incinera-tion (Table 6). About 29 percent were classified as G and 69 percent as P candi-dates. Of the G and P categories, liquid injection incineration may be appropri-ate for approximately 50 percent, fluidized bed incineration for 91 percent and rotary kiln incineration for close to 100 percent of the organic compounds listed as HW (6).

Based on chemical composition information, the data in Table 6 gave a rough indication of the incinerability of the compounds found within each RGN. For

TABLE 6

Summary of Incineration Feasibility for Appropriate Technologies, Based on RGN (Rotz (3)).

RGN	Total HW in Group	Percent Incineration			Feasibility Technology		
		% G	% P	%N	% L	% R	% F
3	9	33	56	11	33	89	56
4	12	58	42	0	75	100	100
5	7	71	29	0	86	86	86
6	26	4	92	4	4	96	92
7	52	25	73	2	38	98	92
8	12	17	75	8	33	92	92
9	7	14	86	0	0	100	100
12	2	0	100	0	50	100	100
13	15	73	20	7	73	93	87
14	24	42	58	0	67	100	100
16	14	100	0	0	36	100	93
17	100	11	88	1	48	99	68
18	2	50	50	0	100	100	100
19	12	75	25	0	58	100	100
20	19	5	95	0	16	100	89
26	8	38	62	0	75	100	88
27	28	11	89	0	36	96	96
28	1	100	0	0	0	100	100
29	1	100	0	0	100	100	100
30	2	100	0	0	100	100	100
31	23	48	52	0	17	96	65
32	16	6	94	0	56	100	100
34	4	75	25	0	75	100	100
TOTAL	396*	29	69	2	51	98	91

KEY—Percentage of substances within the RGN classified

%G-as good candidates for incineration.

%P-as potential candidates for incineration.

%N-as poor candidates for incineration.

%L, R and F-Percentage of substances within the RGN that may be appropriate for liquid injection (L), rotary kiln (R) and fluidized bed (F) incineration, respectively,

RGN missing; incineration not appropriate for RGN, N = 100%

*Several HW have multiple RGN listing.

example, substances falling into RGN's 5, 13, 16, 19, 28, 29, 30 and 34 most likely will be good candiates for incineration because over 70 percent of the compounds in each group were classified as such.

Table 6 also indicated incineration technologies most likely to be applicable to substances within a RGN. For example, for RGN 31, rotary kiln incineration was applicable to 96 percent of the compounds, fluidized bed incineration to 65 percent and liquid injection to only 17 percent of the HW listed in that group.

Looking at each incineration option, the percentage of compounds listed as applicable in each RGN may aid in the selection of a technology for destruction of HW occurring in specific RGN's. Liquid injection incinerators did not appear to be an appropriate choice for destroying a waste containing a large portion of amides (RGN 6) or carbamates (RGN 9), for example. Keeping in mind that other factors not considered in the preparation of Table 6 may be important, rotary kiln incinerators appeared to be the most versatile units for handling substances occurring in the organic reactivity groups listed. Fluidized bed and liquid injection incinerators would appear to be less versatile in their application.

Impact of Thermal Processing on Hazardous Waste Management

Numerous thermal, physical, chemical and biological treatment processes have been identified as candiates to detoxify or to destroy HW (9). Thermal processing tehnologies such as incineration, pyrolysis, calcination and wet air oxidation have received considerable attention in HW management (9). An estimated 60 percent of the Nation's HW load may be processed by thermal treatment techniques (Kiang and Metry (8)). Thermal incineration represents an advanced and proven technology for dealing with HW. About 98 percent of the suitable organic HW represents either good or potential candiates for incineration (Table 6). Yet only six percent of those HW treated or disposed by commercial off-site facilities in 1980 were incinerated (Hill, et al. (10)).

CONCLUSIONS

Focusing on the chemical composition of the HW, approximately 363 hazardous substances were classified into 35 RGN according to molecular functional groups and 7 RGN by chemical classes. Based on available information, the suitable organic HW were classified into RGN and were assessed as to which of several well developed incineration technologies (rotary kiln, fluidized bed and liquid injection) may be applied optimally for destruction of the HW occurring within the RGN (Table 6).

The information presented in this chapter may facilitate the selection of an appropriate thermal processing system for a specific HW stream based on its chemical composition. It also may provide general guidelines as to which chemical classes of HW may be successfully detoxified by specific incineration processes. Based on the information reported, rotary kiln incinerators appeared to be the most versatile process for handling the organic HW listed. Although fluidized bed and liquid injection incinerators appeared to be less versatile in their applications, all three incinerator types represent proven technologies for HW destruction and should experience more widespread application in the future.

REFERENCES

1. Environmental Protection Agency (EPA). 1980. Identificaiton and Listing of Hazardous Waste. *IN* Environmental Protection Agency Hazardous Waste Management System. 40 CFR 261.24.
2. Pennsylvania Department of Environemtal Resources (DER). 1980. *The Pennsylvania Bulletin*, Vol. 10, No. 48, p. 4509.
3. Rotz, C. A. 1982. Thermal Treatment Options for Organic Hazardous Waste. Unpublished M. EPC report, The Pennsylvania State University, University Park, PA.
4. Bonner, T. A. et al. 1981. Engineering Handbook for Hazardous Waste Incineration. U.S. Environmental Protection Agency, SW-889.
5. Hatayama, H. K., J. J. Chen, E. R. de Vera, R. D. Stephens and D. L. Storm. 1980. A Method for Determining the Compatibility of hazardous Wastes. U.S. Environmental Protection Agency, 600/2-80-076.
6. Environmental Protection Agency (EPA). 1980. Background Document. Section 261.24-EP Toxicity Characteristic. Office of Solid Waste, Environmental Protection Agency, Washington, D.C.
7. Environmental Protection Agency (EPA). 1979. 40 CFR 151, National Interim Primary Drinking Water Regulations.
8. Kiang, Y. and A. A. Metry. 1982. *Hazardous Waste Processing Technology.* Ann Arbor Science Publishers, Inc., Ann Arbor, Michgian.
9. Berkowitz, J. B., J. T. Funkhouser and J. I. Stevens. 1978. *Unit Operations for Treatment of Hazardous Industrial Wastes.* Noyes Data Corporation, Park Ridge, New Jersey.
10. Hill, R. D., N. B. Shomaker, R. E. Landreth and C. C. Wiles. 1981. "Four Options for Hazardous Waste Disposal," *Civil Engineering, ASCE*, Vol. 51, No. 9.

Hazardous and Toxic Wastes: Technology, Management and Health Effects. Edited by S.K. Majumdar and E. Willard Miller. © 1984, The Pennsylvania Academy of Science.

Chapter Two

A Sequenced Industrial Waste Incineration Research Program

Richard Carnes[1] and E. Timothy Oppelt[2]

[1]Environmental Scientist
U.S. Environmental Protection Agency Combustion Research Facility
Jefferson, Arkansas 72079

[2]Chief, Incineration Research Group
U.S. Environmental Protection Agency
26 West St. Clair Street
Cincinnati, Ohio 45268

The United States is generally recognized as an industrially based society. The unusually high standard of living enjoyed in the United States can be directly attributed to our industry base. We all recognize the clear fact of chemistry, in that, no chemical reaction goes to completion. In any industrial chemical process there is by nature a waste stream that requires safe and efficient disposal.

Past industrial waste disposal practices have been primarily directed toward the landfill or land disposal options rather than destruction technology. As a result, more and more incidents of poor disposal practices or illegal acts are being discovered. These result in groundwater contamination, surface water pollution, and large land masses basically unuseable for any purpose.

A viable alternative to the use of the land as a storage sink for a wide variety of industrial wastes is the process of incineration. When discussing incineration in the context of waste disposal, we are talking of high temperature processes, those capable of sustained operation in excess of 1000 °C. We are also usually referring to oxidative processes, those processes occuring in an excess of oxygen, rather than the pyrolysis type operations, those occuring at or near the absence of oxygen. For purposes of this paper, we will limit our discussion to two distinct technologies, rotary kiln and liquid injection (1) and how laboratory techniques are being developed to advance our understanding of the incineration process.

The Rotary Kiln Incineration System

Operation:

Rotary kiln incinerators are generally refractory-lined cylindrical shells mounted at a slight incline from the horizontal plane. The speed of rotation may be used to control the residence time and mixing with combustion air. They are generally used by industry, the military, and municipalities to degrade solid and liquid combustible wastes, but combustible gases may also be oxidized. Recently, rotary kiln incinerators have been used to successfully dispose obsolete chemical warfare agents and munitions. Figure 1 is a schematic of what a general rotary kiln system involves. This schematic is typical of most rotary kilns, including small portable units currently being used in hazardous waste disposal site restoration and demilitarization projects.

Two types of rotary kilns are currently being manufactured in the United States today, cocurrent (burner at the front end with waste feed) and countercurrent (burner at the back end). For a waste which easily sustains combustion, the positioning of the burner is arbitrary from an incineration standpoint; both types will destroy a waste. However, for a waste having low combustibility (such as a high volume sludge), the countercurrent design offers the advantage of controlling temperature at both ends, which all but eliminates problems such as overheating the refractory lining. The countercurrent flow technique has been reputed to carry excessive ash over into the air pollution control system due to the associated higher velocities involved, however, this condition also increases the turbulence during combustion which is generally a desirable factor.

Optimal length to diameter (L/D) ratios have ranged from 2 to 10, and rotational speeds of 1 to 5 RPM at the kiln periphery are common, depending on the nature of the waste. Residence times vary from a few seconds for a highly combustible gas, to a few hours for a low combustible solid waste. A typical feed capacity range is 600 kg/hr to 2,000 kg/hr for solids, and 630 L/hr to 2,250 L/hr for liquids at temperatures ranging from 800 °C to 1600 °C. Since rotary kilns are normally totally refractory-lined and have no exposed metallic parts, they may operate at high incineration temperatures while experiencing minimal corrosion

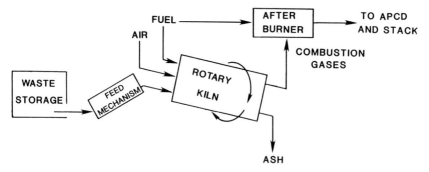

FIGURE 1. Rotary kiln incinerator schematic.

effects. Solid wastes, sometimes packed in fiber drums, are generally fed to the kiln by conveyor. Liquids and sludges are pumped in, with liquids usually being strained, then atomized with steam or air. The kiln and liquid burner are equipped with auxillary fuel burners for initial refractory heating, flame stability, and supplemental heat if necessary.

Afterburners are commonly used to ensure complete combustion of flue gases prior to treatment for air pollutants. Resource recovery (depending on the waste) and heat recovery are also common practices as initial steps to treatment of flue gases.

Types of Wastes:

Numerous hazardous wastes which previously were disposed of in potentially harmful manners (ocean dumping, landfilling, and deep-well injection) are currently being safely and economically destroyed using rotary kiln incinerators combined with proper flue gas handling. Included in this list of primarily toxic wastes are polyvinyl chloride wastes, PCB wastes from capacitors, obsolete munitions, and obsolete chemical warfare agents such as GB, VX, and mustard. Beyond these specific wastes, the rotary kiln incinerator is generally applicable to the destruction and ultimate disposal of any form of hazardous waste material which is combustible at all. Unlikely candidates are noncombustibles such as heavy metals, high moisture content wastes, inert materials, inorganic salts, and the general group of materials having a high inorganic content.

Advantages:
  (1) Will incinerate a wide variety of liquid and solid hazardous wastes.
  (2) Will incinerate materials passing through a melt phase.
  (3) Capable of receiving liquids and solids independently or in combination.
  (4) Feed capability for drums and bulk containers.
  (5) Adaptable to wide variety of feed mechanism designs.
  (6) Characterized by high turbulence and air exposure of solid wastes.
  (7) Continuous ash removal which does not interfere with the waste oxidation.
  (8) No moving parts inside the kiln (except when chains are added).
  (9) Adaptable for use with a wet gas scrubbing system.
  (10) The retention or residence time of the nonvolatile component can be controlled by adjusting the rotational speed.
  (11) The waste can be fed directly into the kiln without any preparation such as preheating, mixing, etc.
  (12) Rotary kilns can be operated at temperatures in excess of 2500 °F (1400 °C), making them well suited for the destruction of toxic compounds that are difficult to thermally degrade.
  (13) The rotational speed control of the kiln also allows a turndown ratio (maximum to minimum operating range) of about 50 percent.

Disadvantages:
(1) High capital cost for installation.
(2) Operating care necessary to prevent refractory damage; thermal shock is a particularly damaging event.
(3) Airborne particles may be carried out of kiln before complete combustion.
(4) Spherical or cylindrical items may roll through kiln before complete combustion.
(5) The rotary kiln frequently requires additional makeup air due to air leakage via the kiln end seals.
(6) Drying or ignition grates, if used prior to the rotary kiln, can cause problems with melt plugging of grates and grate mechanisms.
(7) High particulate loadings.
(8) Relatively low thermal efficiency.
(9) Problems in maintaining seals at either end of the kiln are a significant operating difficulty.
(10) Drying of aqueous sludge wastes or melting of some solid wastes can result in clinker or ring formation on refractory walls.

The Liquid Injection Incineration System
Operation:
Liquid injection incinerators are currently the most commonly used incinerator for hazardous waste disposal. A wide variety of units are marketed today, with the 2 major types being horizontally- and vertically-fired units. A less common unit is the tangentially-fired vortex combustor. Figure 2 is a schematic of a horizontally-fired unit. As the name implies, the liquid injection incinerator is confined to hazardous liquids, slurries, and sludges with a viscosity value of 10,000 SSU* or less. The reason for this limitation being that a liquid waste must be converted to a gas prior to combustion. This change is brought about in the combustion chamber, and is generally expedited by increasing the waste surface area through atomization. An ideal size droplet is about 40μ or less, and is attainable mechanically using rotary cup or pressure atomization, or via gas-fluid nozzles and high pressure air or steam.

The key to efficient destruction of liquid hazardous wastes lies in minimizing unevaporated droplets and unreacted vapors. Just as for the rotary kiln, temperature, residence time, and turbulence may be optimized to increase destruction efficiencies. Typical combustion chamber residence time and temperature ranges are 0.5 to 2 seconds and 700 °C to 1650 °C, respectively. Liquid injection incinerators are variable dimensionally, and have feed rates up to 5600 L/hr.

---

*To obtain the Saybolt universal viscosity equivalent to a kinematic viscosity determined at $t°F$, multiply the equivalent Saybold universal viscosity at 100°F by $1 + (t-100)$ 0.000064; e.g., 10 centistokes at 210°F are equivalent to 58.8 x 1.0070, or 59.2 SSU at 210°F. (Handbook of Chemistry and Physics, 45th edition).

The combustion chamber is a refractory-lined cylinder. Burners are normally located in the chamber in such a manner that the flames do impinge on the refractory walls. The combustion chamber wall can be actively cooled by process air prior to its entry into the combustion zone, thus preheating the air to between 150°C and 370°C.

Liquid waste fuel is transferred from drums into a feed tank. The tank is pressurized with nitrogen, and waste is fed to the incinerator using a remote control valve and a compatible flowmeter. The fuel line is purged with N_2 after use. A recirculation system is used to mix the tank contents. Normally a gas (for example, propane) preheats the incinerator system to an equilibrium temperature of approximately 815°C before introduction of the waste liquid.

Of the three types of units discussed earlier, the horizontal and vertical are basically similar in operating conditions. The tangentially-fired unit is known to have a much higher heat release and generally superior mixing than the previous two units, making it more attractive. However, these conditions lend to increased deterioration of the refractory lining from thermal effects and erosion.

As is obvious from the previous description of the rotary kiln and liquid injection incineration systems, much design and operational information is generally

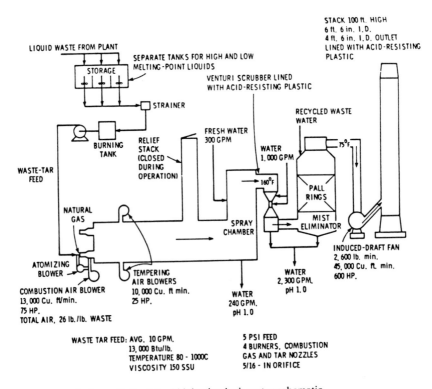

FIGURE 2. Horizontally-fired liquid injection incinerator schematic.

available. However, what is particularly needed is the basic information on how organic compounds act in a high temperature system and what can happen if the 3Ts (time, temperature, and turbulence) are out of balance. Why do some compounds exhibit much greater thermal stability than others? What does incinerability of a compound mean when considering a permit application? How can the controlled experiments of the laboratory be properly and correctly translated to pilot-scale technology and on to full scale for a much more thorough understanding of the truly complex phenomenon we call incineration?

Laboratory Thermal Decomposition System Development

The following is a presentation of the historical development of laboratory procedures used in thermal decomposition studies.

The early thermal decomposition research activities at the University of Dayton were aimed at developing special instrumentation to study the thermal decomposition properties of organic polymeric materials. In 1974, the emphasis shifted to developing special thermal systems which could address the thermal decomposition behavior of industrial organic wastes, toxic organic materials, and other environmentally sensitive substances. The first such system was the Discontinuous Thermal System (DTS). This was a relatively simple instrumentation assembly consisting of a heatable sample inlet region, a high-temperature quartz tubular reactor, an effluent trap for collection of unreacted parent compounds along with various reaction products, and a detached programmed temperature gas chromatograph. The DTS provided valuable information with respect to the thermal decomposition of various pesticides, e.g., Kepone, Mirex, and DDT (2).

From an instrumentation standpoint, this particular system was important for another reason. It laid the groundwork for more sophisticated studies by identifying those variables which are of concern to laboratory-scale thermal decomposition studies. These variables are listed in Table 1. Also, while conducting experiments using the DTS the importance of the formation of products of incomplete combustion (PICs) was clearly revealed. This important finding with respect to the thermal decomposition behavior of organic substances led to the development of the Thermal Decomposition Analytical System (TDAS).

In a general sense the TDAS (3) is similar to the DTS in that it consists of a thermally programmable inlet region that is connected to a quartz tubular reactor with a cryogenic in-line trap for capture of the effluent products from the

TABLE 1

Gas Phase Thermal Decomposition Variables

Exposure Temperature
Residence Time
Composition of Atmosphere
Chamber Pressure
Residence Time Distribution

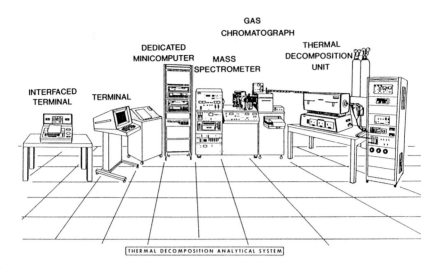

FIGURE 3. Artist's rendering of the TDAS.

reactor. In reality, however, the TDAS is a far more sophisticated system and has been changed to the point that it bears little resemblance to its predecessor, see Figure 3. The TDAS, unlike the DTS, is a truly continuous system. Connected downstream of the cryogenic trap is an LKB 2091 GC/MS which is used for analyzing the effluent products. This unit also has a dedicated minicomputer which proceses the data and aids in the interpretation of the mass spectra.

With its GC/MS and dedicated minicomputer, the TDAS proved to be a powerful tool for the investigation of gas phase thermal decomposition behavior. Over the past four years, the unit has been used to analyze numerous pesticides, industrial organic wastes, and pure compounds. The TDAS is the most sophisticated system thus far designed and by its nature it is also a relatively expensive system to assemble and maintain. With that in mind, the USEPA sponsored research to design an instrument with capabilities similar to the TDAS but which incorporated recent technological advances at a much reduced cost. This led to the development of the Thermal Decomposition Unit-Gas Chromatographic (TDU-GC) systems (4).

Just as the TDAS was similar to the DTS, the TDU-GC was very similar to the TDAS. As with the TDAS, the TDU-GC incorporated several improvements. In particular, the TDU-GC's sample inlet region was designed to be more versatile with respect to the different types of samples to be analyzed. The largest change was replacing the LKB 2091 with a Varian Vista series programmed temperature high-resolution gas chromatograph. The versatility and operating characteristics of this particular GC, particularly its ability to operate at cryogenic temperatures expanded the analytical capabilities of this system significantly (see Figures 4 and 5).

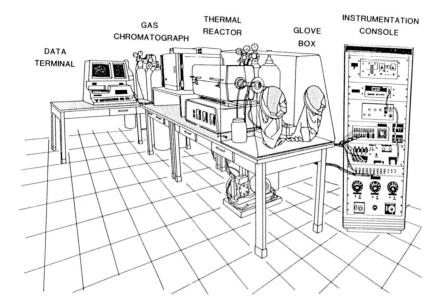

FIGURE 4. Artist's sketch of the TDU-GC system.

THERMAL DECOMPOSITION UNIT

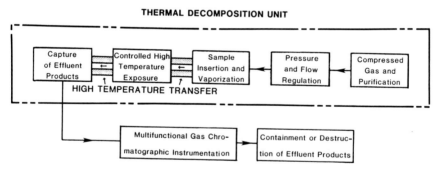

TDU-GC SYSTEM

FIGURE 5. TDU-GC block diagram.

Since it became operational in late 1981, the TDU-GC has shown that it can perform many of the same tasks as the TDAS. The major difference between the two instrumentation assemblies is that while the TDU-GC can detect the presence of PICs, to actually identify them would require further analysis. As far as cost, a TDU-GC system may be assembled for about one-third the cost of a TDAS. The TDU-GC has proven it capabilities by developing data on the basic oxidation kinetics for a series of compounds. These basic data are presently being accumulated in an attempt to provide researchers, permit writers, operators, and others a potential incinerability ranking index with direct applications to PIC formation prediction.

As work with the TDAS and the TDU-GC systems progressed, the enormous demand for data concerning the thermal stability of both pure compounds and complex industrial organic waste mixtures became critical. Events and discussions at a recent hazardous waste incineration conference (5) clearly pointed out the urgent need to determine thermal decomposition behaviors for a very large number of toxic organic compounds. Eventually hundreds, and possibly thousands, of different organic compounds need to be tested. Accordingly, analytical instrumentation is desperately needed which can rapidly determine the thermal decomposition properties of various principal organic hazardous constituents (POHCs) prior to subjecting these organic substances to controlled high-temperature incineration. In view of the urgency of this situation, researchers conceptualized the design of a relatively simple Packaged Thermal Reactor System (PTRS).

The objective of the controlled high-temperature incineration disposal method is to convert all incoming hazardous organic materials to substances which are environmentally acceptable, or easily contained (scrubbed) and eventually rendered harmless. Attention must be focused upon both the hazardous organic substances present in the feed (the POHCs) and the various intermediate and stable products of incomplete combustion (PICs), which in many cases may be more toxic than the parent substances.

It is currently the opinion of many that the characteristic best describing incinerability is thermal stability. This particular term is itself not a highly descriptive term; that is, one can not readily attach dimensions to thermal stability as is done with some other physical parameters. It is basically a characteristic established through experimentation. Even so, much is already known about the thermal stability of different classes of chemical compounds.

The family of aromatic compounds appears to have the greatest thermal stability, with the exception of some very small molecules, e.g., ammonia and hydrogen cyanide, etc. In addition, many organic toxicants are aromatic in nature. Specifically, the halogenated aromatic compounds as a class are probably some of the most toxic compounds.

Concept of the Packaged Thermal Reactor System:

The design goals for the PTRS were to develop a system which could rapidly and safely determine the DE (Destruction Efficiency) for an organic material at a given residence time, exposure temperature, and flowing gaseous atmosphere. The instrument should also be small, compact, and relatively portable so that it can be used as part of a mobile field unit or as part of a permanent installation. Finally, it should be an inexpensive unit to assemble and maintain. The concept of such an instrument is presented in block diagram form in Figure 6 and is shown schematically in Figure 7.

The schematic shown in Figure 7 describes one version of the originally proposed PTRS (5). The various components within the dashed-line boundary

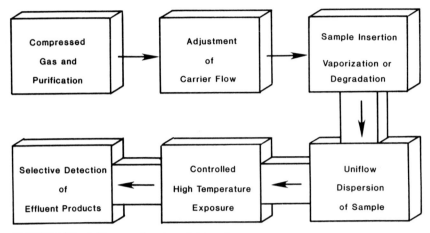

FIGURE 6. Block diagram of packaged thermal reactor system.

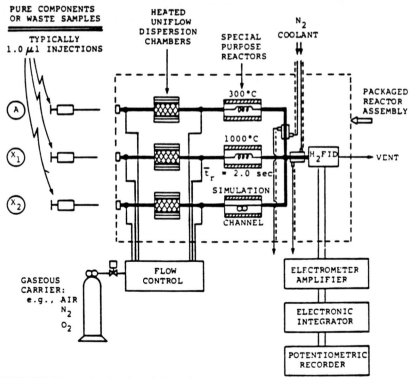

FIGURE 7. Schematic of packaged thermal reactor system.

represent the packaged reactor assembly. The components outside the boundary are typically found in almost every chemical instrumentation laboratory. An analyst who has been conducting gas chromatographic analyses could readily operate this unit. In addition, once this unit has been installed in a laboratory, the thermal decomposition data can be obtained very quickly. A single determination should take less than ten minutes, and an extremely small sample can be used to minimize safety problems.

In Figure 7, it is seen that localized cooling is an option that can be used for condensing and concentrating organic substances that have passed through the high-temperature reactors. The same localized cold regions of the transport path can also he heated rapidly to release the trapped organic substances and thereby pass them on downstream to the exit location as narrow concentrated zones.

The design of the packaged reactor will also permit the use of interrupted flow techniques. Such as arrangement permits the use of atmospheres with differing oxygen concentrations. Also, with this reactor assembly, tests can be made with almost any type of flowing gas, e.g., air, oxygen, steam, hydrogen, or even levels of chlorine.

The USEPA Combustion Research Facility (CRF)

The CRF located in Jefferson, Arkansas, is the only facility of its kind known to exist for the sole purpose of hazardous waste incineration research and development. The CRF has on-site waste characterization laboratories, including the aforementioned TDU-GC, for precombustion analysis of incoming waste streams to insure complete safety of the planned pilot-scale incineration experiments. An aditional aspect of the characterization process is to make sure the waste in question is compatible with all aspects of the incineration technology. The CRF also has on-site detailed analytical chemistry capability excluding mass spectrometry. In this way, samples collected during the research can be analyzed on-site thus reducing the time each must be handled and will expedite the results of each test so the report can be developed in a reasonable time frame. Figure 8 presents a floor plan of the CRF. The TDU-GC is presently stationed in Room 005.

The original purpose of the research program, which eventually led to the concept of the CRF, was to carry out a number of pilot-scale test burns on materials that had previously been studied either in part or completely at the laboratory scale. The explicit purpose of these experiments was to better allow the extrapolation of laboratory-scale results to large-scale systems (albeit still pilot scale from an engineering design standpoint) and to indicate differences between the idealized laboratory system and the real-world equipment.

The basic function of the CRF is to conduct detailed parametric studies for a number of industrial wastes using rotary kiln technology and liquid injection incineration technology. The results will augment the data base on which permit writers can draw for their decision making process and to further the state-of-

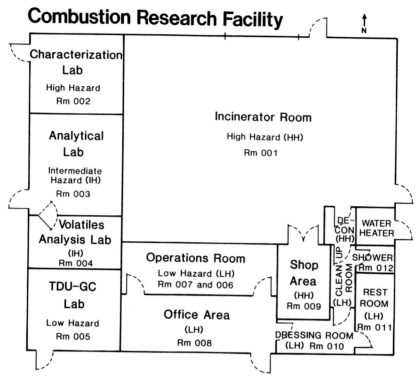

FIGURE 8. CRF Floor Plan.

the-knowledge in high temperature combustion. Much thought has been given to just how experiments will be planned and conducted. The following section presents the experimental design for a waste at CRF.

Experimental Design:

The philosophy that will govern the design and carrying out of the incineration experiments at the CRF is based on the following assertion: any experiment that does not result in a detectable emission is not useful in the determination of the necessary operating conditions for the acceptable disposal of the target waste stream. On the other hand, emissions of hazardous materials from the CRF are unacceptable. These apparently incompatible requirements can be made compatible by the observation that the transfer line between the kiln and the afterburner has been fitted with suitable sampling ports. In general, it is anticipated that the afterburner will be operated under conditions such that there will be no detectable emissions; the kiln conditions may be adjusted to a variety of temperatures. This mode of operation allows the determination of the thermal stability of the POHCs while at the same time providing protection of the facility and its surroundings. Sometime in the course of conducting experiments at the

CRF, a measurable discharge of pollutant must be detected at the stack in order for the total system performance evaluation to be meaningful. During these times, it is proposed to take these stack gases and pass them through an activated carbon filter for ultimate collection of trace organics.

In a very general sense, it can be envisioned that the function of the rotary kiln is the separation of the volatile organics from the non-combustibles in the waste matrix and the subsequent mixing of the voltilized materials with combustion air for ultimate presentation to the thermal environment of the afterburner. The function of the latter is the final complete oxidation of the organics. The proposed method of use of the kiln serves to enlarge somewhat the function of the kiln, in that, primary and even some secondary oxidation of the organics will occur in the kiln. Under the conditions that are expected, it will be possible to determine the DE for the identified POHCs as a function of temperature and of oxygen content while at the same time maintaining a DRE (Destruction and Removal Efficiency) at least as high as that required by the Incineration Regulations.

To illustrate the above we select a waste that contains, as POHC, hexachlorobenzene. It has been shown that a DRE of 99.99 percent can be obtained for this material at a temperature of 1000°C, a residence time of 2 seconds, and 20 percent excess air. Therefore, the afterburner will be operated at 1100°C, 2 seconds residence time and 20 percent excess air. The waste will be introduced into the kiln which will operate on the following schedule:

Initial temperature   500°C
Intermediate temperatures   600°C, 700°C, 800°C, 900°C
Final temperature   1000°C

At each intermediate temperature a sample will be taken from the transfer duct. The result of this series of experiments will be a DE as a function of exposure time. This will not be a complete series since the actual residence time of the mixed volatilized HCB and the combustion air will be unknown.

When the basic nature of the thermal stability of HCB has been determined as above, it will be of importance that the effect of exposure time be determined. With the feed to the kiln as before and the kiln temperature adjusted at some intermediate temperature, the afterburner will be operated according to the following schedule:

Initial temperature   1100°C
Intermediate temperature   1050°C, 1000°C, 950°C and 900°C

At each of the selected afterburner temperatures, samples will be taken at the sampling ports that have been provided in the output port of the afterburner (while still monitoring the stack to assure acceptable emissions from that source).

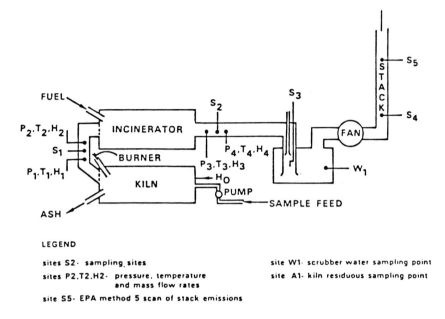

LEGEND

sites S2- sampling sites

sites P2.T2.H2- pressure. temperature
and mass flow rates

site S5- EPA method 5 scan of stack emissions

site W1- scrubber water sampling point

site A1- kiln residuous sampling point

FIGURE 9. Schematic Layout of incinerator showing sampling points.

In this manner, the full thermal cross-section may be obtained while at the same time providing full control over emissions from the system.

In addition to the obvious safety factors associated with the proposed method of operation, an additional benefit should accrue in that sampling of the kiln transfer duct should allow the determination of the nature and quantities of PICs that can be formed by the incineration process. Further, since the kiln can be operated in either a reducing atmosphere (pryolytic mode) or an oxidative atmosphere, it should be possible to not only identify the PICs but to begin to infer something about the formation processes. This latter study can also be conducted under conditions that prevent the emission of the hazardous PICs or their precursors, the hazardous POHCs.

Although the discussion has involved only HCB, it will be this general approach that will be applied to all waste streams investigated at the CRF.

Major Equipment Items:

Initially the incineration technology that will be available at the CRF will be a rotary kiln/afterburner system that is fitted with conventional wet combustion gas scrubbing (Figure 9). The general features of the incinerator are listed in Table 2 which follows.

142,522

TABLE 2
Characteristics of Incinerator

KILN
   Kiln Dimensions - 4 ft in diameter by 8 ft in length
   Kiln Rotation Speed - from 0.1 rpm to 1.2 rpm
   Burner - Iron Fireman, Model C-120-G-SMC fired by natural gas with a
        rated heat output of 1 x 10^6 Btu/hr.
   Structural Information - constructed of ¼ inch plate steel and lined with
                  cast-in-place refractory usable to 2500 °F.
   Feed System - equipped with liquid and semi-liquid feed equipment which
            will include Moyno pumping systems as well as high pressure
            liquid feed systems.

AFTERBURNER
   Dimensions - 3 ft in diameter, 10 ft in length
   Burner - Iron Fireman, Model C-120-G-SMC fired with natural gas with rated
        heat output of 1 x 10^6 Btu/hr.
   Structural Information - made of ¼ inch plate steel lined with cast-in-
                  place refractory useful to 2700 °F.

SCRUBBER
   Conventional wet system supplied by Andersen 2000.

CONCLUSIONS

One of the disadvantages of the use of trial burns in existing equipment lies in the difficulty involved with intercomparison of such data. The present difficulties with the available data base derive principally from this problem. It was for the specific purpose of circumventing this build-in uncertainty that the CRF was conceived, designed, and constructed. With the availability of a dedicated facility and a full time dedicated staff, the rate of return of comparable and reliable data on thermal characteristics of a variety of industrial waste streams will be forthcoming.

SUMMARY

By its continuous support of the research efforts at the University of Dayton (UD), the EPA is committed to a better understanding of the basic principal determining thermal decomposition or organic compounds. The broad systems evolution process at UD has demonstrated a step wise and thorough research approach to grasping this understanding and translating it to the technical community.

The CRF concept is the next logical step in the process we loosely call scaling. Through detailed, well-planned and conceived research at pilot scale, the interpretation of laboratory data can be directly implemented. Armed with the results of pilot-scale research, we can then turn to the next challenge facing us—a better understanding of combustion on a full-field scale system.

Researchers throughout the country are investigating various scales between laboratory and pilot, and perhaps, the results of their efforts will be incorporated into the CRF operations. The Agency has recently awarded a major contract effort aimed at a more thorough understanding of the scale-up phenomenon, no doubt the CRF will supply a major portion of the data for this effort.

A colleague of ours recently made a profound statement that will have a lasting impact on the entire research effort into hazardous waste incineration. It follows as a direct quote.

"Most advances that are made in the physical sciences are associated directly or indirectly with the ability to measure something new or different. The observation of new phenomena and the ability to accurately and repetitively characterize such observations are of paramount importance with respect to experimental inquiry. Therefore, it is readily apparent that instrumentation and special measurement techniques play very important roles in research activities. The capability of developing special instruments to measure and record select phenomena is accordingly of vital importance."

W.A.R. 1982

REFERENCES

1. "Engineering Handbook for Hazardous Waste Incineration," SW-889, September 1981.
2. Duvall, D. S. and Rubey, W. A., *Laboratory Evaluation of High-Temperature Destruction of Kepone and Related Pesticides,* Report for U.S. Environmental Protection Agency, EPA-600/2-76-299, December 1976.
3. Rubey, W. A., *Design Considerations for a Thermal Decomposition Analytical System (TDAS),* Report for U.S. Environmental Protection Agency, EPA-600/2-80-098, August 1980.
4. Dellinger, B., Duvall, D. S., Hall, D. L., Rubey, W. A., and Carnes, R. A., *Laboratory Determination of High-Temperature Decomposition Behavior of Industrial Organic Materials,* Paper presented at 75th Annual Meeting at Air Pollution Control Association, New Orleans, June 1982.
5. Hazardous Waste Incineration Conference, Jointly Sponsored by American Society of Mechanical Engineers and U.S. Environmental Protection Agency, Williamsburg, Virginia, May 1981.
6. Rubey, W. A., *A Packaged Thermal Reactor System for Characterizing Thermal Stability of Organic Substances,* University of Dayton Research Institute Technical Memorandum, UDR-TM-81-30, August 1981.

Hazardous and Toxic Wastes: Technology, Management and Health Effects. Edited by S.K. Majumdar and E. Willard Miller. © 1984, The Pennsylvania Academy of Science.

Chapter Three

Hazardous Waste Incinerator Design Considerations

David M. Pitts, M. Ch.E. and James J. Cudahy, M.S.Ch.E.

IT Enviroscience
312 Directors Drive
Knoxville, Tennessee 37923

With the advent of the Resource Conservation and Recovery Act (RCRA) and the recent focus of national attention on hazardous waste (HW) and hazardous waste problems such as those at Love Canal, the combustion process (incineration) is emerging as an important processing step for hazardous waste destruction. While relatively expensive compared to landfill, incineration is essentially a destruction process which removes the HW from the environment. The landfilling of HWs, however, is in many cases only a very long-term storage process during which hazardous components can potentially diffuse into the air and leach into ground water. In recognition of this it is anticipated that the EPA will soon start prohibiting certain liquid HWs in landfill, and eventually develop regulations prohibiting the landfilling of certain other non-liquid HWs.[1]

The publication of the first RCRA incineration regulations in 1978[2] brought industrial incineration from a qualitative study of black smoke, carbon monoxide, and total hydrocarbon to the new and relatively unknown quantitative area of destruction removal efficiency (DRE). The new RCRA DRE performance standard has been one of the significant occurrences in the history of industrial incineration.

When designing a hazardous waste incinerator, an understanding of the fundamental physical and chemical processes taking place in the incineration process needs to be translated into practical hardware designs. This chapter will examine the significant considerations associated with the design of a hazardous waste combustion system that can achieve DREs of $\geq 99.99\%$ for liquid HWs.

The scope of this chapter is limited to the considerations associated with the design or up-grading of a liquid injection incinerator, primarily the burner and combustion chamber. These considerations and relationships were developed for the design of a pilot liquid injection incineration system for the EPA's Combustion Research Facility in Jefferson, Arkansas.

RCRA INCINERATION HISTORY

Because of the importance of the RCRA incineration regulations to the subject of HW incineration a brief summary of the history of these regulations follows. The proposed Section 3004 incineration regulations, first published in the December 18, 1978, *Federal Register,*[2] contained specific incineration technical performance and design standards. These performance and design standards mandated criteria such as combustion temperature, retention time, carbon monoxide, combustion efficiency, toxic component destruction efficiency, and exhaust gas removal requirements for halogens and particulates. During the allowable comment period, industry responded strongly to some of the proposed performance and design standards, indicating that the combined EPA and industry incineration data base used to set the standards was poor and in need of further development and documentation. Consequently, when EPA published the revised Section 3004 regulations in the May 19, 1980, *Federal Register,*[3] the incineration performance and design standard approach had been replaced with a proposed three-phase regulatory program.

The May 19, 1980, Phase I standards were interim status standards which regulated the operation of existing HW incinerators. The general requirements for incineration in 40 CFR 265 basically involved improving existing incineration facility operating procedures.

The Phase II incineration standards were published in the January 23, 1981, *Federal Register.*[4] The January 23, 1981, 40 CFR 264 incineration regulations set forth three performance standards that must be met by new HW incinerators during a trial burn test. The three performance standards involved principal organic hazardous constituent (POHC) destruction, exhaust gas HCl removal, and maximum particulate emission.

The most recent interim final RCRA incineration regulations (40 CFR 122, 264, and 265) were published in the June 24, 1982 *Federal Register.*[5] These regulations apply to both new and existing incinerators and while changed somewhat from the January 23, 1981, version still included the 99.99% DRE performance standard.

RCRA INCINERATOR PERFORMANCE STANDARDS

The main HW incineration performance standards for DRE, HCl, and particulate are listed in 40 CFR 264.343. The key points of these standards, particularly the 99.99% DRE, have had significant impact on the design of HW incinerators. The particulate standard requires [3]0.08 grains/dscf corrected to 7% O_2, and the HCl standard requires an HCl removal of 99% or 4 lb/hr HCl stack emission, whichever is greater.[6]

Continuous measurement of the concentrations of POHC, HCl, and particulates in a stack gas is currently beyond the state of the art. Therefore, compliance with the performance standards given in 40 CFR 264.343 can only be measured during a trial burn by using EPA stack sampling techniques.

POHC DESTRUCTION REMOVAL EFFICIENCY

The concept and selection of POHCs is an important part of the incineration regulations. POHCs, which are to be sampled and determined during permit trial burns, are to be selected by the EPA permit writers from the RCRA Appendix VIII constituents present in the wastes to be incinerated.[4] Appendix VIII is a list of about 370 organic and inorganic hazardous chemicals first published in Part 261 of the May 19, 1980, *Federal Register*.[3] The Appendix VIII constituents have been defined by the EPA as substances that scientific studies have shown to have toxic, carcionogenic, mutagenic, or teratogenic effects on humans or other life forms.[7] The actual number of Appendix VIII chemicals is higher then 370, because more than 30 of the Appendix VIII constituents are chemical groups such as isomers or salts. For example, the constituent listed as chlorinated benzenes represents all the possible isomers (monochlorobenzene through hexachlorobenzene). The most recent version of Appendix VIII appears in the May 20, 1981, *Federal Register*.[8]

The POHC selection process is based on the concentration of Appendix VIII constituents in the incineration waste feed and on the degree of difficulty of incinerating the Appendix VIII constitutents. Those Appendix VIII constitutents that represent the greatest degree of difficulty of incineration (highest thermal oxidation stability) will be the ones most likely to be designated as POHCs.[4]

The POHC DRE standard (40 CFR 264.343a) requires that a HW incinerator must achieve a DRE of 99.99% for each POHC designated by the EPA permit writer. The DRE during the trial burn is determined for each POHC according to the following equation:

$$DRE = \frac{(W_{in} - W_{out})}{W_{in}} \times 100\% \qquad (1)$$

where W_{in} = mass feed rate of a POHC in the waste stream feeding the incinerator

W_{out} = mass emission rate of the same POHC present in exhaust emissions

As can be seen from Equation 1, each DRE is based on the feed rate of each POHC into the incinerator and the emission rate of each POHC in the HW incineration combustion gas measured after going through the incineration system's air pollution control (APC) equipment. The DRE calculation does not include any POHC present in either the incinerator ash or any APC effluents such as scrubber water. The DRE calculation includes only the stack emission of the original POHC that was fed into the HW incinerator and does not consider combustion by-products or products of incomplete combustion (PIC).

The DRE is also a function of the APC system. In incineration systems equipped with wet scrubbers such as venturis, uncombusted POHCs with high aqueous solubility and low vapor pressures will have high potential for being removed. With wet scrubbers or dry APC equipment (such as electrostatic precipitators or baghouses), uncombusted POHCs that are adsorbed on soot and particulate will be removed from the combustion gas along with the particulate collected by the APC equipment.

IT Enviroscience designed the pilot HW liquid injection incinerator at the EPA's Combustion Research Facility to achieve DREs of 99.99% or greater for any organic Appendix VIII constituent and also as a research tool that would be used to elucidate the importance of various fundamental chemical and physical processes occurring during the combustion of liquid HWs. A description of the pilot liquid incinerator follows.

EPA PILOT LIQUID INCINERATOR DESCRIPTION

The EPA's pilot liquid incinerator system was conceived as a flexible research tool with the purpose of carrying out pilot-scale test burns on hazardous liquid and sludge wastes. The effect of important incineration parameters, such as temperature, turbulence, residence time, burner type and configuration, atomization, etc., on the destruction efficiency of waste constituents will be studied on a scale that will yield results translatable to full-scale systems.

Important features of the combustion chamber relative to its use as a flexible research tool include the following:

- A variable length combustion chamber that will facilitate the study of overall residence time relative to destruction efficiency at constant turbulence (mixing)
- The capability to study different burner systems and the impact of different burner systems on different categories of wastes
- The control and accurate measurement of primary and secondary combustion air flow
- The control and accurate measurement of atomization air or steam
- The capability for alternative temperature control by water injection
- The capability for aqueous waste injection
- Hot gas sampling and monitoring ports located strategically along the length of the combustion chamber

- A combustion chamber configuration that will allow the study of the turbulent effects associated with a non-linear gas path
- A combustion chamber configuration which allows potential two-chamber operation as well as the future implementation of secondary combustion
- The use and design of insulating refractory to limit heat losses so that useful fundamental and scaleup data can be obtained
- High alumina service refractory which will allow the study of halogenated organic and salt-laden wastes while reducing the corrosive impact of these wastes on the combustion chamber refractory

The overall pilot liquid waste system consists of the combustion burner and chamber and an air pollution control system. The APC includes a quench chamber, a packed bed acid gas absorption tower, an ionizing wet scrubber for particulate removal, a carbon adsorption system for removal of trace organics during non-optimum conditions, and an induced draft fan.

The flexibility designed into the pilot unit would not normally be required for a full-scale HW liquid injection incinerator designed to handle a specific waste profile. The important considerations and parametric relationships, however, would be the same, and these are the tools engineers would use in the design of an industrial combustion system.

GENERAL INCINERATION CONSIDERATIONS

The ultimate goal of an incineration system for the disposal of waste liquids containing hazardous constituents is, of course, to efficiently burn the waste and destroy the hazardous constituents by thermal oxidation. The type of equipment normally used to carry out the thermal oxidation consists of a burner and a combustion chamber. The combustible wastes, liquids, slurries, and/or in some cases sludges are introduced through the burner or burners along with auxiliary fuel if required. Noncombustible or aqueous wastes often bypass the burner and are atomized directly into the combustion chamber.

Air pollution control equipment is required to remove contaminants from the thermal oxidation system effluent gas. In general, scrubbing equipment such as packed or tray tower absorbers is required for removing acid gases such as HCl and SO_2. The liquid effluents from the absorption equipment may require further processing. For solid particulates, dry collection methods can be used, such as bag filters or electrostatic precipitators. In hazardous waste incineration, however, wet collection methods are normally required because of the wet processing requirements for the removal of acid gases. Venturi scrubbers, wet electrostatic precipitators, and ionizing wet scrubbers are typically used.

Although common in Europe, energy recovery from the combustion of hazardous wastes is not generally practiced in the United States. Proven European energy recovery technology is, however, available in the U.S. for the recovery of energy from corrosive hazardous waste combustion gases.[9]

The following sections of this chapter will focus on the design considerations associated with the burner and combustion chamber for a liquid incineration system.

BURNER AND NOZZLE DESIGN CONSIDERATIONS

The terms burner, burner gun, and burner nozzle are often used interchangeably. In this discussion the term burner refers to the assembly wherein liquid waste and/or fuel is atomized either mechanically, hydraulically, or pneumatically by air or steam, mixed with combustion air, and ignited so that a continuous and stable flame is supported within the burner assembly. The burner gun is a critical part of the burner assembly. The purpose of the burner gun is to atomize or subdivide the liquid feed into fine droplets in order to enhance vaporization and mixing with the combustion oxygen. The combustion reactions only take place in the vapor phase, and the liquid waste must therefore be vaporized in order for the combustion process to start and continue. The burner nozzle or burner tip is the term normally used to describe the end of the burner gun where the waste liquid is atomized by mixing with high-pressure air or steam and the vaporized liquid flame is started.

In the burner, heat from the flame must be radiated into the atomized spray from the nozzle. This requirement is accomplished by the burner tile which is a refractory block with a conical or cylindrical hole through the center. The flame passes through the burner tile. The tile serves to maintain stable ignition by radiating heat to the fuel air mixture and aiding in the liquid vaporization process. While flames can be generated in hot combustion chambers without burner tiles, flames tend to be much more stable over a wide range of fuel and air inputs if a burner tile is used.

A special case of burner nozzle use is when an aqueous injection nozzle is used to atomize and introduce a waste liquid (usually a waste with a low heat content such as an aqueous waste) into the main combustion chamber. Depending on the heat content of the waste, this type of nozzle may or may not support a flame.

NOZZLE DESIGN CHARACTERISTICS

For hazardous waste incineration the purpose of the burner nozzle or the aqueous injection nozzle is to subdivide or atomize the liquid into droplet sizes as small as possible to provide the maximum surface area for vaporization and heat and mass transfer. This atomization is often limited by certain characteristics of the wastes including specific gravity, viscosity, and solids content. Nozzles have been developed to produce mists with particle sizes as low as 1 μ.[10] Typical oil burner nozzles yield droplets in the 10 to 50 μ range.[11]

Atomization of Organics

Atomization in nozzles is accomplished either pneumatically or hydraulically by the following means:

- Low pressure air or steam atomization (1 to 10 psig)
- High pressure air or steam atomization (25 to 100 psig)
- Mechanical (hydraulic) atomization through specially designed orifices (25 to 450 psig)

In nozzles where atomization is effected by mechanical means, hydraulic pressure forces the liquid through orifices that cause unstable sheets or jets to form and subsequently collapse into ligaments which break up into droplets of various sizes. Droplet size normally increases as capacity requirements increase because the orifice sizes must be larger for larger flow rates at a given pressure drop. The droplet size is also a strong function of the viscosity and surface tension of the liquid in mechanically atomized nozzles.

External Atomization

In pneumatic nozzles, the shearing force required to break up the liquid sheet is attained by air or steam pressure. In the operation of a typical externally atomized pneumatic nozzle, the liquid is pumped out a center orifice and steam or air is delivered through an annulus and directed at the external surface of the liquid stream which breaks up the liquid into fine droplets. A hollow cone spray can be obtained if the liquid is delivered through the annulus and the atomizing fluid through the center nozzle and directed outward against the liquid sheet. Liquid pressure requirements in systems such as these range from 5 to 25 psig. A good rule of thumb for the ratio of atomizing fluid mass rate to liquid mass rate is 0.3 lb of steam or air per lb of atomizing liquid. External mix nozzles are generally a good choice for liquids containing high suspended solids concentrations.

Internal Atomization

Nozzles that internally mix the liquid and atomizing fluid use a combination of hydraulic and pneumatic forces to effect atomization. The advantage of these nozzles is the relatively high turndown ratio of about 4:1 that can be obtained at reasonable hydraulic pressures ($\leq$ 150 psig). At the maximum liquid flow rate the nozzle is essentially a hydraulic nozzle. As the flow rate of liquid and associated pressure is decreased, the atomizing steam or air flow is increased to maintain a constant differential pressure across the outer tip openings.

Large Droplets

It is extremely important that the atomization nozzle system be designed and maintained to produce a narrow droplet size distribution and few very large drops. Large liquid droplets while few in number can contain a disproportionate amount of organic relative to smaller droplets. A recent theoretical evaluation of droplet size[12] indicated that for liquid droplets having an evaporation rate constant of 0.25 mm²/sec, a 50 μ droplet outside the flame envelope would take about 0.01 sec to evaporate in an 1800 °F combustion chamber. A 500 μ droplet, however, under the same conditions would take about 1.0 sec to evaporate. The 50 μ droplet in an incinerator with a 1.0-sec residence time would therefore be vaporized and subjected to high temperature for essentially all of the available residence time. The 500 μ droplet, however, would not be subjected to the same high temperature and residence time distribution as the 50 μ droplet. It is apparent from this example that evaporation of large liquid droplets could be a thermal oxidation rate limiting step in systems that produce too many liquid droplets. Too many large liquid droplets could be a significant factor in the failure of an incineration system to meet the 99.99% DRE.

Aqueous Atomization

Aqueous wastes, while typically not highly viscous, are generally more difficult to effectively atomize than low viscosity organic liquid wastes because of the significantly higher surface tension of the water. This means that atomization of aqueous wastes will generally result in larger diameter drops than for most organic liquid wastes under the same atomization conditions. Higher heat of vaporization of water relative to most organic liquids also increases the aqueous droplet evaporation time in the combustion chamber. As will be indicated in a later section, the atomization, droplet size, and evaporation of organic aqueous wastes appear, in some cases, to be very important to the DRE. These factors are possibly rate limiting steps in the combustion of high water solubility organics present in aqueous wastes. The design of an atomization system for high solubility organically contaminated aqueous wastes therefore requires special considerations. These considerations include the production of minimal large droplets, the best possible atomized droplet size distribution, and a high droplet vaporization rate. After evaporation, sufficient mixing with oxygen and a high temperature residence time are important to achieve 99.99% DREs.

BURNER CONSIDERATIONS

The combustion of liquids entails the application of heat to vaporize the fuel. Heat flux can come from the existing flame, from recirculating hot combustion products, from radiating refractory surfaces, or from a pilot burner.

Turbulent Diffusion Flames

The type of flames generally occurring in industrial incineration systems are called turbulent diffusion flames. In turbulent diffusion flames the fuel vapors diffuse from the liquid surface and mix with the oxygen in the air. As the temperature of the flame increases, the rate of the exothermic oxidation reactions increases and associated heat is released until full ignition and combustion ensues. The combustion rate for this type of flame process is always mixing or diffusion limited.

The greatest part of the destruction of the hazardous constituents of an atomized liquid waste fired through a burner occurs inside the turbulent diffusion flame emanating from the burner. The temperature, shape, and characteristics of that flame dictate the relative degree of destruction efficiency that can be expected. These flame characteristics are determined by the design of the burner and the specific combustion characteristics of the waste.

Reaction Kinetics

Three types of kinetic rates occur in an incineration system: flame oxidation, vapor phase thermal oxidation outside the flame, and pyrolysis both in and outside the flame envelope. The flame oxidation kinetic rate is several orders of magnitude greater than the flameless vapor phase oxidation because of the much higher temperature in the flame envelope, and the flameless oxidation kinetic rate is several orders of magnitude higher than that of the flameless pyrolysis reaction rate.[13] Flameless vapor phase thermal oxidation data done by the University of Dayton Research Institute (UDRI) in low oxygen gases and inert gases also confirms the lower kinetic rates under pyrolysis conditions.[14]

Flame Shapes

In general, a higher destruction efficiency can be expected from a short bushy non-luminous flame than from a long slender flame. Good mixing accomplished by high turbulence and high velocities produces a high heat release rate per unit volume in a short bushy flame.[15] In a long slender flame, poor or delayed mixing and low velocities produce a longer, more slender, less intense, and more luminous flame. Long slender flames can be produced only if the rate of mixing of the vaporized fuel and air is low enough so that the two fluids can travel some longitudinal distance from the burner before mixing and burning. Intense heating of the vapor in the absence of air, by radiation from the combustion occurring at the flame-air-vapor interface, thermally cracks some of the gaseous hydrocarbons into free carbon and hydrogen. It is the opaque carbon particles which become luminous and emit radiant energy.[16]

Heat Release Rates

Typical oil and waste burners with long slender flames exhibit combustion heat releases in the range of 100,000 Btu/hr/ft^3. Short bushy flames such as those produced by gas burners with considerable refractory surface, high mixture pressures, and thorough mixing can release as much as 40 MM Btu/hr/ft^3 of combustion volume.

For hazardous waste application, high intensity or vortex type burners have been developed to produce highly mixed intense short flames with heat release rates in the range of 500,000 to 1 million Btu/hr/ft^3 of combustion volume.[15] In this high intensity burner design liquid fuel is introduced at the centerline of the burner chamber through a high pressure steam or air atomized nozzle. The air for combustion is brought in from the side of the burner and passes through swirl vanes so that the high velocity rotating air mixes with the fuel at the entry point to the burner tile or chamber and causes a slight vacuum just downstream from the entry point. This vacuum causes the hot combustion gases from the flame to be pulled back into the turbulent air-fuel mixture entering the flame zone. Thus the mixture is preheated, vaporized, and ignited almost instantaneously and a tangentially rotating short bushy non-luminous flame is produced inside the burner chamber.[15]

Waste Characteristics/High Intensity Burner
A number of waste characteristics determine whether a waste liquid is suitable for such a high intensity burner. In order to maintain stability for the high intensity short flame, the adiabatic flame temperature in the burner must be maintained at some minimum value. The adiabatic flame temperature is a function of the heat of combustion of the waste and the burner excess air level. Thus, at a minimum specified excess air at which the burner can operate, (typically 15 to 20% excess air for hazardous waste incineration application) the heat of combustion will dictate the flame temperature. If the heat of combustion of the waste is less than about 4500 Btu/lb, the adiabatic flame temperature will fall below 2200 °F, the cutoff point for maintaining the stable short flame.[15] These burners can also have dual fuel capability, however, so an auxiliary gas fuel can be burned simultaneously with the liquid waste to provide flame stability and to increase the overall flame temperature.

Flame stability can be maintained in high intensity burners with a low heat of combustion waste without the use of auxiliary fuel if a somewhat longer than normal flame is generated. For hazardous waste destruction, however, a higher flame temperature is desirable. The use of auxiliary fuel or blending to a higher Btu/lb would therefore be appropriate to accomplish a higher flame temperature and increase the probability of attaining a 99.99% DRE.

Other characteristics of the HW liquid which impact the suitability of the high intensity short flame burner include the solids concentration, the viscosity, and the salt concentration of the HW liquid. The solids concentration and viscosity of the liquid affect the design of the burner nozzle. For this type of high intensity burner, mixing is very important and the liquid droplets resulting from the atomization in the nozzle must therefore be as small and as uniformly sized as possible. Obviously viscous fluids and large particles act against this.

High viscosity liquids (> 500 SSU) typically generate longer more luminous flames than those generated by low viscosity liquids depending on the specific design of the burner and nozzle and the type of atomization. Typical oil burners for high viscosity Bunker C fuel oil have large orifices and generate longer less intense flames than the same type burner for less viscous No. 2 fuel oil.

High viscosity and/or high solids concentration wastes can be fired in a cyclonic burner which applies centrifugal forces to atomize and mix the fluids. Sludges have been burned in cyclonic type burners firing auxiliary fuel. The flame in a cyclonic burner, however, will not be high intensity.

For wastes containing alkali metal salts, the high temperature intense flame inside a vortex burner chamber will cause the salts and alkali oxides to vaporize, react with and rapidly destroy the burner chamber refractory.[15]

Custom-Designed Burners

If an industrial HW has a high viscosity, a low heat of combustion, and contains alkali metal salts and/or suspended solids, a compromise generally will be required relative to the temperature, shape, and intensity of the flame that can be achieved by a given burner design. Custom-designed burners are common for hazardous waste incineration application.

The burner design and resulting flame characteristics significantly affect the destruction of the hazardous constituents that can be achieved in the burner, and therefore have a signficant impact on the subsequent combustion chamber design and residence time requirements. For example, residence time requirements and thus the length of the combustion chamber would likely be less for a system employing a short intense flame burner than for a system that had to use a long slender flame-type burner.

Burner Research

At present no specific correlations exist that relate the DRE with burner design details and flame characteristics. An important aspect of the pilot liquid injection incineration system designed for the EPA's Combustion Research Facility is that it can accommodate interchangeable burner and waste injection nozzle configurations so that these relationships can be studied.

COMBUSTION CHAMBER DESIGN

Assuming that the bulk of the destruction of the POHCs of a liquid waste has taken place in the burner flame envelope, the residual destruction to meet a specified DRE must be carried out in the combustion chamber. To accomplish this the residual POHC must be exposed to a sufficient amount of oxygen at a high enough temperature for a sufficient amount of time. The definition of these requirements for a specific waste in a system with a specific burner and combustion chamber configuration will be studied in the EPA's pilot liquid incineration system. After this research HW combustion chamber design can then be dictated by experimentally determined correlations for the important process parameters of incinerator temperature, residence time, excess air, and mixing.

The following sections will examine the relationship between these important process parameters and the combustion chamber design and sizing constraints.

Capacity Rating - Heat and Material Balance

The design of a combustion chamber requires the knowledge and analysis of a number of interrelated parameters, including the process parameters previously mentioned as well as the composition and flow of materials and combustion products through the system. Ultimately it is the combustion gas flow rate that will dictate the size of the combustion chamber and the mixing and residence time requirements. Combustion gas flowrate is proportional to the firing rate or heat duty in the system in Btu/hr. The heat duty is directly related to the composition of the waste and/or fuel to be fired, incinerator bulk gas temperature, excess air level, and heat loss of the system.

Thus, the first step in designing a combustion chamber is to carry out the heat and material balance calculations required to define the process and specify the sizing constraints. Computerized material and energy balances can be used to calculate the following process parameters:

> Total system heat duty (MM Btuh)
> Combustion gas composition
> Combustion gas flowrate
> Required auxiliary fuel and/or
> Required excess air

The important inputs to these calculations include:

> Flowrate, composition, and heat of combustion of waste fuel
> Required average temperature in combustion chamber
> Heat losses from combustion chamber
> Composition of auxiliary fuel

Once the heat and mass balances have defined the relationships of the important system parameters, their effect on the combustion chamber size and requirements can be determined.

Mixing

Three of the four important process design parameters for an incineration system (temperature, time, and excess air) are most commonly used to define a combustion process. It is possible, however, that two incineration systems with identical temperature, residence time, and excess air levels could exhibit different DREs if one incinerator had better mixing. In general, a better mixed combustion chamber will perform better in terms of combustion and destruction efficiency than one with less mixing even though other process parameters are identical.

The degree of mixing in a combustion chamber is related to the flow pattern and velocity of the fluids moving through the chamber. Mixing in a combustion chamber can be enhanced by a number of mechanical means including introducing the secondary air through swirl vanes to provide rotational energy or providing baffles and restrictions at different points in the chamber to change flow pattern. Most industrial liquid incineration combustion chambers, however, are cylindrical vessels with a straight through gas flow pattern.

If mechanical means are not used to enhance mixing, the mixing characteristics will be defined by the turbulent flow patterns and velocity profiles established in the system due to the flow rate and temperature gradients inside the chamber. A historical rule of thumb is that combustion gas velocities in the range of 10 to 20 ft/sec are required to promote turbulence and therefore good mixing. This rule of thumb, however, is not always conducive to good mixing.

It is the dimensionless Reynolds number (N_{Re}) that indicates whether a flow pattern is actually turbulent. The Reynolds number is defined as:

$$N_{Re} = \frac{DpV}{\mu} \tag{2}$$

where D = the diameter of the chamber
   p = the density of the fluid
   V = the velocity of the fluid through the chamber
   μ = the viscosity of the fluid

For $N_{Re} \leq 2200$ flow is laminar. In the laminar regime fluids tend to flow without lateral mixing, and there are neither crosscurrents nor eddies. At $N_{Re} \geq 10,000$ flow is turbulent with a mass of eddies of various sizes coexisting in the flowing stream and resulting in significant lateral mixing. Transition region flow at N_{Re}s between 2,200 and 10,000 is characterized by the formation of eddies with some lateral mixing. Combustion chambers should therefore be sized to yield a N_{Re} greater than 10,000 to ensure that the gas flow is in the turbulent regime so good mixing can occur. The length of the combustion chamber is then calculated based on the required residence time and the volumetric flow.

Heat Loss

The heat loss to the surroundings from a combustion chamber is a very important parameter of the incineration system which must be considered in the design of the combustion chamber. The actual heat loss from a system can have a significant impact on what happens in the combustion chamber and therefore on the DRE.

For example, if a combustion system design was based on adiabatic considerations, the actual heat loss from the system could cause deviation from design operation and therefore the expected performance. For a given waste flow rate, the required excess air level to maintain a specified temperature would decrease.

This would change the combustion gas flowrate which might reduce mixing but would increase the residence time. These changes could have an impact on the DRE. In the extreme, if the heat of combustion of the waste was low, the specified temperature might not be obtainable at a specified minimum excess air level, causing either a lower temperature profile than design or requiring auxiliary fuel firing to maintain temperature. A lower temperature profile could have an obvious deleterious effect on DRE. Firing auxiliary fuel would increase the combustion gas flowrate, yielding potentially better mixing but a lower residence time.

Insulating Refractory

In general, heat loss should be minimized to the extent practical. This is especially critical for small systems because the surface to volume ratio is so much greater for a small system than for a large system. Heat losses in a full-scale system can range from 2 to 5% of the heat duty. Heat loss from a small pilot unit, however, can range from 5 to 50% unless special design steps are taken to reduce this heat loss.

Heat loss from a combustion chamber is normally controlled by placing insulating refractory between the inside service refractory and the steel shell. Insulating refractories are light-weight porous materials with much lower thermal conductivity and heat-storage capacity than service refractories, which must resist abrasion and corrosion and/or erosion by molten metal or slag. The insulating refractories serve a twofold purpose in an incinerator since they limit heat loss as well as prevent the metal shell temperature from becoming excessively hot. Typical practice is to design for steel shell temperatures below 450 °F but above acid gas condensation temperatures. Insulation on the outside of the shell in order to limit heat loss from the combustion chamber is typically not done because of the potential corrosion that would occur to the steel shell from acid gas condensation.

The heat loss is mainly a function of the thermal conductivity and thickness of the refractory system and the inside and outside surface temperatures (hot and cold face temperatures) and the temperature of the ambient air to which the heat is discharged. Typical service refractories exhibit values of thermal conductivity ranging from 5 to 25 Btu/ft²hr °F/in. Insulating refractories have k values in the range of 1 to 10 Btu/ft²hr °F/in. Thermal conductivity is a function of temperature.

Service Refractory Considerations

Another important aspect of combustion chamber design is the selection of the service refractory system. The service refractory is continuously exposed to the high temperature oxidizing atmosphere inside the combustion chamber and must have the right physical properties to withstand this environment. Failure of the refractory system could potentially do significant damage to the combustion chamber.

The significant properties of any refractory including its high temperature strength depend on its mineral makeup and the way these minerals react at high temperatures with combustion chamber gases. A very large number of different refractory compositions exist to provide the optimum service for a particular application. In addition, either brick or monolithic castable refractory systems can be employed depending on the specific system requirements.

For these reasons, the refractory requirements in terms of combustion temperature, halogen and salt slagging potential, etc. are usually specified by the designer who then consults a refractory specialist for detailed specification of the refractory system.

Thermal Oxidation Stability (TOS)

Laboratory studies in flameless vapor phase thermal oxidation by UDRI and Union Carbide Corporation[13,14,17-21] have clearly indicated that some compounds such as chlorinated aromatics are more difficult to destroy in a thermal oxidizer than other compounds. The concept of thermal oxidation stability (TOS) therefore also needs to be considered during the design or upgrading of a pilot or full scale incineration system.

TOS is a key concept of the June 24, 1982, HW incineration regulations.[5] EPA has stated that those Appendix VIII constituents that are the most difficult to incinerate are most likely to be designated as POHCs. Any designer of HW incinerators must therefore have some idea of the TOS of different Appendix VIII and other chemicals during the design process.[17]

EPA has recommended heat of combustion as a TOS indicator,[22] apparently because heat of combustion data and estimation techniques are available for all Appendix VIII constituents. In the EPA heat of combustion model, TOS is inversely proportional to the heat of combustion. The higher the heat of combustion, the lower the TOS of the Appendix VIII constituent.

Two publications, however, indicate that the concept of heat of combustion as a TOS indicator may not be a very good one. Tsang and Shaub[23] point out that, according to the heat of combustion TOS indicator, acknowledged high-TOS chemicals such as benzene, polynuclear aromatics, and polychlorinated biphenyls (PCB) have lower thermal oxidation stabilities than highly unstable chemicals such as nitroglycerine and trinitrotoluene (TNT). Cudahy et al. statistically correlated flameless vapor-phase thermal oxidation DRE data for 15 chemicals with various parameters such as the autoignition temperature (AIT) and the heat of combustion.[24] The laboratory DRE data were expressed as T99.99/2, the temperature necessary to achieve a 99.99% DRE for a particular chemical at a 2-sec residence time. The T99.99/2 was calculated based on laboratory-derived, first-order kinetic constants, assuming Arrhenius temperature dependence for the reaction rate constant. They found that the AIT had a correlation coefficient (R) with the T99.99/2 of 0.94, and that heat of combustion had an R of 0.39 when correlated with T99.99/2.

These correlation coefficients indicate that the AIT is a far better indicator of TOS than heat of combustion in flameless combustion. However, because of the lack of T99.99/2 data under flame combustion conditions, the validity of AIT as a TOS indicator for flame combustion incineration systems is still in question.

Lee et al. clarified the confusion surrounding the EPA's use of heat of combustion as a measure of compound incinerability be defining two new terms, compound destructibility and waste incinerability.[13] According to Lee the heat of combustion is one of the indicators of the overall destruction of a waste blend (waste incinerability) in a particular incinerator and the AIT is a good indicator for ranking the TOS of individual compounds (compound destructibility)[13].

As we shall see in the next section, waste related factors other than TOS must also be considered when designing a combustion system to achieve 99.99% DRE.

DRE DATA

Because the regulatory DRE concept is relatively new and DRE stack gas analytical techniques are still in the developmental stage, very little full-scale DRE data have been published. Two trial burns recently conducted are informative and shed some light on the attainability of the 99.99% DRE. A summary of these trial burns follows.

METROPOLITAN SEWER DISTRICT (MSD) TRIAL BURN

The MSD incinerator in Cincinnati was the site of perhaps the first full scale RCRA trial burns in July and September 1981. The trial burns were funded by the EPA to verify the EPA's RCRA HW trial burn protocols and recommended procedures.[25]

The MSD incinerator is a rotary kiln followed by a fired secondary combustion chamber. Liquid wastes containing eleven different POHCs were fed to the rotary kiln and the secondary combustion chamber. The following eleven POHCs were used:

chloroform	trichloroethane
carbon tetrachloride	tetrachlorethane
tetrachloroethylene	bromodichloromethane
hexachloroethane	pentachloroethane
hexachlorobenzene	dichlorobenzene
hexachlorocyclopentadiene	

During the nine tests, secondary combustion chamber temperatures varied from 899 °C (1650 °F) to 1316 °C (2400 °F) and secondary combustion chamber residence times varied from 1.5 to 3.7 sec. The rotary kiln temperature and residence times are not included in these values. Almost all the DRE values obtained were 99.99% or greater. There was only one test where the 99.99% was clearly not achieved. During this test a DRE of 99.99% was obtained for bromodichloromethane at 1650 °F and 2.3 sec. At 2400 °F and 1.5 sec, however, the bromodichloromethane had a DRE of 99.995%.

The wastes that were incinerated contained water at concentrations ranging from 4.65 to 65.3%. As can be seen from the list of wastes, all the POHCs were low aqueous solubility, high volatility POHCs which would tend to volatize rapidly from aqueous droplets formed during the waste atomization. As discussed previously, it is important to get the POHC into the vapor phase as quickly as possible to maximize the temperature, residence time, and oxygen contact of the POHC. The next trial burn clearly illustrates this concept.

EASTMAN KODAK (EK) TRIAL BURN

The Eastman Kodak Company (EK) in Rochester, New York, as part of an internal research project, conducted extensive DRE testing in their 90 MM Btuh rotary kiln (RK) secondary combustion chamber (SCC) incinerator.[26] Wastes were fed to the RK or the SCC at temperatures ranging from 1300 to 1850 °F and residence times from 1.2 to 3.3 sec. Two waste mixtures were used, one with 40% water and 6,000 Btu/lb and another with 18% water and 10,000 Btu/lb. A total of nine POHCs were studied.

The study attempted to determine the impact on DRE of the following parameters:

Bulk gas temperature   Heat of combustion of POHCs
Residence time   Autoignition temperature
Heat content of waste mixture Aqueous solubility of POHCs

EK concluded that several factors affected the POHC DREs. Kiln fed POHCs that had longer residence times than SCC fed POHCs had higher DREs. Waste mixtures with higher heat contents had higher DREs. Increasing the RK temperature from 1300 to 1600 °F caused higher DREs, but a further increase to 1850 °F caused no significant change in DREs. Finally, EK found that water-soluble compounds in aqueous wastes had lower DREs than water-insoluble compounds. The nine compounds had the following DREs and aqueous solubilities:

Compound	Solubility (wt%)	DRE	AIT (°F)
Heptane	0.005	>99.995	420
Hexane	0.010	>99.995	440
Chloroform	0.80	>99.995	>1200
Methylene chloride	2.0	>99.995	1320
Acetone	∞	99.99	870
Isopropanol	∞	99.99	750
Ethanol	∞	99.88	690
Methanol	∞	99.83	730
1,4-Diethylene dioxide	∞	99.60	—

As can be seen from these data, the DREs are not related to the TOS (AIT), but are apparently related to the compounds' aqueous solubility and Henry's Law constants.

Since highly water soluble organics typically have low volatility in aqueous solutions and since atomized water droplets tend to be larger and take longer to evaporate than organic droplets, the highly soluble organic present in the aqueous droplets apparently takes longer to volatilize and be subjected to the combustion conditions. As previously discussed, high solubility organics in aqueous wastes will require careful design attention to atomization, droplet size, nozzle type, droplet evaporation rate, extended residence time considerations, and possibly waste preheating outside the burner assembly.

SUMMARY

The new RCRA HW incineration regulations have ushered in a new era which will result in a better understanding of the significant combustion design parameters relative to HW destruction. New pilot incineration systems such as the EPA's pilot-liquid incinerator at the Combustion Research Facility in Jefferson, Arkansas, have been designed and will be used to study these important design parameters. This chapter has described various combustion system design considerations that IT Enviroscience considers important in the design of a new HW incinerator or the up-grading of an existing incinerator to achieve 99.99% DREs. These design areas included discussions of nozzles, burner assemblies, flame patterns, atomization of organic and aqueous wastes, large droplets, mixing based on Reynolds numbers, heat loss, refractory, waste characteristics, and thermal oxidation stability of POHCs. A summary of two significant RCRA DRE trial burns is also included.

REFERENCES

1. *Hazardous Waste Report,* May 2, 1983 (4) NO 19, page 1.

2. U.S. EPA Office of Solid Waste. Hazardous Waste, Proposed Guidelines and Regulations and Proposal on Identification and Listing, *Federal Register* 43(243, Part IV):58946-59028, December 18, 1978.
3. U.S. EPA Hazardous Waste Management System: Identification of Listing of Hazardous Waste, *Federal Register* 45(98, Part III):33084-33137, May 19, 1980.
4. U.S. EPA, Incinerator Standards for Owners and Operators of Hazardous Waste Management Facilities; Interim Final Rule and Proposed Rule, *Federal Register* 46(15, Part IV): 7666-7690. January 23, 1981.
5. U.S. EPA, Standards Applicable to Owners and Operators of Hazardous Waste Treatment Facilities, *Federal Register* 47 (122, Part V): 27516-27535, June 24, 1982.
6. Ibid, 27532.
7. U.S. EPA Hazardous Waste Management Sytem: Identification and Listing of Hazardous Waste, *Federal Register* 45(98, Part III):33121. May 19, 1980.
8. U.S. EPA Hazardous Waste Management System; Corrections, *Federal Register* 46(97):27473-27480. May 20, 1981.
9. Novak, R. G., Troxler, W. L., and Dehnke, T. H. 1982. The Technology and Economics of Energy Recovery from Hazardous Waste Incineration. *Presented at the American Institute of Chemical Engineers Summer National Meeting,* Cleveland, Ohio, August 29 through September 1.
10. Kiang, Y. H. and Metry, A. A. 1982. *Hazardous Waste Processing Technology,* Ann Arbor Science, Ann Arbor, Michigan, 58.
11. Reed, R. D. 1981. *Furnace Operations,* 3rd ed. Gulf Publishing Co., Houston, Texas, 122.
12. Miller, D. L., Cundy, V. A., and Matula, R. A. 1983. Incinerability Characteristics of Selected Chlorinated Hydrocarbons, *9th Annual Research Symposium Land Disposal, Incineration and Treatment of Hazardous Waste,* U.S. EPA, May 2-4.
13. Lee, K. C., Morgan, N., Jansen, J. L., and Whipple, G. M. 1982. Revised Model for the Prediction of Time-Temperature Requirements for Thermal Destruction of Dilute Organic Vapors and its Usage for Predicting Compound Destructability, *75th Annual Meeting of the Air Pollution Control Association,* New Orleans, June 20-25.
14. Duval, D. S., Rubey, W. A., and Mescher, J. A. 1980. High Temperature Decomposition of Organic Hazardous Waste, *Proceedings of the Sixth Annual Research Symposium: Treatment and Disposal of Hazardous Waste;* U.S. EPA, MERL, EPA 600/9-980-010; March.
15. Kiang, Y. H. 1977. Total Hazardous Waste Disposal through Combustion, *Industrial Heating* 14 (12), 9.
16. *North American Combustion Handbook,* 1965. North American Mfg. Co., Cleveland, OH, 177.

17. Rubey, W. A. 1980. *Design Considerations for a Thermal Decomposition Analytical System,* EPA-600/2-80-098, August.
18. Duvall, D. S. and Rubey, W. A. 1976. *Laboratory Evaluation of High-Temperature Destruction of Kepone and Related Pesticides,* EPA-600/2-76-299, December.
19. Duval, D. S. and Rubey, W. A. 1977. *Laboratory Evaluation of High-Temperature Destruction of Kepone and Related Pesticides,* EPA-600/2-76-299, December.
20. Lee, K. C., Jahnes, H. J., and Macauley, D. C. 1978. Thermal Oxidation Kinetics of Selected Organic Compounds, *Proceedings of 71st Annual Meeting of the Air Pollution Control Association,* Houston, TX, June.
21. Lee, K. C., Hansen, J. L. and Macauley, D. C. 1979. Predictive Model of the Time-Temperature Requirements for Thermal Destruction of Dilute Organic Vapors, *Proceedings of 72nd Annual Meeting of Air Pollution Control Association,* Cincinnati, OH, June.
22. *Guidance Manual for Hazardous Waste Incinerator Permits,* 1983. U.S. EPA Office of Solid Waste, Washington, D.C., March draft.
23. Tsang, W., and Shaub, W. 1981. Chemical Processes in the Incineration of Hazardous Materials, *paper presented at the American Chemical Symposium on Detoxification of Hazardous Wastes,* New York, NY, August.
24. Cudahy, J. J., Troxler, W. L. and Sroka, L. 1981. *Incineration Characteristics of RCRA Listed Wastes,* U.S. EPA Contract No. 68-03-2568, Work Directive T-7021., Industrial Environmental Research Laboratory, Cincinnati, OH, July.
25. Gorman, P. and Anath, K. P. 1982. *Trial Burn Protocol Verification, Performance Evaluation, and Environmental Assessment of the Cincinnati MSD Hazardous Waste Incinerator,* draft final report, Midwest Research Institute, March.
26. Austin, D. S., Bastian, R. E. and Wood, R. W. 1982. Factors Affecting Performance in a 90 Million Btu/hr Chemical Waste Incinerator: Preliminary Findings, *Technical Conference Middle Atlantic States Section APCA,* April 29-30, 29-36.

Hazardous and Toxic Wastes: Technology, Management and Health Effects. Edited by S.K. Majumdar and E. Willard Miller. © 1984, The Pennsylvania Academy of Science.

Chapter Four

Destruction of Toxic Chemical Wastes By Incineration At Sea

Donald G. Ackerman Jr., Ph.D.[1] and George VanderVelde, Ph.D.[2]
[1]Project Scientist
TRW Inc., Energy and Environmental Division
Redondo Beach, CA 90278
[2]Technical Director
Chemical Waste Management, Inc.
Oak Brook, IL 60521

The United States and other industrialized nations are confronted with the serious and major problem of disposing of ever-increasing quantities of hazardous wastes in an environmentally acceptable manner. Incineration is one of the most environmentally acceptable methods of disposing of hazardous wastes currently available. It is a well developed and well understood technology which provides an optimum solution to the disposal of combustible wastes because it reduces large volumes of such wastes to non-toxic gases and residual amounts of liquids and solids. Incineration is an ultimate disposal technique and is, therefore, particularly useful for disposing of hazardous wastes because it accomplishes essentially total destruction. The purpose of this chapter is to describe incineration as practiced at sea in contrast to incineration practiced on land.

Incineration has long been a practice on land, and the U.S. Environmental Protection Agency (EPA) has sponsored testing of a number of thermal destruction technologies (1,2). EPA considers that emissions from the proper incineration of hazardous wastes pose a minimal long term ecological burden (3).

Incineration at sea onboard appropriately equipped vessels is a relatively recent development. Oceanic incineration of hazardous wastes was first performed in European waters, and industrial organochlorine wastes have been routinely incinerated at sea since 1969. Six vessels have been used to incinerate such wastes: MATTHIAS I, MATTHIAS II, MATTHIAS III, VESTA, VULCANUS, and VULCANUS II. (The MATTHIAS I and MATTHIAS III are no longer operational.)

Oceanic incineration is not yet routine practice in American waters, but it will be soon. In 1980, Chemical Waste Management, Inc., Oak Brook, IL, purchased Ocean Combustion Service BV, Rotterdam, The Netherlands, which operates the VULCANUS and VULCANUS II. Ocean Combustion Service (OCS) intends to operate the VULCANUS II in U.S. waters after international certification is completed and American permits are issued by EPA.

A second American organization is entering the American market. Tacoma Boatbuilding Co., Inc. (TBC), Tacoma, WA, has received an order for two incineration ships from Apollo Co., LP, Lake Success, NY. Also, TBC has acquired At-Sea Incineration, Inc., Greenwich, CT, which will operate the vessels. At-Sea Incineration, Inc., plans to have the two ships and a land-based processing facility operating during the second quarter of 1984 (personal communication, D. G. Ackerman, TRW Inc., with M. Melhoff, TBC, 27 May 1983).

Ocean incineration possesses several advantages over land-based incineration (4):

- Construction or acquisition of fixed sites on land is not required.
- Incineration is performed in remote, uninhabited areas under predictable atmospheric conditions.
- Scrubbing of effluent gas is unnecessary because a major product of combustion, HCl, is dispersed over broad areas of ocean which readily neutralizes the acid.
- Generation and subsequent disposal of solid wastes (i.e., scrubber solids) and of liquids with high dissolved solids content is avoided.
- Catastrophic failure with loss of cargo, while extremely unlikely, presents a substantially lower risk than a similar occurrence on land.

The major difference between oceanic and land-based incinerators is that there is no scrubber on an incineration vessel. Combustion products go directly from the incinerator into the marine environment. Combustion products from a land-based incinerator go through a scrubber or other pollution control device and then into the environment after treatment primarily to reduce the corrosive effects of the emitted acid. When the wastes are organochlorines, the major products of combustion are CO_2, H_2O, and HCl. Not having to treat (neutralize) the HCl leads to significant reductions in capital and operating expoense (5). Other species (e.g., CO, H_2, and Cl_2) will be present at low levels (e.g., 10^{-5} to 10^{-8} mole fraction). Phosphorus, sulfur, and nitrogen in the waste will appear as P_2O_5, SO_x and NO_x. Trace metals that may be present will appear as particulate matter. There may be trace concentrations of unburned constituents of the waste and organic compounds synthesized during the process.

REGULATORY BACKGROUND

Oceanic incineration is regulated both internationally and nationally. Internationally, regulations (6, Appendix C) have been promulgated under the authority of the Convention on the Prevention of Marine Pollution by Dumping of Wastes or Other Matter (London Dumping Convention, LDC). The United States is a signatory to the LDC, and these regulations have force of law in the U.S. Nationally, the Marine Protection, Research, and Sanctuaries Act of 1976, as amended, gave EPA the lead Federal agency authority for regulating ocean dumping. EPA regulates oceanic incineration as form of dumping under promulgated regulations governing ocean dumping (7). These regulations adopt the international standards as minimum national standards until final regulations are promulgated. The regulations are performance-based, and several of the more important performance requirements are:

- A minimum destruction efficiency of 99.9 percent
- A minimum combustion efficiency of 99.95 ± 0.05 percent
- A minimum flame temperature requirement of 1250 °C

TYPICAL WASTES

Currently, only pumpable and combustible liquid organic wastes are suited to oceanic incineration, as existing vessels are not designed to burn solids (e.g., as in a rotary kiln). Organic solids may be suitable if they are soluble or can be dispersed in organic liquids. Wastes may be "wet", emulsions in water, or even contain layers of water. Wastes burned at sea have been almost exclusively organochlorine because they are more costly to dispose of on land (5). Other classes of compounds such as hydrocarbons or oxygenated organics (e.g., alcohols, acids, or ketones) are usually disposed of on land. However, there is no reason to exclude wastes containing these compounds from incineration at sea if they are combustible and do not contain prohibited matter.

INCINERATION VESSELS

Extensive information is available on the VULCANUS (8-22) and VULCANUS II (23,24). Only limited information is available on the MATTHIAS I, MATTHIAS II, MATTHIAS III, and VESTA (20). The VULCANUS and VULCANUS II will be described at some length because the VULCANUS has been operated in U.S. waters and may be again, and the VULCANUS II will be operated in U.S. waters once permits have been issued. Table 1 presents characteristics of these vessels (16,20,24).

THE VULCANUS

The VULCANUS was converted in 1972 from a cargo vessel to a chemical

TABLE 1

Characteristics of incineration ships to date

Charactertistic	Matthias I	Matthias II	Matthias III	Vesta	Vulcanus	Vulcanus II
			Name of Vessel			
Length, m	40	72.8	176.5	72	102	94
Breadth (max.) m	8	10.8	21.9	11	14.4	16
Draft (max.) m	—	5.2	9.2	4.3	7.4	6.2
Capacity, metric tons	780	1,200	15,000 (liquids)	1,300	4,260	4,120
			1,500 (solids)			
Type of incinerator	Cup	Cup	Cup	Funnel	Funnel	Funnel
Number of incinerators	1	1	1	1	2	3
Height, m	6	8.8	15.8	10.7	10.5	11
Combustion chamber I.D., m	5	8.8	15.4	6	4.8	5.3
Number of burners	6	8	20	3	6	9
Feedrate (max.) mt/hr	4.5	16	50	10.5	25	39
Feedrate of air, m³/hr	45,000	100,000	332,000	100,000	180,000	270,000
Residence time, sec	—	—	3-4	~1	~1	~1

tanker fitted with two large incinerators at the stern. The vessel meets all applicable requirements of the International Maritime Organization (IMO) for transportation of hazardous cargo by tanker. She has operated world-wide although she is normally operated in European waters. The VULCANUS is a double-hull, double-bottom vessel. This construction provides double containment of the cargo. Waste is carried in 13 cargo tanks ranging in volume from 117 to 574 m³. Overall capacity is 3,277 m³ (866,000 gal.). Tanks are filled through a manifold on deck using a dockside loading pump. The arrangement of cargo piping controls prohibits discharge overboard of the cargo intended to be incerated during normal handling operations.

Waste is burned in two identical incinerators, each consisting of three sections (combustion chamber, converging section, and stack) through which the combusting gases pass sequentially. Three rotary cup burners are located on the periphery of each incinerator 1.7 m above the base. The burners are directed tangentially to the vertical axes of the incinerators, thereby imparting swirl to and added mixing of the burning gases.

A three-way valve is used on each burner to provide any one of the three conditions: waste feed, fuel oil feed, or shut-off. Although waste and fuel cannot be fed simultaneously into a burner, alternate burners can be operated with fuel and waste to achieve higher or lower combustion temperatures.

The maximum feed rate of waste is 12.5 metric tons (mt)/hr per incinerator, and feed rate is varied inversely with the heat content of the waste in order to maintain an inner wall temperature of about 1200 °C. Feed rate has been measured by sounding the depth of waste in each tank. There is also a closed gauging system and instruments for measuring waste feed rate.

Air for the combustion process is supplied by six large fixed-speed fans, one per burner, each having a rated maximum capacity of 30,000 m³ per hour. Feed rate of air to the incinerators is controlled by an adjustable vane in each air feed line. Typical air feed rates range from 65,000 to 75,000 m³ per hour per incinerator at ambient conditions.

There are three thermocouples (Pt-Pt/10% Rh) in the wall of each incinerator at the level of the burners. Two of these serve the automatic waste shutoff system: one is a primary, and the other is a spare. The third thermocouple serves the temperature readout system. A fourth, "outlet", thermocouple is located in the middle of the converging section between the combustion chamber and the stack and is connected to the temperature readout system. All four thermocouples are mounted flush with the inner wall. Flame temperatures, measured with an optical pyrometer, have been shown to be 250 °C to 300 °C higher than those indicated by the thermocouples (10).

The VULCANUS is equipped, as required by regulations, with an automatic waste feed shut-off system. This system will stop the flow of waste (or fuel oil) to affected burner(s) and shut off power to the metering pump(s) if any of the following problems occur: 1) overload of a metering pump, 2) overload of a combustion air fan, 3) overload of a Gorator (devices in the cargo lines that grinds solids in the waste), 4) failure of feed of primary combustion air, 5) loss of flame, and 6) reduction of wall temperature below the set-point. If the automatic waste shut-off system has functioned, (i.e., shut off the feed of waste or oil), a series of routine manual restart procedures is performed after the fault has been corrected.

Wall temperature is the major regulated control parameter. If wall temperatures fall below a pre-set minimum value, a controller on each incinerator closes valves in the feed pipes and cuts power to the metering pumps. The temperature-set dials on the controllers are mounted in cases which provide visibility but which can be (and are) sealed by cognizant government officials.) The dials are typically set at 1250 °C, which is 50 °C above the value of 1200 °C given in the International Guidelines (6). Waste cannot be fed into the incinerators until the wall temperature reaches that set on the controllers.

There is a system for continuously monitoring concentrations of CO_2, O_2, and CO in the effluent gas. This sytem, designed and installed by Siemens, AG, Federal Republic of Germany, consists of the following components: water-filled gas bubblers to remove acidic gases and particulate matter, refrigeration-based gas conditioners to cool the gas and remove moisture, filters to complete the removal of particulate matter, monitors, pumps, electronics, and calibration gases. The monitors are: 1) CO_2, Ultramat 1, Model M52990, non-dispersive infrared, 0-20% (v/v), 2) CO, Ultramat 2, Model M52014, non-dispersive infrared, 0-500 ppm (v/v), and 3) O_2, Oxymat 2, Model M52010, paramagnetism, 0-15% (v/v). The gas stream from each incinerator is presented alternately to the combustion gas monitoring system every 15 minutes. Readouts from the monitors are located in the "black box" (see next paragraph) and in the control panel in the incinerator room. The system functions largely unattended except for daily calibrations, daily refilling of the gas bubblers, and occasional replacement of the filters.

The "black box" mentioned above is a cabinet mounted on the bridge inside of which are gauges displaying the following information: wall and outlet temperatures in each incinerator; concentrations of O_2, CO_2, and CO in the stack gas and in which incinerator the measurements are being made; day, date, and time; settings of the temperature-set dials; and status (on/off) of the metering pumps. An automatic 35-mm camera is placed within the "black box" and photographs the displays every 11 minutes. The "black box" and/or the camera can be sealed, allowing cognizant government officials to obtain the film for official records at the end of each voyage in order to determine whether operations were in accordance with conditions of the permit for the burn.

THE VULCANUS II

The VULCANUS II is a new vessel, commissioned in November of 1982 and equipped with state-of-the-art technology in navigation, propulsion, and incineration controls. She also is a double-hull, double-bottom vessel and meets all applicable requirements of the IMO relating to the transport of dangerous cargo by tanker. She has an overall length of 94 m, a beam of 16 m, and a maximum draft of 6.2 m. She can operate world-wide. Waste is carried in eight tanks, and her total capacity is 3,170 m³ (837,500 gal.). Table 1 presents characteristics of the VULCANUS II.

The VULCANUS II has three incinerators mounted at the stern. They are virtually identical in size and construction to those on the VULCANUS. There are three rotary cup burners in each incinerator 2.06 m above the base. The maximum feed rate is 13 mt/hr per incinerator, and feed rate is varied inversely with the heat content of the waste in order to maintain stable wall temperatures. Each burner is fed air by a fan capable of supplying up to 30,000 m³/hr. The feed rate of waste is measured by flow meters on each incinerator and a radar-based closing gauging system in the cargo tanks. The feed rate of air to each burner is also measured.

Temperatures are measured with thermocouples (Pt-Pt/10% Rh). There are three thermocouples per incinerator in the wall 4.48 m above the base. There is a fourth thermocouple ("outlet") in each incinerator that is 1.3 m below the exit plane of the stack. There is a combustion gas monitoring system identical to that on the VULCANUS.

The VULCANUS II has a Computerized Data Record System (CDRS) which consists of a micro-computer, program, floppy disc, printer and a land-based reader-printer for the floppy dics. The software is contained in read-only memory and can only be altered by the vendor of the computer. The CDRS acquires data from the incineration process (e.g., temperatures; flow rates of air and waste; concentrations of CO, CO_2, and O_2 and in which incinerator the measurements were made) and other data (e.g., date, time, and position). It calculates combustion efficiency, compares values of certain data with pre-set alarm levels (e.g., for temperatures of the inner wall, concentration of CO, and combustion efficiency), and activates alarms. The CDRS prints all acquired data at 15 minute intervals and stores the same data on the floppy disc. The status of all alarms is also stored and printed. The floppy disc is the official record (corresponding to the film taken in the "black box" on the VULCANUS) and is transmitted to the cognizant regulatory authority after each voyage. The reader-printer is also transmitted to the regulatory authority.

OCEANIC INCINERATION

The First Shell-Waste Burns

The first officially sanctioned oceanic incineration in the U.S. was performed on the VULCANUS between October 1974 and January 1975 (8). The action occurred in an EPA-designated site in the Gulf of Mexico 315 km southeast of Galveston, TX, and 209 km south of Sabine Pass, TX. The site, now the Gulf Ocean Incineration Site, was beyond the continental shelf in water ranging from 914 to 1829 m deep, outside all major shipping lanes, and beyond all commercial fishing and shrimping depths. Figure 1 shows the geographical relationship of the site to various cities in the Gulf of Mexico.

The wastes burned were from Shell Chemical Company's Deer Park, TX, manufacturing complex. It was a mixture of organochlorine compounds consisting chiefly of trichloropropane, trichloroethane, and dichloroethane resulting from the manufacture of a variety of chemicals. Two loads (about 8,400 mt) were burned under a research permit, and EPA coordinated an extensive monitoring, sampling, and surveillance effort during those burns. Stack gas was sampled for HC1 and unburned waste with a train consisting of several impingers to trap HC1, Cl_2, and organic compounds. A bypass in the train enabled the experimenters to direct the sampled gas to a total hydrocarbon analyzer (flame ionization detector) for measurement of low molecular weight hydrocarbons. Concentrations of CO and O_2 also were measured.

Destruction efficiencies averaged 99.95 percent and ranged from 99.92 to 99.98 percent. Wastes were fed at an average rate of 21.2 mt/hr during the first burn and 24.5 mt/hr during the second. Flame temperatures, measured with an optical pyrometer, averaged 1458 °C, while wall temperatures averaged 1255 °C. Concentrations of CO and O_2 averaged 43 ppm and 10.8 percent.

Other testing was performed to assess more fully the environmental suitability of oceanic incineration. Marine sampling and analyses were made by two vessels: the M/V ORCA and the R/V OREGON II. Short-term marine effects were assessed by a battery of tests including determination of pH, chlorinity, organochlorine compounds, and trace metals. Long term effects were assessed by phytoplankton counts, zooplankton counts, and determination of chlorophyll-a and adenosine triphosphate (ATP). The spatial characteristics of the plume were assessed by the R/V OREGON II and by an EPA aircraft. The

OREGON II made sea-level plots of the plume, while the aircraft made cross-wind and axial passes through the plume.

Results of the marine monitoring were (8): Samples of water taken in the vicinity of plume-touchdown showed no significant differences in acidity, chlorinity, or copper (the most abundant heavy metal in the waste) from samples taken in control areas. Organochlorine compounds were not detected (detection limit was 0.5 ppb) in samples of sea water taken in the vicinity of plume-touchdown. Analyses of samples of phytoplankton and zooplankton for chlorophyll-a and adenosine triphosphate showed no obvious or subtle adverse effects. Fish, exposed at the point of plume-touchdown, showed an increase in activity of the liver enzyme system (P-450) involved in metabolizing chemicals foreign to the organism, which indicated that the fish had been exposed to foreign chemicals. However, when the exposed fish were left in clean water for a few days, the level of enzyme activity returned to that of control fish.

The Second Shell-Waste Burns

The second oceanic incineration in U.S. waters occurred during March and April of 1977. Four shiploads of organochlorine wastes from Shell Chemical Company, Deer Park, TX, were burned by the VULCANUS. The first burn (about 4,100 mt of waste) was tested by personnel from TRW Inc. under contract to EPA (10).

Six tests were performed during incineration of waste and one during incineration of fuel oil. Sampling was performed with a high volume EPA Source Assessment Sampling System (SASS train) at a rate of 0.14 cubic meters per minute (4 cfm). Sample acquisition was through a remotely actuated, water-cooled stainless steel proble capable of traversing the starboard incinerator stack diameter of 3.4 meters. Time-composited samples of waste were taken during each test. The effluent gas was simultaneously monitored for total hydrocarbons, CO, CO_2, NO, NO_2, and O_2 concentrations. These parameters were measured in real-time to monitor the overall combustion efficiency of the process.

Combustion efficiency averaged 99.97 percent. Other average results were: feed rate, 11 mt/hr per incinerator; wall temperature, 1261 °C; flame temperature, 1535 °C; concentration of O_2, 10.1 percent; concentration of CO_2, 9.2 percent; concentration of CO, 25 ppm; excess air, 94 percent; and residence time, 0.9 sec (10). The overall destruction efficiency was greater than 99.99 percent.

Herbicide Orange Burns

During the period July - September of 1977, the VULCANUS burned three loads of Herbicide Orange, a total of 10,400 metric tons or 2.31 million gallons. Herbicide Orange consisted of an approximately 50-50 mixture by volume of the n-butyl esters of 2,4-dichlorophenoxyacetic acid (2,4-D) and 2,4,5-trichlorophenoxyacetic acid (2,4,5-T). Certain lots of the herbicide con-

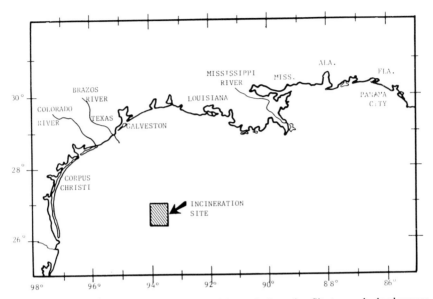

FIGURE 1. Geographical relationship of Gulf Ocean Incineration Site to nearby land masses.

tained the contaminant 2,3,7,8-tetrachlorodibenzo-p-dioxin (TCDD), the concentration of which averaged 1.9 ppm and ranged from < 1 to 47 ppm (11). The operation took place in an EPA-designated incineration site in the Pacific Ocean about 120 miles west of Johnston Atoll. EPA contracted with TRW Inc. to perform testing during the burns.

The three burns were extensively tested and monitored in accordance with test plans approved by EPA and the U.S. Air Force. Stack sampling operations utilized a benzene-filled impinger train and a modified EPA-Method 5 (25) train which incorporated an adsorbent trap to collect organic compounds. The impinger-train was the primary train for acquiring samples for analysis for TCDD, while the modified Method 5 train was used to acquire samples to be analyzed for organic species potentially present in the effluent gas. It also served as a back up to the impinger train. Stack samples were acquired by a remotely activated, water-cooled, stainless steel probe capable of traversing the starboard stack diameter of 3.4 meters. A total of 52 samples of stack gas were taken during the three burns. Time-composited samples of Herbicide Orange as fed to the incinerators also were taken.

During stack sampling operations, incineration effluent products were simultaneously monitored for total hydrocarbons, carbon monoxide, carbon dioxide, and oxygen. Concentrations of these species were measured in real-time to monitor the overall combustion efficiency of the incineration process. Instrumentation for these measurements was housed in a modified shipping container lashed to the ship's deck.

A significant result was derived from on-line monitoring data taken during a

TABLE 2

Destruction efficiencies during burns of Herbicide Orange.

Compound	Concentration in Waste, %	Destruction Efficiency, %
2,4-D, n-butyl ester	50	>99.999
2,4,5-T, n-butyl ester	50	>99.999
TCDD	1.9×10^{-4}	>99.93

traverse across the starboard stack. It was found that wall-effects on the composition of the effluent gas were nonexistent at distances greater than 10 cm from the inside wall. Therefore, sampling of effluent gas and monitoring of combustion efficiency could be determining using a fixed-position probe.

Combustion efficiency averaged 99.99 percent. Averages of other process-related parameters were: feed rate 7.3 mt/hr per incinerator, wall temperature, 1273 °C; concentration of O_2, 8.9 percent; concentration of CO_2, 10.3 percent; concentration of CO, 10 ppm; excess air, 74 percent; and residence time, 1.0 sec. Destruction efficiencies of the major components of Herbicide Orange and TCDD were calculated and are given in Table 2 (12).

Table 2. Destruction efficiencies during burns of Herbicide Orange.

Second PCB Research Burn

During August of 1982, the VULCANUS burned a shipload (3,507 metric tons) of wastes contained polychlorinated biphenyls (PCBs). Testing of emissions from the vessel was performed by personnel from TRW Inc., and EPA (16). Monitoring of ambient air was performed by personnel from JRB Associates (17), and biological monitoring was performed by personnel from TerEco Corp. (18). All work was performed under contract to EPA. Incineration operations occured in the Gulf Ocean Incineration Site.

Ten tests were performed. Time-composited samples of waste were taken during each test. Combustion gases were sampled with a train mandated for use in testing land-based incinerators for emissions of PCBs. It was essentially an EPA Method 5 train, modified to incorporate an adsorbent-filled tube to trap PCBs and other specified organic compounds (chlorobenzenes, (CBs); tetrachlorodibenzofurans (TCDFs); and tetrachlorodibenzo-p-dioxins, (TCDDs)) emitted in the stack gas. Pertinent process-related data, such as wall temperatures and concentrations of CO, CO_2, and O_2 in the combustion gases, were acquired from the Chief Engineer.

Combustion efficiency averaged 99.99 percent. Averages of other process-related incineration parameters were: waste feed rate, 6.22 mt/hr per incinerator, wall temperature, 1303 °C; concentration of O_2, 10.1 percent; concentration of CO_2, 9.1 percent; concentration of CO, 8 ppm; excess air, 95 percent; and residence time, 1.3 sec (16).

Samples from six of the ten tests were analyzed by high resolution gas chromatography - high resolution mass spectrometry. No PCB, CB TCDF, or

TCDD was detected in any of the stack gas samples. Average destruction efficiencies for these compounds are given in Table 3. It was not possible to calculate destruction efficiencies for TCDDs because these compounds were not detected in the wastes (16).

Sampling of ambient air was performed before (baseline) and during the incineration. This sampling was conducted using the research vessel ANTELOPE. Real-time monitoring of HC1 ensured that the plume from the VULCANUS was actually being sampled. Analyses of the baseline samples of ambient air showed no detectable PCBs or other organochlorine compounds (detection limit was 0.3 ng/m³) and only traces of non-chlorinated compounds. Analyses of samples of air from the plume showed no detectable PCBs (detection limit was 0.3 ng/m³) or other organochlorine compounds. Non-chlorinated organic compounds were detected in amounts higher than during the baseline tests. It was theorized that these compounds were generated either by the propulsion engines or by the incineration process (17).

Sampling of surface water and exposure of selected marine organisms were also conducted before and during the incineration (18). Organisms (a species of shrimp and two species of fin fish) were placed in Pelagic Biotal Ocean Monitors (P-BOMs). During incineration operations, P-BOMs were placed in the incineration site so they would be exposed to the plume and in control areas (no exposure) outside the site.

TABLE 3

Destruction efficiencies during burn of PCBs in August 1982.

Compound	Concentration in Waste, % (w/w)	Destruction Efficiency, %
PCBs	27.49	>99.99989
CBs	6.87	>99.99993
TCDFs	4.8×10^{-8}	>99.96

Analyses of surface waters showed that acidity and salinity were the same in control and plume-impact areas. PCBs (Aroclor® 1254) were not detected in baseline and test samples of surface water (detection limit was 0.2 $\mu g/1$ or 0.2 ppb). PCBs were not detected (detection limit was 2 $\mu g/m^2$) in the collectors placed on the P-BOMs. PCBs were not detected in samples of neuston (detection limit was 1 ng Aroclor® 1254 per square meter of surface water). PCBs were not detected (detection limits were 0.025 ppm) in one of the exposed species of fish. PCBs were detected in the other species of fish, but it was found that all fish of that species had been contaminated with PCBs before the testing started. The three species (fish and shrimp) were analyzed for changes in metabolic activity resulting from exposure to the plume, but no effects were found. The investigators concluded that no environmental effects of the incineration were detected.

Test of the VULCANUS II

The most recent test of the VULCANUS II occurred in February 1983 (24). Chemical Waste Management, Inc., contracted with TRW Inc. to measure the emissions of selected volatile organochlorine compounds while a waste containing these compounds was being incinerated and to determine if destruction efficiencies of the incineration process for these compounds were in compliance with current international regulations for oceanic incineration. The testing took place in the North Sea Incineration Site, an area of ocean off the coast of The Netherlands in the vicinity of 52 °16 ′ N longitude and 03 °45 ′ E latitude. The wastes consisted of 245 mt of "light ends" and 797 mt of "heavy ends" resulting from the manufacture of vinyl chloride.

Sampling and chemical analyses were conducted according to a test plan reviewed and approved by EPA. A Volatile Organic Sampling Train (VOST train) was used to acquire samples of the effluent gas. A stainless steel-jacketed, water-cooled probe with a quartz liner was designed and fabricated for conditioning and conducting samples of the stack gas into the VOST train. Four complete tests were performed. Time-compositioned samples of waste also were taken during each test.

Pertinent incineration process-related data were acquired from computer printouts maintained by the Chief Engineer and the navigational log maintained by the Master. These data included temperatures in the incinerators; feed rates of waste and air; and concentrations of carbon dioxide (CO_2), carbon monoxide (CO), and oxygen (O_2) in the stack gas; and ambient temperature, pressure and humidity.

Combustion efficiencies averaged 99.98 percent. Averages of other process-related parameters were: feed rate of light ends, 5.92 mt/hr (one incinerator); feed rate of heavy ends, 9.88 mt/hr (one incinerator); wall temperature, 1166 °C; outlet temperature, 1038 °C; concentration of O_2, 10.6 percent; concentration of CO_2, 9.6 percent; concentration of CO, 22 ppm; excess air, 103 percent; and residence time, 1.1 sec.

Five compounds present in the waste were selected as Principal Organic Hazardous Constituents (POHCs), and all samples of stack gas and waste were analyzed for them. These compounds, their percentage concentrations in the waste, and destruction efficiencies for them are listed in Table 4. It should be noted that CCl_4 and $CHCl_3$, are, respectively, the fourth and tenth hardest-to-destroy compounds in EPA's hierarchy of "incinerability".

In January of 1982, the VULCANUS II was tested by personnel from the Central Research Organization - TNO, Delft, The Netherlands (23). The waste was essentially the same as that burned during the tests in February of 1983 (24). TNO used a modified EPA Method 5 train which incorporated an adsorbent trap for organic compounds. Average results of the test were: combustion efficiency, 99.99; feed rate 12.6 mt/hr/incinerator; wall temperature, 1162 °C; outlet temperature, 941 °C; residence time, 1.2 sec. Destruction efficiencies are given in Table 5.

TABLE 4

Destruction efficiencies during a test of the VULCANUS II in Feburary 1983

Compound	Concentration in Waste, % (w/w)	Destruction Efficiency, %
1,1-dichloroethane	6.8	99.99994
1,2-dichloroethane	7.8	99.99996
1,1,2-trichloroethane	38.9	>99.999995
Chloroform	26.1	99.9998
Carbon tetrachloride	20.5	99.998

TABLE 5

Destruction efficiencies during a test of the VULCANUS II in January 1983

Compound	Concentration in Waste, % (w/w)	Destruction Efficiency, %
1,1,2-trichloroethane	37	>99.9995
Tetrachloroethene	0.78	>99.9934
1,1,1,2-tetrachloroethane	0.80	>99.9990
1,1,2,2-tetrachloroethane	1.68	>99.9983

CONCLUSIONS

Incineration is the most environmentally satisfactory means of disposing or organochlorine wastes currently available and converts these wastes into relatively simple and non-toxic products and produces essentially no environmentally unacceptable by-products. The environmental impacts of oceanic incineration have been extensively addressed (5,6,8,27-34). This research, including the extensive marine monitoring performed during the incineration of two shiploads of organochlorine wastes in 1974 (8) and a shipload of PCBs in August of 1982 (16-18) and results of all of the testing (8-18,20,22-24,34), has led EPA to conclude that oceanic incineration has not produced measureable adverse environmental impacts. EPA also concluded that oceanic incineration was compatible with the intent of the Marine Protection, Research, and Sanctuaries Act of 1972, as amended, and that oceanic incineration was a viable disposal method that should be considered with other methods, including land-based incineration (8).

REFERENCES

1. TRW Inc., Redondo Beach, CA, 1977. Destructing Chemical Wastes in Commercial Scale Incinerators: Final Report, Phase II. U.S. Environmental Protection Agency, Office of Solid Waste. Washington, DC. NTIS PB 278816/3WP.

2. U.S. Environmental Protection Agency. 1981. PCB Disposal by Thermal Destuction. U.S. EDA, Region 6, Dallas, TX. NTIS PB82-241860.
3. U.S. Environmental Protection Agency. 1980. *Federal Register,* 45:33,215 -U.S. Governmental Printing Office, Washington, DC.
4. Venezia, R. A. 1978. Incineration At Sea of Organochlorine Wastes. Paper presented at the Second Conference of the Environment. Paris, France, TRW Inc., Research Triangle Park, N.C.
5. Shih, C. C., J. E. Cotter, D. D. Dean, S. F. Paige, E. P. Pulaski, and C. F. Thorne. 1978. Comparative Cost Analysis and Environmental Assessment for Disposal of OIrganochlorine Wastes. U.S. Environmental Protection Agency. Washington, D.C. Report No. EPA-600/2-78-190.
6. U.S. Department of State and U.S. Environmental Protection Agency. 1979. Final Environmental Impact Statement for the Incineration of Wastes At Sea Under the 1972 Dumping Convention. Office of Environmental Affairs, U.S. Department of State, Washington, D.C.
7. U.S. Environmental Protection Agency. 1981. *Code of Federal Regulations.* Subchapter H - Navigable Waters. Parts 220-229. Pages 178-227. Revised as of 1 July. U.S. Government Printing Office. Washington, D.C.
8. Wastler, T. A., C. K. Offutt, C. K. Fitzsimmons, and P. E. DesRosiers. 1975. Disposal of Organochlorine Wastes by Incineration At-Sea. U.S. Environmental Protection Agency, Washington, D.C. Report No. EPA-430/9-75-014.
9. A. A. Van der Berg, P. M. Houpt, A. G. Keyzer, K. Koopmans, A. C. Lakwijk, P. Van Leeuwen, L. Pot, S. J. Spijk, E. Talman, and B. J. Van Woudt. 1974. On The Occurrence of Organic Chlorides in the Combustion Products of an EDC Tar Waste Burnt by the Incinerator Ship "VULCANUS". A Preliminary Investigation. Central Research Organization-TNO, Delft, The Netherlands. Report No. CL 74/95. October 1975.
10. Clausen, J. F., H. J. Fisher, R. J. Johnson, E. L. Moon, C. C. Shih, R. F. Tobias, and C. A. Zee. 1977. At-Sea Incineration of Organochlorine Wastes Onboard the M/T VULCANUS. U.S. Environmental Protection Agency. Research Triangle Park, NC. Report No. EPA-600/2-77-196.
11. Ackerman, D. G., H. J. Fisher, R. J. Johnson, R. F. Maddalone, B. J. Matthews, E. L. Moon, K. H. Scheyer, C. C. Shih and R. F. Tobias. 1978. At-Sea Incineration of Herbicide Orange Onboard the M/T VULCANUS. U.S. Environmental Protection Agency. Research Triangle Park, N.C. Report No. EPA-600/2-78-086.
12. Ackerman, D. G., R. J. Johnson, E. L. Moon, A. E. Samsonov, and K. H. Scheyer. 1979. At-Sea Incineration: Evaluation of Waste Flow and Combustion Gas Monitoring Instrumentation Onboard the M/T VULCANUS. U.S. Environmental Protection Agency. Research Triangle Park, NC. Report No. EPA-600/2-79-137.

13. Compaan, H., A. A. v. d. Berg, J. M. Timmner, and P. van Leeuwen. 1982. Emission Measurements on Board the Incinerator Ship 'VULCANUS' During the Incineration of Organochlorine and Organofluorine Containing Wastes. Central Research Organization-TNO. Delft, The Netherlands. Report No. CL 81/108. Order No. 30057.

14. Compaan, H. 1982. Monitoring of Combustion Efficiency, Destruction Efficiency, and Safety During the Test Incineration of PCB Waste; Part I: Combustion and Destruction. Central Research Organization-TNO. Delft, The Netherlands. Report No. CL 82/122.

15. Metzger, J. H. and D. G. Ackerman. 1983. A Summary of Events, Communications, and Technical Data Relating to the At-Sea Incineration of PCB-Containing Wastes Onboard the M/T VULCANUS, 20 December 1981-4 January 1982. U.S. Environmental Protection Agency, Cincinnati, OH. Contract No. 68-02-3174, Work Assignment No. 82.

16. Ackerman, D. G., J. F. McGaughey, and D. E. Wagoner. 1983. At-Sea Incineration of PCB-Containing Wastes Onboard the M/T VULCANUS. U.S. Environmental Protection Agency, Research Triangle Park, NC. Report No. EPA-600/7-83-024.

17. Guttman, M. A., N. W. Flynn, and R. F. Shokes. 1983. Ambient Air Monitoring of the August 1982 M/T VULCANUS PCB Incineration at the Gulf of Mexico Designated Site. U.S. Environmental Protection Agency. Office of Water. Washington, D.C.

18. TerEco Corp., College Station, TX. 1982. Biological Monitoring of PCB Incineration in the Gulf of Mexico. U.S. Environmental Protection Agency. Office of Water. Washington, D.C.

19. Ackerman, D. G., Jr., and Ronald A. Venezia. 1981. Research on At-Sea Incineration in the United States. Paper presented at the Third International Ocean Disposal Symposium. Woods Hole Oceanographic Institution, Woods Hole, MA. October 12-16.

20. Compaan, H. 1982. Incineration of Chemical Wastes at Sea. A Short Review. Second Edition. Netherlands Organization for Applied Scientific Research, Central Research Organization - TNO. Delft, The Netherlands. Report No. CL 82/33.

21. Ackerman, D. G., Jr., and J. F. Metzger. 1983. Oceanic Incineration in the United States. Submitted for publication to *Environmental Science and Technology*.

22. Ackerman, D. G., J. H. Metzger, and L. L. Scinto. 1983. History of Environmental Testing of the Chemical Waste Incineration Ships M/T VULCANUS and I/V VULCANUS II. TRW Inc., Energy and Environmental Division. Redondo Beach, CA. Report to Chemical Waste Management, Inc., Oak Brook, IL.

23. Gielen, J. W. J. and H. Compaan. 1983. Monitoring of Combustion Efficiency and Destruction Efficiency During the Certification Voyage of the In-

cineration Vessel "VULCANUS-II", January 1983. Central Research Organization-TNO. Delft, The Netherlands. Report No. R85/53.

24. Ackerman, D. G., R. G. Beimer, and J. F. McGaughey. 1983. Incineration of Volatile Organic Compounds by the M/T VULCANUS II. TRW Inc., Energy and Environmental Division. Redondo Beach, CA. Report to Chemical Waste Management, Inc., Oak Brook, IL.

25. U.S. Environmental Protection Agency. 1980. *Code of Federal Regulations,* Title 40, Part 60, Appendix A, Method 5 - Determination of Particulate Emissions from Stationary Sources - P325. U.S. Government Printing Office. Washington, D.C.

26. U.S. Environmental Protection Agency. 1983. *Federal Register,* 48 (91): 20,985.

27. U.S. Air Force. 1974. Final Environmental Statement for Disposition of Orange Herbicide by Incineration. U.S. Department of the Air Force. Washington, D.C.

28. U.S. Environmental Protection Agency. 1976. Final Environmental Impact Statement, Designation of a Site in the Gulf of Mexico for Incineration of Chemical Wastes. U.S. EPA, Office of Water. Washington, D.C.

29. U.S. Department of Commerce. 1976. Final Environmental Impact Statement: Maritime Administration Chemical Waste Incinerator Ship Project, Vol. 1. U.S. Maritime Administration. Washington, D.C. Document No. MA-EIS-7302-76-041F.

30. Paige, S. F., L. B. Baboolal, H. J. Fisher, K. H. Scheyer, A. M. Shaug, R. L. Tan, and C. F. Thorne. 1978. Environmental Assessment: At-Sea and Land-Based Incineration of Organochlorine Wastes. U.S. Environmental Protection Agency, Office of Water. Washington, D.C. Report No. EPA-600/2-78-087.

31. U.S. Environmental Protection Agency, U.S. Department of Commerce, and U.S. Department of Transportation. 1980. Report of the Interagency Ad Hoc Work Group for the Chemical Waste Incinerator Ship Program. Washington, DC.

32. U.S. Environmental Protection Agency. 1981. Final Environmental Impact Statement for North Atlantic Incineration Site Designation. U.W. EPA, Office of Water. Washington, D.C.

33. U.S. Environmental Protection Agency. 1982. Final Environmental Impact Statement for the Offshore Platform Hazardous Waste Incineration Facility. U.S. EPA. Office of Water. Washington, D.C.

34. U.S. Environmental Protection Agency. 1983. Monitoring Results and Environmental Impact on the Gulf of Mexico Incineration Site from the Incineration of PCBs under Research Permit HQ-81-002. U.S. EPA, Office of Water. Washington, D.C. Also, *Federal Register* 148 (91): 20984.

Hazardous and Toxic Wastes: Technology, Management and Health Effects. Edited by S.K. Majumdar and E. Willard Miller. © 1984, The Pennsylvania Academy of Science.

Chapter Five

Operating Experience in Hazardous Wastes Disposal by Pozzolanic Cementation

Charles L. Smith[1] and Timothy Longosky[2]

[1]Principal Researcher
Envirosafe Services, Inc.
115 Gibraltar Road
Horsham, PA 19044

[2]Marketing Manager
Envirosafe Services, Inc.
115 Gibraltar Road
Horsham, PA 19044

The problem of "hazardous wastes" was first seriously addressed in the last decade. There are two distinct facets:
1. A definition and classification of hazardous wastes.
2. Development of disposal methods, along with enforcement of approved methods.

LEGISLATIVE EFFORTS

Congress has formulated two major pieces of legislation addressing this category of wastes. The Resource Conservation and Recovery Act (1976) defines and establishes classifications of hazardous wastes. It provides guidelines for disposal and establishes a manifest system for following waste from generation through disposal. The act is now about seven years old and is *still* not fully implemented. Enforcement is difficult in many areas, for there is uncertainty in the mechanism of RCRA.

The second significant legislation is the Comprehensive Environmental Response, Compensation and Liability Act of 1980 (Superfund), whose most significant function appears to be establishment of liability for environmental damage through improper handling or disposal. CERCLA has been described as establishing environmental damage liability like a ball of tar, smearing a little bit of liability on anyone who comes in contact with the waste. Generator, transporter and disposer all share in the responsibility for proper final disposal.

WHAT IS ADEQUATE DISPOSAL?

Early estimates placed a quantity of hazardous wastes produced in the United States at about 40 million tons annually. A more recent estimate from the Office of Technology Assessment, an arm of Congress, states that the number is on the order of 255 to 275 million metric tons.[1]

Clearly, the most desirable disposal is through recycling or reclamation. In the real world, however, this is often not a technically achievable, or cost-effective, goal. As the stringency of disposal criteria increases, so will disposal costs, followed by a significant increase in recycling or reclamation; but is unlikely ever to be the dominate means of hazardous wastes disposal.

For disposal, there were originally many options:

TABLE I
Disposal Means Acceptable in the Past

Municipal Sewer
Continguous Water Discharge
Ocean Disposal
Sub-Surface Injection
Lagooning
Incineration
Municipal Landfill
Secure Landfill

Most of these approaches are not acceptable under current regulatory restraints. Lagooning is an intermediate step, not a true disposal. Incineration is clearly limited to combustible materials; primarily organics. The remaining option for the majority of hazardous wastes is—and presumably shall remain into the future—some form of landfilling. This will be the secure landfill in one of several possible forms.

TABLE II
Disposal Means Acceptable for Tomorrow

Incineration
Secure Landfill

The Secure Landfill

Generically, the secure landfill has the primary function of extended waste retention without release of components of that waste into the surrounding environment. If this approach is to provide the final resting place for the majority of solid hazardous wastes in this Nation, *then it must be adequate.*

How is a secure landfill constructed? Throughout most of the United States, secure landfills are typically areas depressed below the normal topology, having a non-permeable barrier or lining. The function of this non-permeable barrier is to resist intrusion of toxic liquids from the landfill into underground aquifers. It has been found that the depth of liquid, and hence the pressure generated, may be significant due to intrusion of rainwater.

The most commonly used non-permeable barrier is clay. Under favorable conditions clay deposits may be hundreds of feet thick; in other scenarios, the clay may have been imported to the site and represent only a few feet of thickness. Synthetic liners are also in use; these are typically asphaltic or synthetic plastics. This non-permeable barrier represents the weak point of a secure landfill, presenting at least three potential failure modes:

1. *Mechanical loss of integrity.* The linings described above can lose integrity (become holed) due to water errosion, subsidence, sysmic activity or through unrelated construction projects.
2. *Vulnerability of clay.* Recent U.S. Environmental Protection Agency -sponsored studies at Texas A and M University have reported that clay, chosen for its low permeability in landfill linings, can develop up to 100 times greater permeability when exposed to basic, neutral polar, and neutral nonpolar organic fluid. Concentrated organic acids and certain inorganic salts may be similarly destructive.[7]
3. *Limits of Synthetic Liners.* Other EPA - sponsored studies, carried out at Matrecon, Inc. have clearly demonstrated that organic liquids commonly present in industrial wastes can attach synthetic plastics or asphaltic linings resulting in vastly increased permeability or in actual generation of holes in the lining.[8].

It would seem that the secure landfill may not be quite as secure as one would hope. Even the common steel drum, once considered an adequate disposal vessel, is no longer so. There was clear evidence that long storage above ground or within landfills can result in some of the environmental mishaps discussed previously.[2] Although diversity of wastes precludes prediction of the life span of steel drums, there is universal agreement that steel drums are not adequate as long-term storage media.[9]

LIABILITY FOR ADEQUATE DISPOSAL

The generator of a hazardous waste seeks to dispose of his problem without incurring any potential liability for environmental damage. CERCLA makes this very difficult to accomplish. In either scenario, self-controlled disposal or a disposal service, the generator maintains some liability for environmental damage. If he chooses to dispose of his own hazardous wastes, he would rationally maintain substantial insurance covering sudden and accidental pollution liability.[3] The need for this insurance is less obvious, but is present, when a disposal service is used. Several attorneys have expressed the opinion that pollution engineers in the employ of the generator have *personal* liability, stating "any person who may have taken part at any time in a private discussion where the outcome was to ignore a possible situation, like properly disposing of hazardous wastes, is liable." "Anyone who withholds information about a dangerous situation is liable."[4]

The transporter of a waste also is confronted with the stigma of liability if something goes wrong, even after he has discharged the wastes into a secure landfill.[5]

Perhaps the ultimate statement is: "under the Comprehensive Environmental Response, Compensation and Liability Act of 1980 (Superfund), any company that arranges for the disposal of hazardous wastes is liable if those wastes are not disposed of properly."[6] Hence, if the generator is depending upon a disposal service, his choice must be a good one.

YESTERDAY, TODAY AND TOMORROW

Yesterday's disposal methods seemed adequate but recently the newpapers have given much attention to Love Canal and The Valley of the Drums.[2] These disposal schemes apparently were adequate in the past. Today, the primary disposal method for inorganic wastes (including incineration residuals) is a secure landfill. However, the discussion above points out that EPA sponsored studies are casting doubts on the security of a secure landfill. Tomorrow it is likely that something better may be required. We know the trend of regulatory constraints continues to grow tighter. Industry's concern is the high cost of remedial action (e.g. Love Canal) if required. The Office of Technology Assessment has estimated that remedial action, retreating a waste disposal area, might cost from 10 to 100 times the cost of adequate initial disposal. An EPA estimate, published in 1983, provides an average cost of disposing of hazardous wastes in compliance with current RCRA regulations at about $90 per metric tons, this is in comparison with the estimated cost of cleaning up the improperly dumped wastes at about $2,000 per metric ton.[1] The message is clear: dispose of the wastes properly the first time.

TOMORROW'S APPROACH

Physicochemical Stabilization

There is an approach presently in commercial practice in North America which provides security above and beyond the chemically secure landfill. This technology, referred to as a belt-and-suspenders approach, is that of placing a physicochemically stabilized mass, containing the hazardous wastes, within an impermeable lining. In this scenario, if lining integrity is breached, the solidified interior virtually eliminates any impact on the environment.[10]

As a simple analogy, consider storing a 55 gallon drum of pesticide in your living room. If the drum fails, the pesticide is released. If, however, the pesticide is cast into a concrete-like lump within the drum, failure of the lining (the drum) has no significant effect. This is the assurance given by the physicochemical stabilization plus impermeable liner.

POZZOLANIC CEMENTATION

The specific stabilization to be discussed is pozzolanic cementation via the Envirosafe process, a process which has been in commercial operation since early 1980. Briefly, hazardous wastes are blended to a pozzolanic cement and placed in a lined landfill. The pozzolanic cement develops a monolithic integrity, generates a clay-like permeability and effectively seals off the waste to provide resistance to leaching. The primary components of the system, beside the waste, are fly ash (collected smoke from coal burning power stations) and lime-based additive.

Chemistry of the Process
Historically, the concept behind the Envirosafe Process is not new. In Italy there still exists an amazing number of nearly 2,000 year old structures using

FIGURE 1. The Colosseum; Pozzolanic Cementation in 80 A.D.

comparable pozzolanic chemistry. Portions of the famed Appian Way, the oldest well-engineered roadway system in the world, still exist. The Colosseum, whose structure is principally pozzolanic conrete, still exists (Figure 1). The Pantheon, in Rome, is a 1,900 year old temple, boasting a *143-foot span of unsupported ceiling dome constructed of pozzolanic conrete*. These and hundreds of other Roman engineering efforts used pozzolanic mortar or pozzolanic concrete based on the same chemistry as the Envirosafe Process. The Romans used volcanic ash as the pozzolan, with lime. The reader will note that there are relatively few volcanic ash sources in North America, but that coal-burning power stations generate another pozzolan - fly ash.

The basic chemistry of the Envirosafe Process is virtually identical to that of portland cement hydration and hardening. Table II lists the basic chemical reactions that occur in both the hardening of portland cement concrete and in the Envirosafe Process; an examination of any good text on portland cement chemistry will confirm these.

TABLE III

Primary Chemical Reactions in The Envirosafe Process

$CaO + SiO_2 + H_2O \rightarrow xCaO = ySiO_2 = zH_2O$

$CaO + Al_2O_3 + H_2O \rightarrow xCaO = yAl_2O_3 = zH_2O$

$CaO + Al_2O_3 + SiO_2 + H_2O \rightarrow wCaO = xAl_2O_3 = ySiO_2 = zH_2O$

$CaO + Al_2O_3 + SO_3 + H_2O \rightarrow 3CaO = Al_2O_3 = 3CaSO_4 = 30\text{-}32\ H_2O$

$CaO + Al_2O_3 + SO_3 + H_2O \rightarrow 3CaO = Al_2O_3 = CaSO_4 = 11\text{-}13\ H_2O$

The properties resulting when the cementitious reactions in the Envirosafe Process proceed are presented in specific examples later in this paper. A generalization of the potential of pozzolanic cementation would be to state that unconfined compressive strengths in excess of 1,000 psi can be achieved, but are not really needed for landfill structures. Permeability coefficients are typically in the 10^{-6} cm/sec range, although designs for 10^{-8} cm/sec ranges are available. EPA leachate analyses conform with Resource Conservation and Recovery Act and/or local regulations.

This is a chemical process. As such, the commercial facility is heavily oriented toward quality control. In the same manner that a structural engineer does not simply throw cement, sand, stone and water indiscriminantly together to create massive dams, highway systems, or buildings; the Envirosafe Process requires design of compositions and monitoring of ongoing operations to provide the environmental integrity which is its mission.

An Operative Facility

At the Envirosafe Services facility in Honeybrook, PA (suburban Philadelphia) the pozzolanic cementation process has been in use since early

FIGURE 2. Envirosafe Process; Proportioning and Blending Facility.

1980. During that time, nearly one-half million tons of hazardous and near-hazardous wastes have been disposed of under Pennsylvania Department of Environmental Resources scrutiny. Incoming waste loads are analyzed and approved before off-loading into a roofed storage building. The waste is preblended with the fly ash reactant to increase handleability and effect dispersion of the fly ash with the waste.

The preblended material is fed into hoppers, then onto a variable speed belt. From here the material is conveyed to an automated weighing system which proportions the required lime addition. Finally, the preblended material, with lime, enters a mixer where constituents are thoroughly dispersed to optimize the pozzolanic reaction (see Figure 2).

The processed material exits from the mixer onto a radial stacker conveyor, from which the product may be stockpiled or loaded directly onto a truck for transport to the adjacent landfill. The consistency of the processed waste at the stacker is similar to damp soil. Regularly taken samples are analyzed to insure that all process specifications are being met.

The processing and landfilling areas are underlain by a minimum of 4 inches of bituminous concrete, as per Pennsylvania DER facility regulations.

Processed waste is conveyed to the active landfill area via dump trucks, using all-weather haul roads. After offloading, the material is spread and graded in layers of about 12 to 24 inches depth. Compaction follows, using steel drum vibratory compactors. The surface grade is maintained at a slope to allow rainwater runoff to sedimentation ponds. In-place, compacted processed waste is tested by quality control personnel; adequate in-place densities after compaction enhance development of environmentally related properties. At this point the landfill operation closely resembles a road-building project rather than a landfill (see Figure 3).

The sedimentation ponds are capable of handling rainfall runoff from the 50-year frequency storm of 24-hour duration from the landfill. Water from the sedimentation ponds may be used to control dust by means of a sprinkler system. The water so applied to the placed, processed waste also serves to maintain surface moisture, aiding the pozzolanic chemical reaction during periods of excessive moisture evaporation.

Performance of the landfill is measured by the quality of rainfall runoff, the quality and quantity of leachate and the quality of monitoring well water. Due to very low permeability of the processed waste as compacted in the landfill and to the engineered slope of the landfill, essentially zero leachate is generated throughout the landfill. No impact has been observed on the quality of monitoring well water, since no leachate has been generated during the landfill operation! Unannounced environmental audits are routinely conducted by our independent corporate Environmental Services Department to monitor compliance with pertinent regulations.

Waste Materials Compatible With the Process

The process described herein is not, repeat NOT, a universal panacea. Pozzolanic cementation cannot be used for all wastes, for the simple reason that some wastes will interfere with the chemistry or "setting" mechanism.

FIGURE 3. Pozzolanic Cementation in Practice.

Historically, high sugar content wastes, for example, are not acceptable. Additionally, others may require disproportionately large amounts of the pozzolanic cement, rendering this approach cost*in*effective. However, about 90% of materials screened in the laboratory are cost-effectively compatible with this system. Table IV exemplifies wastes that have been successfully disposed of by this chemistry in this facility.

Test Data

The best evaluation of any process can be made when real test data are presented. These data are given for:

A. neutralized inorganic metal sludge
B. an electric furnace dust from steel production
C. an oil-bearing sludge from rolling mill operation

TABLE IV

Honey Brook Stabilization Waste Streams

Metal Hydroxide Sludge
Paper Coating Wastes
Electric Furnace Dust
Ceramic/Porcelain Sludge
Aluminum Dross Waste
Calcium Fluoride Sludge
Battery Casings
Foundry Waste
Scrubber Sludge
Rare Earth Metal Sludges
Electroplating Sludges
Chromium Oxide Waste
Grinding Wheel Waste
Titanium Bearing Sludge
Incinerator Ash
Waste Water Treatment Sludges

PAPER COATING MANUFACTURE - PAPER COATING SLUDGE

The generator operates a specialty paper coating process in Pennsylvania. The waste consists primarily of inorganic pigments, filter clays, organic color formers, adhesives and microcapsules. The waste is well over 50% complex organics.

As indicated in Table V, the waste stream was successfully stabilized physically and chemically. Unconfined compressive strength on samples cured for 7 days at 100 °F exceeded 75 psi; permeability of the cured sample was less than 1×10^{-5} cm/sec. Leachate analysis as measured by ASTM Method A indicated substantial reduction in metals, COD, TOC and Phenol content.

TABLE V

Applications of Industrial/Hazardous Waste
Successfully Stabilized by Microencapsulation

General: Paper Coatings Manufacturer
Waste:　Paper Coatings Sludge

Chemical Characteristics

Parameter	Raw Waste DWB Mg/Kg	Stabilized Waste Leachate Analysis, ppm
Total Solids%	52.7	
pH	8.4	11.9
COD	434,000	55
TOC	342,500	16.3
Phenol	13.8	0.6
Oil & Grease	7,800	28.4
Cr	16	0.005
Cu	200	0.05
Zn	610	0.05

Physical Characteristics

Unconfined Compressive Strength: cured 7 days at 100%F (psi)	75 +
Permeability: (cm/sec)	$< 1 \times 10^{-5}$

TYPICAL STEEL PRODUCTION - AIR POLLUTION CONTROL RESIDUE

This air pollution control residue results from particulates generated in the electric arc furnace and are captured in the baghouse collection system. The waste material is defined as hazardous, according to RCRA, and bears the designation "K061".

Analyses indicated zinc, iron and calcium to be the principal constituents, although the analysis clearly shows a multiplicity of metals present, mainly as oxides.

As indicated in Table VI, material was successfully stabilized via pozzolanic cementation. Unconfined compressive strength of the sample (cured 7 days at 100°F) was 100+ psi. Permeability of the cured sample was less than 1×10^{-5} cm/sec. Leachate analysis via ASTM Method A disclosed that the stabilized material met RCRA standards.

TYPICAL ROLLING MILL OPERATION - OIL BEARING SLUDGE

This waste is generated from the dewatering unit operations in the process of milling aluminum. Principal constituents are quartz (about 49%), oil and grease (about 37%) with many heavy metals in hydroxide form as minor constituents.

TABLE VI

Applications of Industrial/Hazardous Waste
Successfully Stabilized by Microencapsulation

Generator: Typical Steel Corporation
Waste: Air Pollution Control Residue, Electric Furnace Dust

CHEMICAL CHARACTERISTICS

Parameter	Raw Waste DWB mg/kg	Stabilized Waste ASTM Method A Leachate Analysis PPM	RCRA Standards PPM
Total Solids %	99.8	-	-
TOC	840	30	-
COD	-	-	-
pH	11.8	11.6	-
Oil/Grease	479	12	-
Ag	235	0.05	5.0
As	34	0.01	5.0
Ba	10	0.75	100
Cd	1680	0.01	1.0
Cr	1155	0.15	5.0
Hg	2.8	0.2	0.2
Pb	49,600	5	5.0
Se	2.0	0.07	1.0

PHYSICAL CHARACTERISTICS

Unconfined Compressive Strength: Cured 7 days @ 100°F (psi)	100 +
Unconfined Compressive Strength: Cured 28 days @ 73°F (psi)	150 +
Permeability, cm/sec^{-1}	< 1 x 10^{-5}

The oil-bearing sludge was sucessfully stabilized, physically and chemically via pozzolanic cementation. Unconfined compressive strengths (cured 7 days at 100°F) exceeded 50 psi. Permeability of the cured sample was less than 5 x 10^{-6} cm/sec. The stabilized material easily met RCRA standards as determined in the ASTM Method A Leachate Analysis. Note specifically that oil and grease content is reduced by a factor of 7,500.

IN SUMMARY. . .

A new waste disposal process has been discussed, in theory, and in actual commercial operation. Since start-up over three years ago, this facility has applied the pozzolanic cementation concept to the environmentally safe disposal of about ½ million tons of hazardous and near-hazardous industrial wastes.

TABLE VII

Applications of Industrial/Hazardous Waste
Successfully Stabilized by Microencapsulation

Generator: Typical Rolling Mill Operation
Waste: Oil Bearing Sludge

CHEMICAL CHARACTERISTICS

Parameter	Raw Waste DWB mg/kg	Stabilized Waste ASTM Method A Leachate Analysis PPM	RCRA Standards PPM
Total Solids %	72	-	-
TOC	90,000	84	-
COD	(348,000)	340	-
pH	8.5	7.1	-
Oil/Grease	(375,000)	50	-
Ag	10	0.05	5.0
As	1.8	0.07	5.0
Ba	10	0.05	100
Cd	10	0.05	1.0
Cr	10	0.05	5.0
Hg	.76	0.002	0.2
Pb	28	0.05	5.0
Se	2.0	0.01	1.0

PHYSICAL CHARACTERISTICS

Unconfined Compressive Strength: Cured 7 days @ 100 °F (psi)	50 +
Unconfined Compressive Strength: Cured 28 days @ 73 °F (psi)	65 +
Permeability, cm/sec^{-1}	$< 5 \times 10^{-6}$

None of us denies the existence of the hazardous waste problem in the United States, but it is clear that there is disagreement on adequacy of disposal means. As implementation of environmental regulations becomes increasingly stringent, the merits of the belt and suspenders approach, as brought to commercial fruition in the Envirosafe Process, may become strikingly evident; this technical approach may easily become state-of-the-art.

REFERENCES

1. _____, *Management Strategies for Hazardous Waste Control - Summary,* Office of Technology Assessment, Congressional Board of the 98th Congress, Washington, D.C., March, 1983.
2. Kiang, Y-H. and Metry, A. A., *Hazardous Waste Processing Technology.* Ann Arbor Science, Ann Arbor, MI, 1982.

3. Powals, R. J., *Protecting Against Gradual Pollution Liability,* Industrial Wastes 27:5, Oct. 1981.
4. Young, R. A., *Who's Liable?,* Pollution Engineering 12:11; Nov. 1980.
5. Cross, F. L., *New Hazards for Hazardous Waste Managers,* Pollution Engineering, 13:8; Aug. 1981.
6. _____, *Chemical Firms Sued Under Superfund Law,* Chemical and Engineering News, Oct. 5, 1981.
7. Brown, K. W. and Anderson, D. C., *Effects of Organic Solvents on the Permeability of Clay Soils,* Report No. EPA-600/2-83-016, March 1983.
8. Haxo, H. E. and White, R. M., *Evaluation of Liner Materials Exposed to Leachate,* (Several Reports Issued under U.S. E.P.A. Contract No. 68-03-2134).
9. Frankel, I., *Estimating the Life Span of A Buried Steel Drum,* Industrial Wastes 28:1, February 1982.
10. Roberts, B. K. and Smith, C. L. *Microencapsulation: Simplified Hazardous Waste Disposal:* Chapter 6 of *Toxic and Hazardous Waste Disposal,* Pojasek, R. B. (Editor); Ann Arbor Science, Ann Arbor, MI 1979.

Hazardous and Toxic Wastes: Technology, Management and Health Effects. Edited by S.K. Majumdar and E. Willard Miller. © 1984, The Pennsylvania Academy of Science.

Chapter Six

Technology of Special Waste Plant in the German Federal Republic

H. G. Rückel

Director
Zweckverband Sondermüllplätze
Mittelfranken
Geschäftsstelle
Rother Straße 56
D-8540 Schwabach
West Germany

Up to about the year 1970, hazardous waste in the German Federal Republic was mostly dumped together with domestic refuse. Before that time no special technology existed for the disposal of special waste. After 1970 industrial production in the GFR increased greatly and hence also the quantity of special waste (in 1982 it was over 2.5 million tonnes). It was recognized that the dumping of these wastes endangers the water supply and harms the rivers in the long term. Therefore in 1972 the law relating to the disposal of waste (1) was enacted by the Federal Government. The law rules that cities and "Land" districts are responsible for the disposal of hazardous waste.

The law further prescribes that:

1. Waste disposal plants require an official permit,
2. Hazardous waste must be monitored by a waybill procedure from the producer via the haulage firm extending to the disposal plant.
3. Hazardous waste may only be transported by qualified transport firms with a permit, and
4. In every Works where particularly hazardous waste occurs, a "Works official responsible for waste" is to be appointed who is responsible for internal Works organisation and monitoring of the waste disposal.

These regulations by the Federal Republic created a network of supraregional special waste treatment plants, as the muncipalities were not in a position financially or technically to construct their own special waste treatment plants. These special waste treatment plants dispose on an average of about 150,000 tonnes per annum of hazardous waste, and are supervised by the respective "Lands" to guarantee they operate properly.

APPLIED TECHNOLOGIES

Input Control

If hazardous wastes are transported to a special waste disposal plant, a declaration from the producer to the haulage firm must accompany the special waste. In this declaration the particular properties must be described, e.g. as to whether the wastes are toxic, have a strong odor, ignite spontaneously, are caustic, contain oil or are likely to cause a fire. Furthermore, a short chemico-physical identification is required, in order to be able to recognise, in broad terms, at input control which disposal method is likely to have to be applied; for example, whether the waste contains cyanogen, nitrate, chromate, is acid or basic, combustible or contains phenol. The producer also has to state whether special anti-pollution measures are necessary or whether special hazards may arise from the waste.

On the basis of these data from the producer of the waste, the waste is subjected to an input control in a special laboratory at the Special Waste disposal plant. In order to accomplish this, as representative a sample as possible is taken of the waste from each delivery. If the waste is delivered in the liquid state in a tank wagon, this is relatively simple. What is more difficult is a representative sample of waste supplied in drums or troughs.

The sample is examined as part of a regular routine at the input laboratory for the following parameters:

pH	zinc
cyanide	chromium
nitrite	iron
chromate	combustibility
nickel	refrangibility of oil emulsions
copper	halogen content of solvents etc.

The analysis should be made in the shortest possible time, as transport firm set great store by rapid handling. From the input laboratory the vehicle making the delivery is sent to one of about thirty unloading points of a Special Waste disposal plant.

The input control is extremely important. Here, on the one hand, it must be detected which noxious substances are contained in the special waste; on the other hand, by correct allocation to the unloading point it is guaranteed that the disposal equipment can operate without problems. If, for example, organic wastes are tipped into an inorganic treatment plant, whole process chains are contaminated and thus become incapable of functioning. Interference with operations on account of comprehensive cleaning procedures then results.

Special Waste Tips

The following may be stored on a definitive basis in a special waste tip (2) in the GFR:

a) De-toxified, neutralised and compact de-watered residues from the chemico-physical special waste treatment plant,

b) Combustion residues from the Special Waste combustion plant

c) Soil which is contaminated with oil, not dripping

d) Production-specific special wastes which are delivered already de-toxified, neutralised and de-watered,

e) Other solid production-specific wastes which cannot be disposed of together with domestic refuse and which can be incorporated in the tip without prior treatment.

Insofar as the special waste delivered to input control cannot be reliably defined, the waste must be parked at an interim storage place until there is an adequate analysis of it.

At a Special Waste tip it is mainly wastes which are a danger to groundwater which are stored. On this account the tip must be designed in such a way that this cannot occur. In the German Federal Republic special waste tips, insofar as this is geologically possible, are located in worked-out claypits. By means of many probe drillings the subsoil is examined to see whether the existing natural strata of clay are massive and impervious. However, one does not rely on the natural impermeability of the soil, but applies additional artificial clay strata.

According to present requirements, three optimally compacted strata of clay each 30 cm thick are constructed. The permeability correction value should reach at least 1×10^{-8} m/sec. Over this clay seal a drainage system is laid which absorbs the seepage water and conducts it away. This is to avoid a build-up of water pressure on the sealing layer. The seepage water is collected, and because of the widely varying ingredients in it has to be purified in the chemico-physical treatment plant of a Special Waste disposal plant before it can be diverted to an efficient settling plant.

The special waste is incorporated in coffers of about 100 x 100 m. area and with a layer thickness of a maximum of 1.5 m. The waste is poured in a gradient of at least 3%, so that rain can be conducted away before it penetrates into the body of the deposit, so far as possible. Care should be taken to compact the substance to be incorporated as much as possible. Substances with a pungent odor should be covered without delay. The same applies to substances which easily blow away, and which in exceptional cases can be bound by wetting with water. By adding lime, the body of the deposit is kept basic, in order to limit metal solubility.

After about 7.5 m. dumping height an intermediate sealing layer of clay (layer thickness approx. 30 cm, compacted to the greatest possible extent) is introduced into the Special Waste tip. This interim sealing stratum is to prevent the penetra-

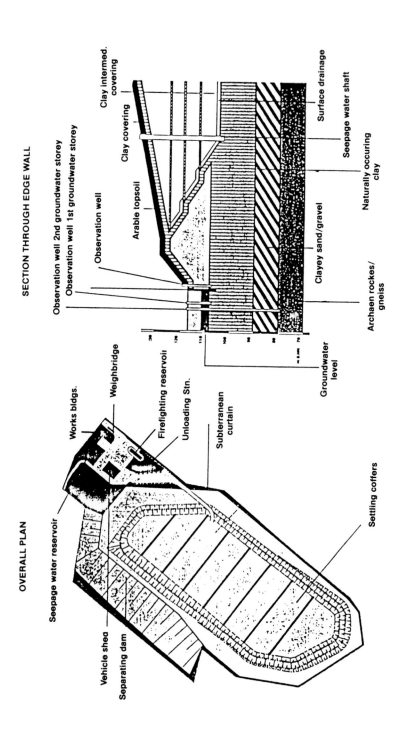

SECTION THROUGH EDGE WALL

Clay intermed. covering
Clay covering
Observation well 2nd groundwater storey
Observation well 1st groundwater storey
Observation well
Arable topsoil
Surface drainage
Seepage water shaft
Naturally occuring clay
Clayey sand/gravel
Archaen rockes/ gneiss
Groundwater level

OVERALL PLAN

Works bldgs.
Weighbridge
Firefighting reservoii
Unloading Stn.
Subterranean curtain
Seepage water reservoir
Vehicle shea
Separating dam
Settling coffers

tion of precipitates into the layers of refuse lying below, in order to reduce the occurrence of seepage water. This layer is covered with 20 cm of soil and grassed over; consequently, drying-out and crack formation are prevented.

Once a Special Waste tip has been filled, the entire surface is covered with a clay stratum 1 m. thick. This clay stratum is designed so that the surface water flows off to the edges of the tip and can be conducted away. Above the clay stratum a soil layer about 2 m. thick is dumped, and planted.

Surrounding the tip, wells are sunk, which extend downward to the groundwater. Samples are taken monthly in order to detect whether the tip is impervious.

A Special Waste tip will probably have to be attended for at least 50 years after it has been sealed. Care must be taken to see that the seepage water is continually conducted away and purified. In a Special Waste tip generally there are no formations of gas, since organic wastes are mainly burnt in the Federal Republic.

The Special Waste tip must be identified for posterity. Generally speaking such areas would no longer be suitable for building. However, they are very good for use as parking lots. When constructing a tip, therefore, landscaping and re-cultivation plans are already compulsorily prescribed by law, and must be carried out as soon as partial sections of the tip have been filled.

For operating a tip, in most cases one push loader, one bucket loader, one dredger and a mechanical sweeper are used. Great store is set by a good visual impression.

Underground Tip

There are highly toxic solid wastes which may not be stored at a Special Waste tip, for example carburising agents containing cyanogen, barium or nitrite, residues or plant protection or pesticide agents. These wastes are stored permanently in a central location in the GFR in a disused salt mine (3), at a depth of 700 m. The stored wastes must fulfill the following pre-requisites:

The waste must not give off gases and must be packed in a hermetically sealed undamaged lead container with welded-on pressure covers or a bolted-on tension ring cover. Drums of uniform size must be placed on disposable pallets an secured firmly so that they cannot shake loose, by means of flat steel strips. Every drum is inscribed with a code number which designates the contents and the supplier. In this way it is possible, even after years, to identify the wastes stored in the salt mine. Substances which are wastes today may become valuable substances in a few years' time.

Chemico-Physical Treatment of Special Waste

Given types of special waste have to be pre-treated because of their chemical, physical or toxicological properties, before they can be dumped, thermodestructed or recycled for re-use.

These types of special waste can be classified basically into two categories:

1) Liquid waste, special waste containing solids and organic matter, for example emulsions, water, containing solids and oil, the contents of oil and gasoline separators, residues from tank-cleaning and the like. These wastes originate predominantly from metallurgical and mineral oil processing industries and also from filling stations and vehicle repair works.

2) Liquid waste, special waste containing solids or free from solids contaminated with inorganic material, for example acids or caustic solutions, effluents and slurries containing cyanide, nitrite, chromate and heavy metals from the metalworking, electrical engineering and electroplating industries.

Wastes are also delivered to repositories which are contaminated both organically and inorganically. The treatment of these wastes is particularly burdensome and often presents the special waste disposal plants with considerable problems.

Emulsion-Splitting-Effluent Purification

The aim of the process described below is to separate the wastes delivered into the phases of solids, oil and water.

The oil phase can either be recovered or be thermodestructed in Special Waste combustion plants; the solid substance phase, still containing considerable fractions of oil and therefore rich in calorific value, is burnt in a Special Waste combustion plant. Finally, the water phase can either be introduced direct into the sewerage system; or, if conditions for its introduction are not met, it must be detoxified by one of the processes described in the following section.

On account of the high degree of pollution of the emulsions and oil/water mixtures delivered, in contrast to what happens in many emulsion separation plants in industry, in Special Waste treatment plants (4) only those processes can be used which are insensitive to high contents of solids.

Before the actual emulsion-splitting, the solid substances are separated. Coarse substances (sand, heavy metal particles) settle in sedimentation bunkers. Floating light substances are retained by screening or sieving devices. Very fine solid particles, known as scum, are separated by mechanical de-watering, e.g. with decanters or vacuum drum filters.

For the separation of the oil and water phase, the actual emulsion-splitting, as a rule chemical processes are used in Speical Waste treatment plants. In these processes the oil/water mixture is split by the addition of acids or inorganic salts, e.g. magnesium chloride of ferric chloride. The mode of action of the chemical processes is based on the fact that by the addition of strong electrolytes the oil/water phase boundary layers are destroyed. If used acids are available in adequate quantity, these can be used for splitting emulsions.

Some Special Waste disposal plants operate thermal emulsion separation installations. These are mutli-stage vaporising plants in which, because of the differing temperatures of evaporation, the oil and water phases are separated from each other. On account of their high energy requirement, thermal emulsion separation installations can only be used economically where, for example, steam is available cheaply from a waste combustion plant.

Chemical De-Toxication of Special Waste

The processes described below are chemical de-toxication methods by means of which the noxious ingredients, mainly inorganic, are decomposed to such an extent that after de-toxication the limit values prescribed by the licensing authorities for introduction into the sewage system or a main outfall are achieved. Dilution, for example with effluents which are not highly contaminated, from other parts of the Works at the Special Waste treatment plant, is expressly not permitted by the licensing and inspection authority.

The de-toxication reactions usual in Special Waste treatment, such as neutralisation of acids and caustic solutions, de-toxication of cyanide, nitrite and chromate, and also precipitation of heavy metals are known processes which are well described generally in the Literature.

Because of the special environmental problems involved and the new developments emerging, only the processes for cyanide de-toxication and heavy metal precipitation are tested in detail.

Cyanide De-Toxication

As a rule the cyanide ion is destroyed by oxidation. In Special Waste disposal plants it is mostly an aqueous sodium hypochlorite solution which is used as the oxidation agent. The reaction takes place in the alkaline range (pH greater than or equal to 11), to form first of all cyanate which is less toxic, and can be concluded, after lowering the pH, by oxidation of the cyanate to form nitrogen and carbon dioxide.

With suitably long reaction periods and high excesses of oxidation agent, even the heavy metal/cyanide complexes, which are generally very stable to attack by oxidation, e.g. those involving copper and nickel, can be decomposed with sodium hypochlorite. In the oxidation of the cyanide ion with sodium hypochlorite, care must be taken in the first stage in particular to maintain a high pH figure, as at too low a pH the highly toxic cyanogen chloride may be formed.

One drawback of this process is the undesired salting of the effluent associated with it. An oxidation agent in which any additional salting is avoided is hydrogen peroxide. The use of H_2O_2 as oxidant for cyanide detoxication, however, is only possible to a restricted extent in Special Waste treatment plants. On account of the generally high concentrations of heavy metal in the effluent containing cyanide to be detoxified, the H_2O_2 is catalytically decomposed to a large extent already before the actual reaction, so that one has to operate at high excesses of oxidant, as a result of which the process becomes uneconomic.

Processes discussed at various times for the thermo-hydrolytic cracking of cyanides at high pressures and temperatures, for example in a reactor, in which

CO_2 and nitrogen or formate and ammonia occur as the reaction products, have proved not to be transferable on a large industrial scale to special waste treatment. In some cases highly concentrated effluents containing cyanide are thermally treated by injecting them into the secondary combustion chamber of a Special Water combustion plant.

Heavy Metal Precipitation

The aim of heavy metal precipitation is to transfer the toxic heavy metals occurring in the effluents of the electroplating industry, especially nickel, copper, cadmium, mercury and lead, into compounds which are sparingly soluble and to separate them by filtration. As a rule, the heavy metal ions are precipitated out as sparingly soluble hydroxides or basic compounds with milk of lime.

Solutions containing complex-formers are being delivered to an increasing extent to Special Waste treatment plants. In these effluents the heavy metals are kept in solution by organic complex-formers, e.g. ethylenediaminetetra-acetic acid (EDTA). Precipitation with milk of lime is here not possible on account of the high stability of the complex compounds. In these cases sodium sulphide is used as the precipitant. The minor solubility product of metal sulphides is usually sufficient to precipitate metals from their complexes.

To separate off the precipitated compounds, hydroxides or sulphides, filter presses have been successful. The filter cake occurring with a dry substance between 20 and 40 per cent by weight has as a rule a compact consistency and can be stored at special waste tips.

Increase in Amounts of Waste and Technology for the Chemico-Physical Treatment of Special Waste

For some years industry has increasingly carried out a part of the necessary chemico-physical pre-treatment at plants in its own works, mainly for cost reasons. For Special Waste disposal firms this means a drop in quantity and bound up with this a smaller utilisation of capacities of their existing plants which are generally of a high industrial standard.

However, the delivery of particularly problematical wastes, which require a high treatment outlay, is increasing. An example of this are the increasing deliveries of effluents containing complex-formers or of wastes which are contaminated from a number of sources, e.g. organically and inorganically. Pre-treatment at a firm's own plant frequently does not compare to the high technical standards of the special waste treatment plants. The result of this, inter alia, is that the sludge occurring with a large number of industrial firms and which is introduced into the municipal settling plant has too high a content of heavy metal.

Special attention has been paid recently to the high salt contamination of industrial effluents. This also applies to effluent from Special Waste treatment. Therefore, in the long term, new thermal and physical processes, such as for example vaporisation and reverse osmosis to reduce the salt content in Special Waste treatment plants, are being used.

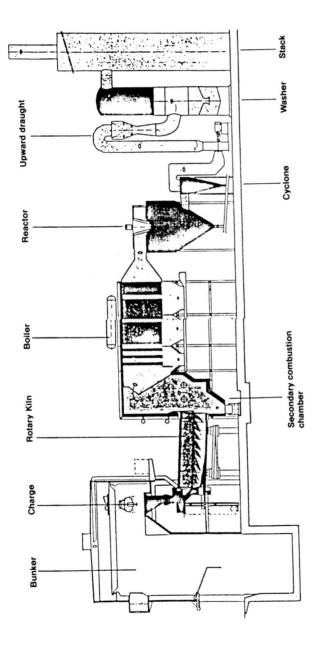

Thermodestruction of Special Waste

In the German Federal Republic organic special wastes are burnt in Special Waste combustion plants (5). Here, rotary kiln furnaces have universally proved to be the most applicable system of burning. This type of furnace can be used for solid and pasty waste substances and for liquids. Fluctuations in the composition of the waste and mechanical properties permit of reliable operation of the rotary kiln furnace in practically all calorific value areas occurring.

In the rotary kiln furnace the following typical wastes can be combusted:

1. *Solid wastes:*
   Soil contaminated with oil (dripping), drums and containers contaminated with Special Waste, hardened pigments and greases, plastic wastes, bleaching earths, organic chemical residues, bitumen, tar wastes.

2. *Pasty wastes:*
   Pigment residues, tank cleaning residues, oily industrial sludges, used greases, grinding residues, refinery wastes.

3. *Liquid wastes:*
   Used oil with differing contents of water, solvents contaminated oils, emulsions containing oil, effluents containing phenol.

Modern Special Waste combustion plants in the German Federal Republic can be described as having the following zones:

a) *Reception zone*
   For the delivery of these wastes, tank stores, drum stores and bunkers are available. Here the wastes are stored separately according to their consistency. As a result of this interim storage of the waste there is the possibility of composing an optimum combustion mixture at any time.

b) *Bunkers:*
   In the bunker solid and sludge-type wastes are stored on an intermediate basis. Here the wastes are largely homogenised by mixing. This is a prerequisite for continuous combustion.

c) *End wall:*
   The wastes are pushed over the end wall into the rotary kiln. For this purpose the following feed-in devices are available:
   Feed hopper for solid wastes
   Feed hopper for sludge wastes which are conveyed by a
   thick matter pump
   Solvent burner
   Effluent lances
   Light oil burners

d) *Rotary kiln:*
   The rotary kiln is the actual combustion unit. In this steel drum, lined with refractory bricks, the wastes are burnt at temperatures of 1100 to 1400 °C and with a high excess of air. The non-combustible substances contained in the wastes are discharged at the end of the rotary drum as liquid slag; they

fall into a waterbath and there granulate.

e) *Secondary combustion chamber:*

In the secondary combustion chamber, by means of additional injection of secondary air, it is ensured that all organic ingredients are completely combusted. In this zone effluent with a high organic contamination can be injected in addition. An additional burner fired with fuel oil automatically ensures that the combustion temperature in the secondary combustion chamber does not drop below 900 °C. If required, for example if polychlorinated biphenyls (PCB's) are present, the temperature in the secondary combustion chamber can be raised to 1200 °C.

f) *Steam boiler:*

In the steam boiler, energy is drawn from the hot flue gases, and the temperature drops to 250 °C. With the steam recovered, the operation can be supplied with heat on a regular basis. Electricity is also produced for the plant's use. From time to time the excess current is supplied to the local network.

g) *Flue gas purification:*

The modern Special Waste combustion plants in the German Federal Republic have efficient flue gas purification installations. They consist for example, of the following complexes:

1. Reactor

The reactor fulfils two tasks. The flue gas is cooled by injecting washing sols from the exhaust air washer from 250 °C to 160 °C. The water injected then evaporates. The salts contained in the washing sol crystallise and other harmful substances are separated off in these crystals. A part of these salts is already discharged in the reactor from the combustion gases.

2. Cyclone

The cyclones connected after the reactor serve to remove dust and salts from the flue gases.

3. Flue gas washing operation

In the flue gas washing operation the flue gas is freed from harmful substances in three stages. The first stage, known as quenching, serves to cool the flue gases by the injection of washing water from 160° C to 65 °C. Hydrogen chloride (HCl) and hydrogen fluoride (HF) are washed out at the same time. In the second stage, known as the precipitant layer, the last residues of hydrogen chloride and fluoride and washed out. In the third stage, known as ring jets, sulphur dioxide (SO_2) and aerosols are washed out. The flue gases are then passed through a mist collector and leave the stack of the combustion plant freed from harmful substances. The washing water from all three stages is jointly recycled to the reactor as colloidal solutions.

In modern installations the following or even lower pure gas values are achieved:

HCl   50 mg/cu.m.
SO$_2$ 150 mg/cu.m.
HF   5 mg/cu.m.
Dust   100 mg/cu.m.

h) *Disposal of residues*

The harmful substances contained in the special wastes are practically removed from the flue gas and occur in solid form as slag, ash and salts. In this form they can be transported without danger and can be stored permanently at Special Waste tips. However, care must be taken to see that they can no longer come into solution.

Combustion at Sea

At present some 50,000 tonnes a year of highly chlorinated hydrocarbons are still being taken from the German Federal Republic and burnt in the North sea by means of special ships. In the future this possibility will be much restricted, as soon as adequate combustion capacities are available on land for these wastes.

New Technologies for the Thermodestruction of Special Wastes

At the ZVSMM a combustion plant has been designed in which Special Waste can be combusted as well as domestic refuse and sewage sludge. This plant should be ready for operation by 1988. If the design proves of value, separate Special Waste combustion plant can be largely dispensed with.

PROSPECTS

In the German Federal Republic, research projects are proceeding, with support from the authorities, which have the following aims:

1. Compacting of non-homogeneous special waste
2. Vaporising of effluent strongly contaminated with salts from Special Waste disposal plants
3. Recovery of metals from electroplating effluents
4. Optimisation of flue gas treatment plants connected up to Special Waste combustion plants.

FOOTNOTES

(1) Bundesgesetzblatt I Seite 873/1972
(2) Zweckverband Sondermüllplätze Mittelfranken (ZVSMM)
    Rother Straße 56
    D-8540 Schwabach

(3) Untertagedeponie Herfa-Neurode
Kali + Salz AG
Werk Wintershall
D-6432 Meeringen 1
(4) Gesellschaft zur Beseitigung von Sondermüll
in Bayern mbH
D-8076 Ebenhausen
(5) Hessische Industriemüll GmbH
Außerhalb 34
D-6081 Biebesheim

Hazardous and Toxic Wastes: Technology, Management and Health Effects. Edited by S.K. Majumdar and E. Willard Miller. © 1984, The Pennsylvania Academy of Science.

Chapter Seven

The Regional Treatment Facility for Organic Hazardous Waste of the State of Hesse in Biebesheim, Germany

Alois Scharsach
Sales Manager, Environmental Engineering Division
Von Roll Ltd.
P.O. Box 760
Zurich, CH-8037 Switzerland

THE HESSIAN HAZARDOUS WASTE DISPOSAL SYSTEM

In 1978 the Government of the State of Hesse issued a decree for the disposal of hazardous waste from industry and trade in order to ensure an organized and environmentally safe disposal of industrial waste in the entire state. This decree, called the Hazardous Waste Decree, stipulates that the waste generators have to leave industrial waste for disposal to the hazardous waste disposal authority as far as they are not in a position to dispose of such waste in approved and permitted own facilities.

The Hazardous Waste Decree declares the Hessische Industriemüll GmbH (H.I.M.) at Wiesbaden responsible for the disposal of industrial waste. The H.I.M., as a result, is bound to accept and safely dispose of all industrial waste generated in the State of Hesse. The fact that the H.I.M. so is actually holding a state monopoly commits this company to equal treatment of all waste producers, large or small, and to charge cost-covering disposal fees.

The State of Hesse holds 26 per cent of the shares of H.I.M. The other 74 per cent are held by the waste generators, 23 larger or smaller private companies.

In order to comply with its obligations, the H.I.M. provides the following disposal facilities:

The Hazardous Waste Treatment Plant at Biebesheim

- two plants for the chemical-physical treatment of liquid anorganic hazardous waste and thin anorganic sludges, one in Frankfurt and one in Kassel
- a plant for the treatment of solid, semi-solid and liquid organic hazardous waste and an emulsions splitting plant in Biebesheim
- a secured hazardous waste landfill in Mainflingen

The headquarters of the H.I.M. are located in Wiesbaden.

The Hazardous Waste Decree also sets rules for close communications between waste generator, supervising authority and H.I.M. by means of a manifest system. (See diagram on next page.)

THE HAZARDOUS WASTE INCINERATION PLANT IN BIEBESHEIM

Within the H.I.M. organization the Hazardous Waste Incineration Plant in Biebesheim is designed to handle all kinds of solid, semi-solid and liquid organic waste, including its intermediate storage and final disposal through thermal treatment.

The plant in Biebesheim consists of:
- material receiving and laboratories
- barrel storage
- storage bunker for solid and semi-solid waste

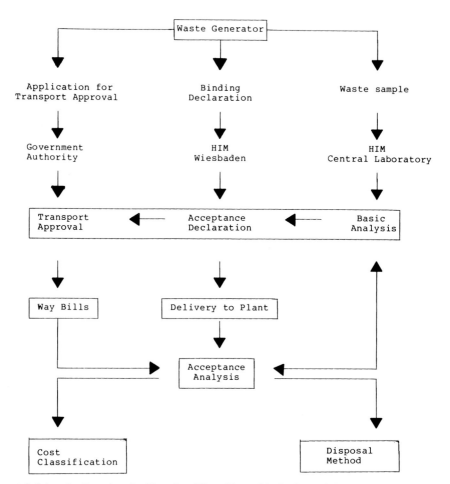

Administrative Procedure for Hazardous Waste Disposal in the State of Hesse

- 2 incineration trains, independently operated
- tank farm
- emulsions splitting plant.

The incineration plant is designed for a capacity of approx. 50.000 metric tons per year, one third of which will be solid waste, one third semi-solid and one third liquid waste. The total capacity is spread over two independent incineration trains in order to ensure the best possible availability.

The contract for the design and the turn-key delivery of the two incineration trains was awarded to Von Roll Ltd. in Zurich, Switzerland. The construction of the plant started in 1980, test operation began at the end of 1981 and since spring 1982 the plant has been in full operation.

HIM BIEBESHEIM
Lageplan / Layout

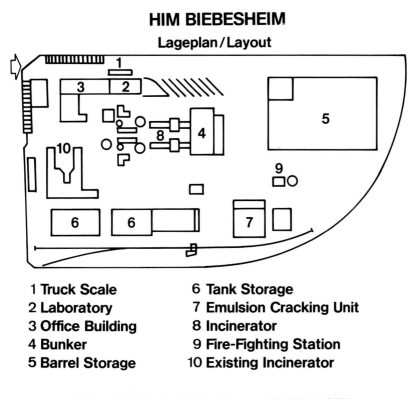

1 Truck Scale	6 Tank Storage
2 Laboratory	7 Emulsion Cracking Unit
3 Office Building	8 Incinerator
4 Bunker	9 Fire-Fighting Station
5 Barrel Storage	10 Existing Incinerator

RECEPTION AND STORAGE OF THE WASTE

All incoming waste has to pass the arrival check. There deliveries are weighed, qualified and registered as to quantity, kind, composition and consistency. From there the waste is directed to several unloading and storage areas.

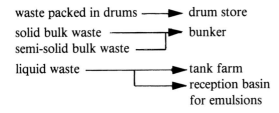

The storage facilities are provided with pretreatment devised for preparing the waste for the main process step.

This essentially applies to waste delivered in drums and liquids arriving in tanks cars. The solid material straightaway tipped into the bunker can be mixed and homogenized with the crane installation for more stable combustion.

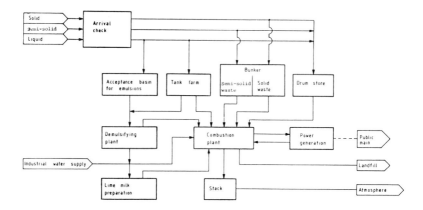

PRE-TREATMENT AND PREPARATION OF WASTE

Waste in drums

It is presumed that the waste in drums is delivered in three different conditions:
- solid
- semi-solid
- liquid.

Accordingly, the following operations can be performed in the drum storage area:
- opening of drums by drum cutting machine
- testing of drum contents if necessary in the laboratory
- extraction of liquids by vacuum into an intermediate tank and from there to the main tank farm
- partial emptying in case of high heat value by a drum tipping device.

All drum handling stations are interconnected by roller conveyors. The entire system enables the most varied manipulations to be performed ensuring optimum drum preparation.

Drums prepared for incineration are then moved on roller conveyors to the drum charging station.

Liquid Waste

This waste is divided into two streams:
- to the tank farm
- to the emulsions splitting plant.

The tank farm is an outdoor set-up. All tanks are placed in a number of sealed concrete pits. In front of the tanks are the reception basins for waste oil, waste water, emulsions, solvents, sludge and heavy oil, together with their pertaining pumps and pipe work. There are 25 reception and storage tanks and 2 mixing vessels, some of them equipped with agitation devices.

As gases may be generated in the tanks through reactions or during filling, the tanks are connected to a common venting pipeline which carries the exhaust gases into the rotary kiln furnaces.

In the event of fire the entire tank farm can either be sprayed or smothered with foam as required with regard to the chemical properties of the tank contents.

At the emulsions splitting plant a collecting basin is in front of the installation. From here the emulsions infiltrated with coagulants are passed to a sludge centrifuge. The extracted sludges are then conveyed to the waste bunker. The stripped liquor is passed over a belt filter press for further separation of solids and subsequently pumped to the emulsions splitting unit. The splitting is effected through thermal treatment. The emulsion is evaporated in consecutive stages and again condensed. The water phase is neutralized and discharged into the sewer, the separated oil constituents are pumped to the tank farm.

Solid and semi-solid bulk waste

The solid and semi-solid waste is unloaded into the bunker. This bunker has a storage capacity of approximately 1,500 m³ and is divided into 3 sections of equal size.

From the bunker the waste is transported by the overhead travelling crane into one of the two feed hoppers of each of the two incineration trains, whereas one hopper is for solid and the other for pumpable semi-solid waste.

THE INCINERATION TRAINS

Each incineration train consists of the following plant units:
- waste feeding installation
- rotary kiln with front wall assembly
- secondary combustion chamber
- steam boiler
- flue gas treatment plant
- stack

and is designed for the following capacities:
- solid and semi-solid waste   13,500 - 20,000 t/a
  (10,470 - 18,830 kJ/kg LHV)
- liquid waste   5,000 - 11,500 t/a
  (16,730 - 39,300 kJ/kg LHV)
- additional organically
  contaminated water without
  significant heat value   16,000 t/a.
- heat generation in the rotary kiln:

  14.0 MW maximum continuous rating
  17.5 MW short time peak load.

- Heat generation in the rotary kiln and the secondary combustion chamber combined:

17.5 MW maximum continuous rating

22.7 MW short time peak load.

HIM-VERBRENNUNGSANLAGE BIEBESHEIM

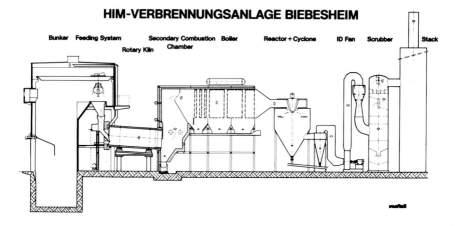

Waste Feeding Installation

The solid waste is loaded by the clam shell grab into the charging hopper and passed into the rotary kiln through the feeding chute. The charging hopper can be closed at its top by a duplex sliding gate and at its lower end by means of a flap gate, both devices together functioning as an air seal. The operation sequence is automatically controlled and interlocked with the drum charging installation.

Semi-solid waste is loaded with the clam shell grab into the tank above the double piston pump which then conveys the material into the rotary kiln through a steam-operated atomizing lance.

An elevator loads the drums into the sluice chamber. Next, a ram device pushes the drums through the feeding chute into the rotary kiln. The sluice chamber is sealed off from the feeding chute by a locking flap which is interlocked with the top level position of the drum elevator. This is to ensure adquate air seal. The entire operating sequence is automatically controlled and interlocked with the charging procedure of solid waste.

Liquid waste is supplied to the respective injection lances and burners through pumps and pipelines.

These are in particular:

in the rotary kiln front wall

- lance for solvents, heavy fuel oil, concentrates from the emulsion splitting plant
- lance for thin sludges
- lance for special wastes.

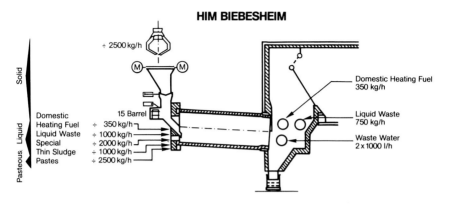

in the secondary combustion chamber
- two lateral opposite lances for waste water
- one burner for waste oil, solvents and concentrates from the emulsion splitting plant

Extra light fuel oil is supplied through a loop pipeline and fired for ignition and support firing purposes
in the rotary kiln front wall
- for starting up the plant or support firing in the rotary kiln

in the secondary combustion chamber
- burner for securing the temperature at 900 °C in compliance with the Clean Air Act.

Rotary Kiln with Front Wall Assembly, Secondary Combustion Chamber and Steam Boiler

The entire combustion system including the rotary kiln with its front wall assembly, secondary combustion chamber and steam boiler is designed to allow for fluctuations in the composition of the semi-solid, liquid and solid waste within a wide range, and to enable direct corrective action to be taken in the kiln.

The flow conditions in the rotary kiln and the secondary combustion chamber are devised to ensure effective admission of secondary air into the secondary combustion chamber.

The rotary kiln assembly together with adjoined secondary combustion chamber constitute the core of the combustion line.

The plant is able to operate continously at an operation temperature of 1300° in the rotary kiln. Furthermore the design allows for a residence time of the flue gas of approximately 4 seconds at temperatures above 1200 °C before the introduction of secondary air.

HIM BIEBESHEIM

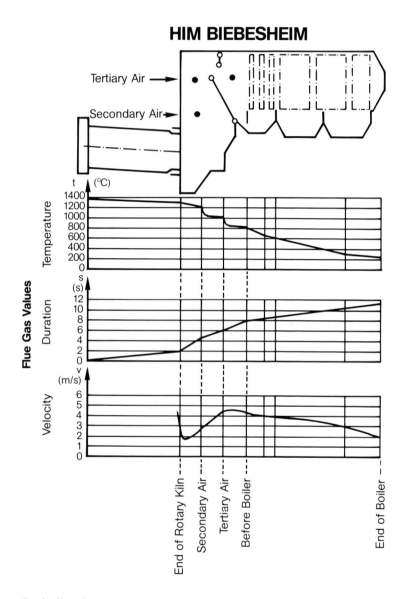

Basically, the rotary kiln can be operated either with solid or liquid slag discharge. In any case, it should be attempted to conserve a solid coat of slag on the lining, and in order to achieve this as best as possible, two conditions are to be met:

- maintaining the steadiest possible combustion chamber temperature in correlation to the average slag temperature

- efficient heat transfer through the refractory lining and drum kiln shell achieved by water-cooling the kiln shell.

The kiln is rotated by means of a hydraulic power unit.

The secondary combustion chamber is immediately connected to the rotary kiln. It ensures a complete burnout of the combustion gases under the following conditions:

- maintaining a minimum temperature of 900 °C. A light oil burner is provided and starts automatically when the temperature drops below this set point.
- long residence time of the combustion gases through low flow velocities.
- additional supply of secondary and tertiary air at the secondary combustion chamber front wall.

The secondary combustion chamber is fully refractory lined up to and including its passage to the boiler. The combustion gas path is lengthened by two tube panels connected to the boiler and functioning as flow reversing baffles. This also ensures efficient gas burnout and furthermore achieves a preliminary separation of fly ash particles.

Secondary and tertiary air supply is arranged at the front side through multiple air ducts. The secondary air promotes complete combustion gas burnout whilst the tertiary air cools the combustion gas before its entry into the boiler. This temperature limitation is required to avoid boiler sooting-up by deposits of sticky ash particles.

The burners and also the waste water lances are placed lateral in the lower section of the secondary combustion chamber. Thus their input takes place before the injection of secondary air, i.e. the organic constituents are incinerated before reaching tertiary air.

The steam boiler is placed downstream of the secondary combustion chamber. In this way, a separation is established between the actual combustion process and the gas cooling. New in connection with the steam boiler is the tertiary air inlet assembly as intergral part of the boiler system. This inlet is located in the top section of the secondary combustion chamber. The tertiary air injection effects a shock-like cooling of the completely burnt-out gases from 1,200 °C down to 800 °C.

Also, the liquid and fused ash particles and salts carried in the gas stream are cooled down, and by that means it is possible to sufficiently reduce the sticking properties of the fly ash, which are responsible for the soot build-up on the boiler surfaces. The boiler cleaning during operation itself is accomplished by rapping, generating a vibration in the boiler tube panels. The rapping device is located at the side of the boiler and operated automatically.

The boiler configuration provides a gas flow perpendicular to the heating surfaces without a change of direction. The flow velocity is as low as 4 m/sec. This very low gas velocity prevents boiler metal erosion to a very great extent.

The heating surface is divided into evaporator, superheater and economizer sections. The steam parameters were deliberately selected low to encounter boiler corrosion which may occur in waste combustion plants operating with steam data of higher magnitude.

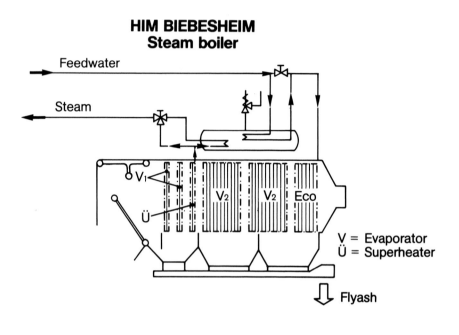

Flue Gas Treatment

Due to the heterogenous composition of hazardous waste, the content of noxious particles in the exhaust gases will constantly fluctuate. On the other hand, very strict official regulations have to be met, and it therefore became imperative to put very strict demands on the design and performance of flue gas treatment plants.

Taking into account the experiences gathered in actually operating incineration plants for wastes from industry and trade, together with values computed from an assumed composition of industrial waste material, as far as the load in the raw gas is concerned, the following noxious substance contents were used as the base for the design of the flue gas treatment plant. For maximum allowable stack emissions the German Clean Air Act as well as the official stipulations were decisive, and the limits as below had to be guaranteed:

	Raw Gas mg/Nm³	Guaranteed Limit at Stack mg/Nm³
HCl	4000	100
HF	250	5
SO₂	2000	200
Dust	8000	75
Cd	31	0.14
Pb	143	0.63
Cr	56	0.25
Cu	33	0.14
V	0.55	0.0024
Zn	219	0.96

It was known that the absorption of acids, gaseous noxious substances and also dust particles of not-too-small grain size can be effectively accomplished with relative ease in wet gas scrubbing plants. Considerably more difficult to solve was the separating of aerosols from the gas. In order to achieve this, a "conditioning" of the aerosols was required, linked with an enlargement of the particles, which leads to a simplification of the separation procedure. This conditioning was brought about by

• agglomeration of several aerosol particles to larger particles
• adherence and fixation of water to these single larger particles

The separation of aerosols makes it possible to remove fine dust and metal oxides with an efficiency rate of at least 99.5%, and by virtue of the very fine droplet formation in the waste gas stream after the scrubber, the subsequent reheating of the gas can be forgone as no water fallout has to be feared. This means a significant advantage in terms of process economy and operation.

A simplified flow sheet illustrates the process basics of the effluent-free flue gas cleaning system (next page).

After leaving the boiler at a temperature of 250 - 280 °C, the hot flue gases pass through the reactor (1) from the top down in parallel flow with the injected thin sludge which comes from the circulating tanks of the two gas scrubbing stages (7 and 8).

In the drying process the thin sludge is combined with the primary dust of the flue gases and the dried product is discharged via airlocks from the reactor and the cyclones installed upstream (2). The dust-freed gases reach the quench (3) and a layer of packing (4), which constitute the first gas scrubbing stage. The circulating liquid of this stage contains free hydrochloric acid, the concentration of which is regulated by admixture of Ca(OH)₂. From the sludge separating device (7) thin sludge is continuously passed into the reactor (1) via the mixing tank (9). Located above the first gas scrubbing stage is set a second one which is equipped with ring-jet elements (5) for the separation of aerosols. Each scrubbing stage is followed by a droplet separating unit (6). The fresh water inflow serves to flush the droplet separators. To enable the second washing stage to absorb SO_2, liquid NaOH is added to the washing liquor drawn from the circulating vessel (8).

HIM BIEBESHEIM
Process Principle

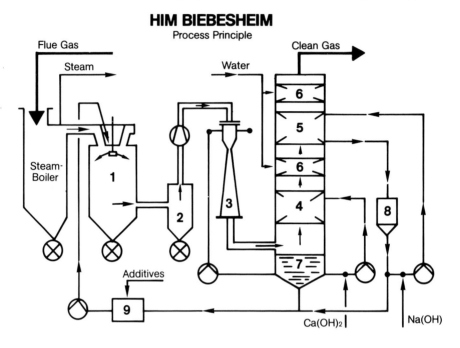

In line with the promotion of this project by the German Federal Ministry for Research and Technology an extensive test and analysis program will be carried out in cooperation with the USEPA.

A material and mass balance will indicate possible effects on the environment by the hazardous waste treatment plant, which will be minimized by an appropriate optimizing process.

Official measurements to determine the degree of the separation of the noxious substances and dust species have not been conducted yet. However, some preliminary tests have brought the following results:

	Guaranteed Limits mg/Nm³	Actually Measured Mean Values mg/Nm³
SO_2	200	150
HCl	100	17
Dust	75	52.6
• therein:		
Pb	0.63	0.126
Cd	0.14	0.054

Zn	0.96	0.956
Cu	0.14	0.298
Cr	0.24	0.014

(at 11% O_2, wet gas)

STEAM UTILIZATION

The steam raised in the two waste boilers is utilized for power generation and as process steam at the plant itself. A condensing-extraction turbo generator produces 3.5 MW at maximum load. Running the entire plant requires approximately 1.8 MW, the surplus is sold to the public utility company.

The extracted steam is used for internal process purposes such as heating of storage and blending tanks, heating of the building, as atomizing medium at the liquid waste and sludge lances and for the emulsions splitting plant.

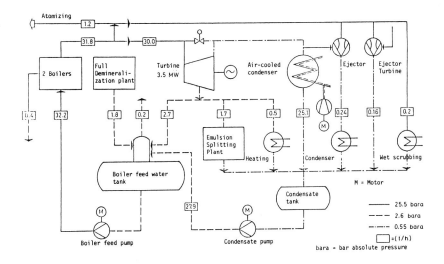

CONCLUSION

The successful operation of the Biebesheim plant in its first year of operation demonstrates in an excellent manner that a proven technology for a safe disposal of hazardous waste under observation of the strictest environmental standards is available. The well functioning of the H.I.M. organization also makes clear that regional disposal facilities should be backed up by appropriate legislation as a kind of guarantee of its function. The active involvement of the waste generators finally gives to those the security of having access at the lowest possible cost to a waste disposal organization that follows the strict and extensive environmental standards of the State of Hesse.

REFERENCES

1. Erbach, Dipl.-Ing. G., 1982. Hessische Industriemüll-Beseitigungsanlage (HIM) - Erfahrungen des ersten Betriebsjahres at *VGB-Fachtagung "Müllverbrennung"*, Bielefeld October 1, 1982 and München on October 22, 1982.

2. Hessische Industriemüll GmbH, 1982, Sonderabfallverbrennungsanlage Biebesheim, Konzept und Betriebserfahrungen *Report*

3. Fattinger, Dr. V. 1979, Abwasserfreies Rauchgaswaschverfahren zum Abscheiden des Feinstaubes und der Schadgase nach Abfallverbrennungsanlagen at *International Recycling Congress, Berlin, October 1-2, 1979*.

PART 2

Sites—Distribution, Selection and Geological Considerations

The amount of hazardous and toxic wastes has increased tremendously in the past 40 years. As a consequence many of these hazardous materials have been improperly discarded creating sites that pollute the surrounding environment. The problems of removing material from a hazardous site is complex for these sites consist of a diverse group of solids, liquids and gases.

The initial step in gaining a perspective to the problem is to determine the spatial pattern of the hazardous waste sites. Chapter eight presents the current distribution of hazardous waste sites, including treatment, storage and disposal in Pennsylvania. This is followed by an analysis of their distribution relating the spatial patterns to selected demographic, land use, and economic factors.

The geological considerations are of major importance in the selection of a waste disposal site. The diversity of waste materials usually increases the difficulty of selecting a satisfactory site. This complex problem requires extensive study and evaluation of a site before it can be shown that it is geologically sound as a waste storage and disposal site. This problem may be further complicated by the political requirement that each region or state dispose of its own hazardous materials. There simply may not be a geologically satisfactory site within the political area.

The geological time perspective is also a fundamental consideration. The movement of surface, soil, and ground water may be at an extremely rapid rate or so slow that they appear unaffected by land use activities and waste disposal practices. There is, however, strong evidence that due to the presence of hazardous materials in many areas there is a degradation of the water quality. Chapter ten treats the time perspective in the subsurface management of waste. Such aspects are considered as potential contributing and hydrologically active areas, waste characteristic, hydrogeologic setting of waste disposal sites, surficial deposits, bedrock strata, and groundwater flow systems.

The final paper in Part Two treats theory versus reality in the siting of hazardous waste facilities. It is recognized that in the past, siting divisions did not consider safety as a major concern. By utilizing accepted criteria for siting hazardous waste dumps, the study revealed the most suitable and least suitable areas in the region. There is nevertheless public opposition for new hazardous waste facilities motivated by fear of impact on the resident's health, degradation of the natural resources, decreased property values, and a general decline in the quality of life in the region.

Hazardous and Toxic Wastes: Technology, Management and Health Effects. Edited by S.K. Majumdar and E. Willard Miller. © 1984, The Pennsylvania Academy of Science.

Chapter Eight

A Geography of Hazardous Waste Sites in Pennsylvania

George A. Schnell[1] Ph.D. and Mark Stephen Monmonier Ph.D.[2]

[1]Department of Geography
The College at New Paltz
State University of New York
New Paltz, New York 12561

[2]Department of Geography
Syracuse University
Syracuse, New York 13210

Among the aspects of the hazardous waste issue, one seems to grab headlines regularly—the "discovery" of sites in different parts of the United States. Obviously, those living within or near the affected place are vitally interested in conditions which may well be life threatening. Moreover, the authorities at a variety of levels—from local to national—often become adversaries as the problem gains definition and requires immediate as well as long term solutions. As sites are "discovered" by agencies involved or, more frequently it is hoped, firms or individuals apply for permits to transport, store, or dispose of hazardous materials, a collection of individual sites evolves into a distributional pattern with important ramifications for settlement and migration, economic development (including commercial and residential as well as industrial and agricultural), and a range of political-social activities at several levels.

In this chapter, we deal with the current distribution of hazardous waste sites, including treatment, storage, and disposal, in Pennsylvania. After describing the distribution of sites, we seek reasons for the distribution by relating the spatial pattern to selected demographic, land use, and economic factors. Our approach is quantitative-cartographic and we hope to stir the reader's interest in the geography of hazardous waste as a fundamental aspect of the greater topic.

DATA

Data on hazardous waste sites for this study were provided by the U.S. Environmental Protection Agency, Region III, in Philadelphia, and consist of the names and addresses of notifiers in conformance with the Resource Conservation and Recovery Act. Specifically, firms that require a permit based upon the statutes are identified by a process code indicating whether they treat, store, or dispose of hazardous waste, as defined by the Commonwealth's Department of Environmental Resources' Solid Waste and Hazardous Waste Management Regulations (1), and describing the method of treatment, storage, or disposal. Thus, our primary data concern the sites at which solid or hazardous waste may "cause or significantly contribute to an increase in mortality or morbidity. . ." and "pose a substantial present or potential hazard to human health or the environment when improperly treated, stored, transported, disposed of, or otherwise managed." (2) Of significance, especially in a paper on Pennsylvania, coal refuse is not included among hazardous waste as defined here. Rather, the Coal Refuse Disposal Control Act (3) regulates that industry and allows us to concentrate on other activities, especially manufacturing. Treatment, storage, and disposal facilities are examined in lieu of the more general term "generator," or producer, of hazardous waste because of the definitive data base used here. The terms defined for use in the Commonwealth of Pennsylvania are:

"Treatment—A method, technique, or process. . .designed to change the physical, chemical, or biological character or composition of any waste. . . so as to render it neutral or nonhazardous.

"Storage—The containment of waste on a temporary basis in such a manner as not to constitute disposal of such waste. It shall be presumed that the containment of waste in excess of one year constitutes disposal.

"Disposal—The incineration, deposition, injection, dumping, spilling, leaking, or placing of solid waste into or on the land or water in a manner that the solid waste or its constituent. . . enters the environment, is emitted into the air, or is discharged to the waters. . ." (4)

GEOGRAPHIC PATTERNS

The spatial patterns of hazardous waste sites, shown in Figures 1, 2, 3, and 4 and listed by category by county in Table 1, represent respectively all sites, storage sites, treatment sites, and disposal sites. Except for disposal sites, these patterns reflect population distribution, with the most prominent concentration in the southeast, around Philadelphia and Delaware counties, and a secondary concentration in Allegheny County. For both storage and treatment sites, minor clusters can be seen around the industrial cities of Reading, Lancaster, and York and, for storage sites alone, similar clusterings can be observed around Bethlehem, Allentown, Harrisburg, Erie, as well as along the Ohio River in Allegheny and Beaver counties. Other noteworthy clusters of storage sites are evident in a north-south belt across the center of Montgomery County, and in

All Sites: Storage, Treatment, Disposal

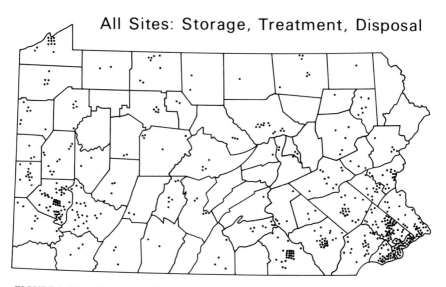

FIGURE 1. Hazardous waste sites of all types.

Storage Sites

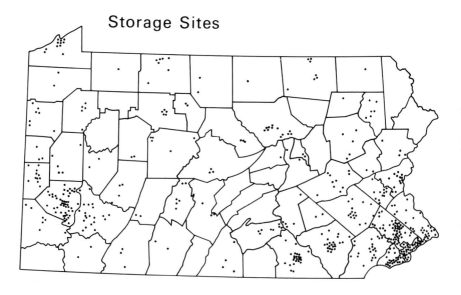

FIGURE 2. Hazardous waste sites—storage.

Treatment Sites

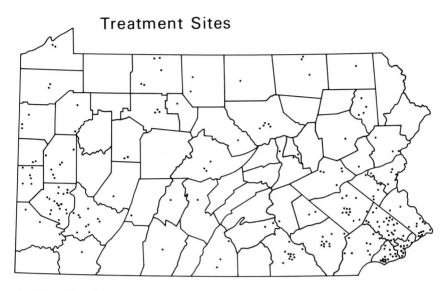

FIGURE 3. Hazardous waste sites—treatment.

Disposal Sites

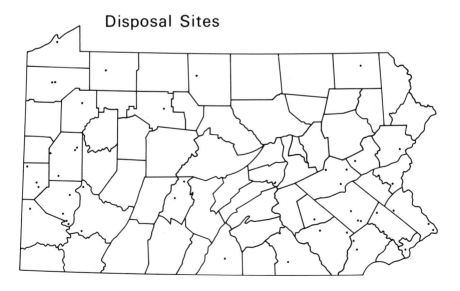

FIGURE 4. Hazardous waste sites—disposal.

TABLE 1

Treatment, Storage, and Disposal Sites. Pennsylvania Counties, 1983

County	All Sites	Storage Sites Total	Storage Combined With Treatment	Storage Combined With Disposal	Treatment Sites Total	Treatment Combined With Storage	Treatment Combined With Disposal	Disposal Sites Total	Disposal Combined With Storage	Disposal Combined With Treatment	3-Function Sites
Adams	0	0	0	0	0	0	0	0	0	0	0
Allegheny	50	46	16	1	17	16	0	4	1	0	0
Armstrong	2	2	0	0	0	0	0	0	0	0	0
Beaver	12	11	5	3	6	5	2	3	3	2	2
Bedford	2	2	1	0	1	1	0	0	0	0	0
Berks	19	17	12	2	13	12	2	3	2	2	2
Blair	6	6	2	2	2	2	1	3	2	1	1
Bradford	7	7	3	0	3	3	0	0	0	0	0
Bucks	33	31	13	2	16	13	3	3	2	3	2
Butler	11	5	2	0	7	2	3	4	0	3	0
Cambria	2	2	1	0	1	1	0	0	0	0	0
Cameron	2	2	1	0	1	1	0	0	0	0	0
Carbon	3	2	1	0	1	1	0	1	0	0	0
Centre	4	3	2	0	3	2	0	0	0	0	0
Chester	22	21	8	2	9	8	0	2	2	0	0
Clarion	0	0	0	0	0	0	0	0	0	0	0
Clearfield	5	5	0	0	0	0	0	0	0	0	0
Clinton	3	3	1	0	1	1	0	0	0	0	0
Columbia	0	0	0	0	0	0	0	0	0	0	0
Crawford	5	2	0	0	2	0	1	2	0	1	0
Cumberland	8	8	2	0	2	2	0	0	0	0	0
Dauphin	5	5	1	0	1	1	0	0	0	0	0
Delaware	26	24	12	0	14	12	0	2	0	0	0
Elk	5	5	3	1	3	3	1	1	1	1	1
Erie	12	11	1	1	2	1	1	1	1	1	1
Fayette	3	2	0	0	1	0	0	0	0	0	0

County											
Forest	0	0	0	0	0	0	0	0	0	0	0
Franklin	4	4	3	1	3	3	1	1	1	1	1
Fulton	0	0	0	0	0	0	0	0	0	0	0
Greene	0	0	0	0	0	0	0	0	0	0	0
Huntingdon	4	3	0	0	1	0	0	0	0	0	0
Indiana	4	4	1	0	1	1	0	0	0	0	0
Jefferson	2	2	2	0	2	2	0	0	0	0	0
Juniata	0	0	0	0	0	0	0	0	0	0	0
Lackawanna	6	6	4	0	4	4	0	0	0	0	0
Lancaster	23	23	10	2	10	10	1	2	2	1	1
Lawrence	3	3	2	0	2	2	0	0	0	0	0
Lebanon	3	2	2	0	2	2	0	1	0	0	0
Lehigh	12	12	5	0	5	5	0	0	0	0	0
Luzerne	5	4	1	0	2	1	0	0	0	0	0
Lycoming	13	12	4	0	5	4	0	0	0	0	0
Mc Kean	8	7	4	0	5	4	0	0	0	0	0
Mercer	7	7	3	0	3	3	0	0	0	0	0
Mifflin	0	0	0	0	0	0	0	0	0	0	0
Monroe	5	4	3	0	3	3	1	1	0	0	0
Montgomery	52	49	20	0	23	20	0	1	0	0	0
Montour	2	2	1	0	1	1	1	0	0	1	1
Northampton	14	13	5	1	6	5	0	1	1	0	0
Northumberland	3	2	1	0	2	1	0	0	0	1	1
Perry	0	0	0	0	0	0	1	0	0	0	0
Philadelphia	31	30	8	1	9	8	0	1	1	0	1
Pike	0	0	0	0	0	0	1	0	0	0	0
Potter	2	2	2	1	2	2	1	1	1	1	1
Schuylkill	10	10	5	3	5	5	3	3	3	3	3
Snyder	0	0	0	0	0	0	0	0	0	0	0
Somerset	3	3	0	0	0	0	0	0	0	0	0
Sullivan	0	0	0	0	0	0	0	0	0	0	0
Susquehanna	1	1	1	1	1	1	1	1	1	1	1
Tioga	1	1	1	0	1	1	0	0	0	0	0

TABLE 1 Continued

County	All Sites	Storage Sites			Treatment Sites			Disposal Sites			3-Function Sites
		Total	Combined With Treatment	Disposal	Total	Combined With Storage	Disposal	Total	Combined With Storage	Treatment	
Union	1	1	0	0	0	0	0	0	0	0	0
Venango	5	5	2	1	2	2	1	1	1	1	1
Warren	2	1	0	0	0	0	0	0	0	0	0
Washington	7	7	3	1	3	3	1	1	1	1	1
Wayne	0	0	0	0	0	0	0	0	0	0	0
Westmoreland	22	21	10	1	10	10	1	2	1	1	1
Wyoming	1	1	0	0	0	0	0	0	0	0	0
York	25[1]	23	10	1	11	10	1	1	1	1	1
Pennsylvania Total	528[2]	487	200	28	230	200	28	45	28	28	22

Source: Compiled by authors from data provided by the United States Environmental Protection Agency.
[1]Figures for York County include one site with activity not specified.
[2]Rows may not sum to totals shown because of multiple activities at some sites.

Lycoming and northern Bradford counties. In at least a general sense, the patterns of storage and treatment sites correspond to those of population, urban and built-up land, manufacturing employment, and value added in manufacturing.

Although a more thorough study requires correlation analysis, employed later in this chapter, a careful examination of the maps of waste sites reveals not only a spatial association with the explanatory factors but also a tendency, even in remote, nonmetropolitan areas toward clusters of two, three or more sites. These nonmetropolitan clusters suggest that industries generating hazardous waste and possibly nonhostile attitudes toward both industry and its waste are more concentrated than data aggregated by county unit would suggest. Thus, although sites are heavily concentrated in the metropolitan sections of the Commonwealth, a perusal of the original E.P.A. list (in which sites are identified by address) reveals some clusters of note in nonmetropolitan counties—Bradford City in McKean County, East Stroudsburg in Monroe County, Meadville in Crawford County, St. Marys in Elk County, and Towanda in Bradford County. Each of these clusters appears clearly on the "all sites" map, Figure 1, across the state's northern reaches.

Disposal sites, by far the more dangerous to those living and working nearby, exhibit a pattern very different from those of storage and treatment sites. Although there are a few disposal sites present, Philadelphia and surrounding counties have proportionately fewer disposal sites than they have storage and treatment sites. This observation is valid also for most if not all of the other clusters noted for storage and treatment sites: Pittsburgh, Erie, Allentown, Bethlehem, Reading, Lancaster, and York.

The spatial clustering of hazardous waste sites is evident in Figure 5, a plot of the cumulative percentage of sites (scaled along the vertical axis) and the cumulative percentage of counties (scaled along the horizontal axis). This graph indicates that almost three-quarters of the state's 528 hazardous waste sites are concentrated in that quarter of the Commonwealth's counties with the highest frequency of sites. The absence of hazardous waste sites in 13 counties underlies much of this clustering, as do the noteworthy concentrations of 52 sites in Montgomery County, 50 sites in Allegheny County, 33 sites in Bucks County, and 31 in Philadelphia. Not surprisingly, the pattern of storage permits, which characterizes 487 sites, or 92 percent, is highly similar to that for all sites. Thirteen counties have no storage sites, whereas Montgomery and Allegheny counties have 49 and 46, respectively. The state's 230 treatment sites, 44 percent of the total, have a similar pattern of concentration—not surprising insofar as 200 of these are also storage sites and only 19 counties lack treatment sites. More noticeably clustered, though, are Pennsylvania's disposal sites, only 22 of which also include both storage and treatment functions. Only 9 percent of all hazardous waste sites have (or have applied for) a disposal permit. Forty-one counties have no disposal sites, whereas Allegheny and Bucks have 4 each.

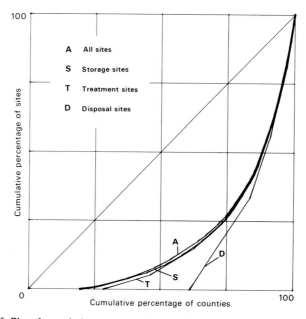

FIGURE 5. Plot of cumulative percentage frequencies for number of hazardous waste sites and counties, which are rank-ordered from those with no sites (on the left) to the county with the most sites (on the right). Degree of departure from the diagonal illustrates the degree of concentration: if all counties had an equal number of sites, the curve would be a straight, diagonal line, whereas if all sites were concentrated in a single county, the curve would be a reverse-L.

RELATED GEOGRAPHIC DISTRIBUTIONS

Since more populous counties are likely to have more industrial generators of hazardous waste than relatively less populated areas, we examine next the distribution of the Commonwealth's population. Figure 6, a dot map of Pennsylvania's population distribution in 1980, depicts a pattern which is easy to describe. There are two foci. The first, centered on Philadelphia, extends outward to include clusters as far as Harrisburg and York to the west and Scranton and Wilkes Barre to the north. The second is a dense area centered on Pittsburgh and Allegheny County, extending into the periphery and outward along the numerous valleys. The only cluster of any size along the northern tier is Erie and its suburbs and satellites in northwestern Pennsylvania. Although clusters of dots exist elsewhere on the map, they represent smaller settlements than those described above and almost all are restricted to locations within Standard Metropolitan Statistical Areas (SMSAs). Indeed, as Table 2 illustrates, almost 82 percent of the nearly 12 million inhabitants are residents of the 30 counties comprising the 15 SMSAs of the Commonwealth. Philadelphia SMSA and Pittsburgh SMSA account for half of the total, housing 31 and 19 percent, respectively. Moreover,

TABLE 2

Population in Pennsylvania's SMSAs and Central Cities. 1980. in Thousands

SMSA/City	No.	Percent	SMSA/City	No.	Percent
Allentown-Bethlehem-Easton[1]	552	4.7	Philadelphia[1]	3,683	31.0
Allentown city	104	0.9	Philadelphia city	1,688	14.2
Bethlehem city	70	0.6	Pittsburgh	2,264	19.1
Easton city	26	0.2	Pittsburgh city	424	3.6
Altoona	137	1.2	Reading	313	2.7
Altoona city	57	0.5	Reading city	79	0.7
Binghamton[1]	38	0.3	Sharon	128	1.1
Erie	280	2.4	Sharon city	19	0.2
Erie city	119	1.0	State College	113	1.0
Harrisburg	446	3.8	State College Borough	36	0.3
Harrisburg city	53	0.4	Williamsport	118	1.0
Johnstown	265	2.2	Williamsport city	33	0.3
Johnstown city	35	0.3	York	381	3.2
Lancaster	362	3.1	York city	45	0.4
Lancaster city	55	0.5	All 15 SMSAs[1]	9,720	81.9
Northeast Pennsylvania	640	5.4	All 18 Central Cities	3,010	25.4
Hazelton city	27	0.2			
Scranton city	88	0.7	The Commonwealth	11,867	100.0
Wikes Barre city	52	0.4			

Source: Computed by authors from 1980 Census data.
[1]Includes only those parts in Pennsylvania.

the 18 central cities listed in Table 2 house more than 3 million Pennsylvanians, the City of Philadelphia alone accounting for almost 1.7 million of the aggregate. At the other extreme, the populations of Cameron, Forest, and Sullivan counties barely sum to 18 thousand, and exemplify the relatively empty northern and north-central areas of Pennsylvania.

For this study, we measure the effect of urban development not only by place of residence of the population but by its direct effect upon the landscape. Specifically, we have computed by county the percentage of land area classified by the U.S. Geological Survey in its nationwide inventory of land use/cover as urban or built up (5). Included are all parcels, 4 hectares or larger, of residential, commercial, industrial, transportation, or mixed urban land, whether within or outside an urbanized area or urban place as defined by the Bureau of the Census (6). Similarly, tracts within cities or towns but not built up or otherwise involved in urban activities are excluded from the numerator. When mapped by county units, as in Figure 7, urban and built-up land as a percentage of all land shows clearly the effects of terrain, accessibility, and metropolitan development. Counties surrounding Philadelphia and Pittsburgh record the highest percentages, whereas counties in rugged, remote parts of the Appalachian Plateau and the Ridge and Valley province register the lowest values. Locally high percentages, contrasting with more rural neighboring counties, occur in Erie County, in the northwest; Lackawanna and Luzerne counties, in the northeast; and Dauphin

Population Distribution

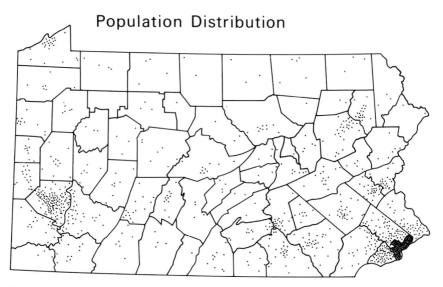

One dot represents 0.1 percent of the total state population.

FIGURE 6. Population distribution, 1980.

Urban and Built-up Land as Percentage of All Land

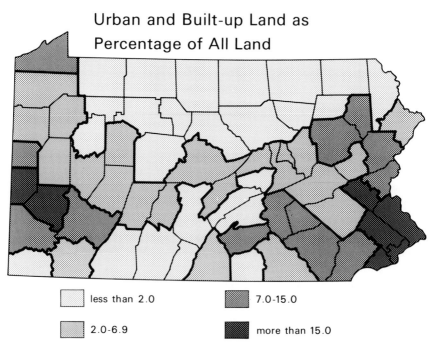

less than 2.0

2.0-6.9

7.0-15.0

more than 15.0

FIGURE 7. Urban or built-up land as a percentage of all land.

Manufacturing Employment as Percentage of the Labor Force

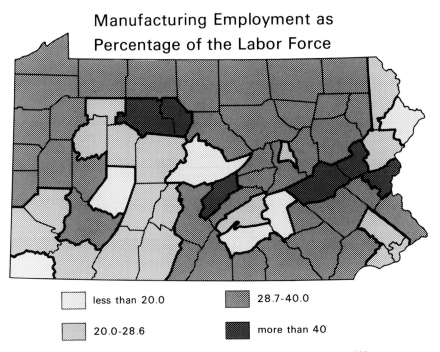

less than 20.0

20.0-28.6

28.7-40.0

more than 40

FIGURE 8. Manufacturing employment as a percentage of the labor force, 1980.

and Lancaster counties, in the southeast.

A companion choropleth map, Figure 8, illustrates the geographic pattern of manufacturing employment. Defined here as persons 16 years of age or older employed in manufacturing in 1980 as a percentage of all persons in the labor force in that year (7), this measure of the relative prominence of manufacturing within the county has a geographic pattern very different from that of urban and built-up land: the metropolitan regions centered on Philadelphia and Pittsburgh have proportionately less manufacturing employment than many far more rural counties. Indeed, manufacturing employment as a percentage of all employment is most prominent in Elk (55.0 percent), Cameron (47.2), Carbon and Northampton (43.8 percent each), Schuylkill (40.8), and Mifflin (40.7), only two of which might be considered urban-industrial. The lowest percentages occur in Greene (10.4), Pike (15.8), Indiana (17.7), Centre (17.8), Dauphin (19.4), and Cumberland (19.7), which are more heavily involved in various combinations of agricultural, mining, or service activities.

A map of the proportion of the labor force employed in manufacturing ignores, of course, the size of a county's labor force as well as its gross productivity. The spatial patterns of these factors are portrayed in Figures 9 and 10, maps that focus attention on the magnitude of manufacturing employment and value added, respectively (8). As might be expected, both distributions are heavily

Employment in Manufacturing

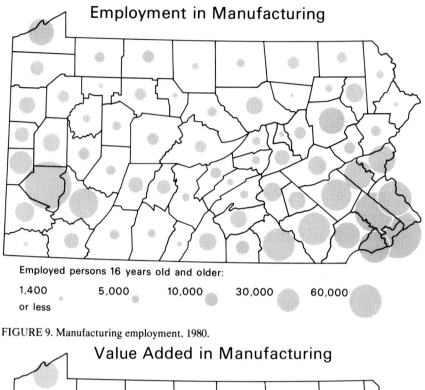

Employed persons 16 years old and older:

1,400   5,000   10,000   30,000   60,000
or less

FIGURE 9. Manufacturing employment, 1980.

Value Added in Manufacturing

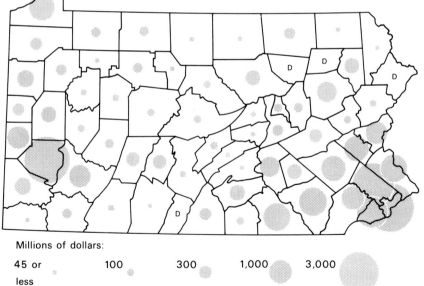

Millions of dollars:

45 or   100   300   1,000   3,000
less

FIGURE 10. Value added in manufacturing, 1977. D indicates county for which nondisclosure provision required that the value not be reported.

TABLE 3

Correlation between Waste-site/Area Rates and Independent Variables

	Site Density			
	All	Storage	Treatment	Disposal
Product-moment Correlation Coefficients				
Population density	0.904	0.911	0.753	0.521
Ratio of population/urban and built-up land	0.565	0.577	0.423	0.233
Value added in manufacturing/urban and built-up land	0.299	0.304	0.277	0.349
Percentage of labor force in manufacturing	−0.166	−0.168	−0.129	0.091
Value added per manufacturing employee	0.234	0.229	0.295	0.095
Urban and built-up land as percent of all land	0.977	0.974	0.943	0.557
Coefficient of Multiple Determination	0.973	0.972	0.924	0.413
Rank-Order Correlation Coefficients				
Population density	0.801	0.780	0.736	0.468
Ratio of population/urban and built-up land	0.054	0.060	0.007	−0.176
Value added in manufacturing/urban and built-up land	0.599	0.592	0.546	0.292
Percentage of labor force in manufacturing	0.105	0.099	0.142	0.228
Value added per manufacturing employee	0.494	0.476	0.407	0.236
Urban and built-up land as percent of all land	0.802	0.780	0.747	0.521

Source: Computed by authors from Census data and data provided by the Environmental Protection Agency.

focused on the southeastern part of the Commonwealth, with noteworthy secondary concentrations centered on Pittsburgh and Erie. As with total population, manufacturing employment and value added have very low densities in the northern and central regions of Pennsylvania.

CORRELATION ANALYSIS

To provide consistent, size-independent intensity measurements for correlation and regression analysis, the counts of all sites, as well as storage, treatment, and disposal sites were converted to ratios by dividing by land area. A similar adjustment was made for county population, converting that to population density. Urban and built-up land was treated as a percentage of total county land area, and manufacturing employment was measured as a percentage of the total county labor force. Value added per manufacturing employee provided a measure of employee productivity. Value added and total population size served as numerators of two additional intensity ratios obtained by dividing by the amount of urban and built-up land. Pearson product-moment and Spearman rank-order correlation coefficients between each dependent, hazardous waste variable and each independent variable are shown in Table 3, which also reports coefficients of multiple determination for each dependent variable.

An inspection of the product-moment correlation coefficients in Table 3 suggests two significant observations: The distribution of disposal sites is far more difficult to explain than the pattern of all, storage, and treatment sites, and urban and built-up land as a percentage of all land is the single most powerful independent variable in accounting for all four hazardous waste distributions. Urban and built-up land accounts for 95 percent of the distribution of all sites and storage sites, 89 percent of treatment sites' geographic pattern, and 31 percent of the distribution of disposal sites. Also noteworthy as a predictor is population density, which accounts for 82 percent of the pattern of all sites and storage sites, 57 percent of the pattern of treatment sites, and 27 percent of the pattern of disposal sites. The ratio of population to urban land is only moderately correlated with the density of hazardous waste sites; it accounts for 32 percent of the pattern of all sites, 33 percent of the pattern of storage sites, 18 percent of the pattern of treatment sites, and 5 percent of disposal sites' pattern. Least significant are the three independent variables related to manufacturing. For all sites, storage sites, and treatment sites, the correlation is even negative, suggesting very weakly that counties with higher-than-average employment in manufacturing tend to have slightly lower-than-average densities for all but disposal sites. Yet, one manufacturing index, the ratio of value added to urban and built-up land, has the third highest association with disposal sites, accounting for 12 percent of the variation in disposal-site density.

Rank-order correlation generally supports the results of product-moment correlation. Again, the percentage of urban and built-up land is the most powerful predictor, followed closely by population density. The statistical association is higher once again for the densities of all and storage sites than for the less numerous treatment sites, with the lowest correlations and the weakest relationships generally recorded for the least numerous disposal sites. With the rank-order correlations, however, the ratio of population to urban land is virtually insignificant, whereas the ratios of value added to both urban land and manufacturing employment have moderately high correlations with the county-unit densities of all, storage, and treatment sites. These differences originate, at least in part, in the large number of zero densities for counties without hazardous waste sites of any type.

Correlation analysis, as shown in Table 4, verifies the general similarity in the patterns of county-unit densities for all, storage, and treatment sites noted in our discussion of the dot maps of hazardous waste sites. Not surprisingly, the highest correlation between sites is for all sites and storers, a category which comprises 92 percent of all sites. The patterns of storage and treatment sites are closely related, though, with product-moment and rank-order correlations of 0.95 and 0.92, respectively. Neither storage nor treatment sites has more than a moderate correlation between its density distribution and that of disposal sites.

Disposal sites are more hazardous than either treatment or storage facilities. That their distribution differs from those of the other two types suggests that dif-

TABLE 4

Similarity Matrices of Correlation Coefficients for the Waste-site/Area Rates

	Product-moment Coefficients				Rank-order Coefficients			
	All	Storage	Treatment	Disposal	All	Storage	Treatment	Disposal
All	1.000	0.999	0.953	0.556	1.000	0.987	0.937	0.602
Storage	0.999	1.000	0.946	0.547	0.987	1.000	0.919	0.558
Treatment	0.953	0.946	1.000	0.493	0.937	0.919	1.000	0.615
Disposal	0.556	0.547	0.493	1.000	0.602	0.558	0.615	1.000

Source: Computed by authors from Census data and data provided by the Environmental Protection Agency.

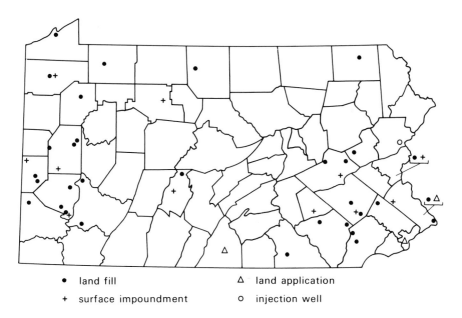

•	land fill	Δ	land application
+	surface impoundment	o	injection well

FIGURE 11. Disposal sites by type.

ferent processes might have produced these patterns. To be sure, local opposition can be strong, and many communities have objected vehemently to serving as depositories for the waste of other areas (9). Yet, whatever combinations of ignorance and political weakness might have promoted the evolution of the pattern in Figure 11, it is likely to change, given recent revelations of heretofore unknown disposal sites and repeated warnings by health authorities and the National Academy of Sciences that hazardous waste is more likely to kill than stink.

CONCLUDING REMARKS

Our concluding remarks reflect a regret that information on the comparative hazardousness of the sites mapped is unavailable. Rating schemes have been developed and tested, and an appropriate comparative index might soon be available (10). For now, though, we present in Figure 11 the geographic pattern of the disposal methods reported for the Commonwealth's 45 (disposal) sites. That the clear majority are landfills, a low-cost option, is reassuring to neither local residents nor public health authorities, who consider landfills and other underground or buried storage a hazard with a definite time penalty—ill effects might not surface until the solution is too costly or the biochemical effects are irreversible. The other disposal methods shown—surface impoundment, land application, and injection well—are not much more reassuring given that they merely hide, shift, or disperse the problem, and perhaps delay a more nearly permanent solution that could convert the waste to less harmful substances. The current situation, especially regarding disposal tactics, might well affect the death and tax rates of future generations, as present distributional patterns play a vital, prominent role in the geography of Pennsylvania in the future.

ACKNOWLEDGEMENT

The authors gratefully acknowledge the assistance of Paul J. Pocavich, Division of Hazardous Waste Management, the Department of Environmental Resources of the Commonwealth of Pennsylvania, in acquiring the primary data used in this study.

REFERENCES

1. These regulations and all definitions which follow in this section are from Commonwealth of Pennsylvania, Legislative Reference Bureau. 1982. *Pennsylvania Bulletin*. 12: 2981-3084.
2. *Pa. Bull.*, 2984.
3. *Pa. Bull.*, 2984.
4. *Pa. Bull.*, 2983 and 2986.
5. The state and county summary tabulations used to compute this variable were compiled with GIRAS software, described in Mitchell, W. B. 1977. GIRAS: A Geographic Information Retrieval and Analysis System for Handling Land Use and Land Cover Data. *U.S. Geological Survey Professional Paper 1059.*
6. Anderson, J. R. and others. 1976. A Land Use and Land Cover Classification System for Use with Remote Sensor Data. *U.S. Geological Survey Professional Paper 964.*
7. These percentages were calculated from data in U.S. Bureau of the Census. 1983. *1980 Census of Population and Housing, Advance Estimates of*

Social, Economic, and Housing Characteristics, Supplementary Report PHC80-52-40, Pennsylvania. U.S. Government Printing Office, Washington, D.C.

8. Data on value added are from U.S. Bureau of the Census. 1980. *1977 Census of Manufactures, Geographic Area Series, MC77-A-39, Pennsylvania.* U.S. Government Printing Office, Washington, D.C.

9. See Morell, D. and Christopher Margorion. 1982. *Siting Hazarous Waste Facilities.* Ballinger Publishing Company, Cambridge, Mass., 21-46; and Duberg, J. A. and others. 1980. Siting of Hazardous Waste Management Facilities and Public Opposition. *Environmental Impact Assessment Review.* 1:84-88.

10. See, for example, Kufs, C. and others. n.d. Rating the Hazard Potential of Waste Disposal Facilities. *Proceedings of the U.S. EPA National Conference on Management of Uncontrolled Hazardous Waste Sites, October 15-17, 1980.* Hazardous Materials Control Research Institute, Silver Spring, Md., 30-41.

Hazardous and Toxic Wastes: Technology, Management and Health Effects. Edited by S.K. Majumdar and E. Willard Miller. © 1984, The Pennsylvania Academy of Science.

Chapter Nine

Geological Factors in Disposal Site Selection

Robert F. Schmalz, Ph.D.

Department of Geosciences
The Pennsylvania State University
University Park, PA 16802

During the years following World War II, waste production in the United States increased dramatically. For the first time, the quantity of waste generated exceeded the amount that the environment could accommodate without deleterious effects. This increase was caused by a rapid growth in population and a sharp rise in the annual per capita generation of all forms of waste (U.S. EPA, 1980-a). The resulting problems of waste disposal were aggravated by the concentration of population and industry in urban-industrial complexes, particularly along the nation's coastlines, and by the disproportionately rapid rise in the quantity of hazardous materials in the waste stream (U.S. EPA, 1980-b).

The term "hazardous waste" commonly evokes an image of leaking barrels of foul-smelling liquids inspired, perhaps, by news photographs of Love Canal or industrial wastes dumps in the New Jersey Meadows. In fact, hazardous wastes comprise an extremely diverse group of solids, liquids and gases, ranging from exotic organic compounds to substances as commonplace as the salts used for highway de-icing (Nemerow, 1978; Parizek, 1970). Some of these chemicals are toxic, others are carcinogenic, teratogenic, or mutagenic; in some cases the deleterious effect is immediately evident, while in others, the impact may not become apparent until years after exposure or until the concentration in the environment (or in an organism) accumulates to some critical threshold level (Pye, et al. 1983; Claus, et al. 1972). This diversity sometimes makes it difficult to identify hazardous constituents in the waste stream, and may provoke vigorous debate about their maximum acceptable levels of concentration in the environment (Hohenemser, et al, 1983). These points are raised here only to help to explain the uncertainties and inconsistencies which sometimes seem to characterize our efforts to protect the environment from hazardous waste contamination.

HAZARDOUS WASTE MANAGEMENT

Many constituents of hazardous waste break down spontaneously by oxidation, hydrolysis or biodegradation in the environment (Metry, 1980-a). Others can be detoxified by chemical reaction or destroyed by high temperature incineration (Metry, 1980-a; Sittig, 1979; U.S. EPA, 1980-c, 1981). Still others, including a surprising variety of industrial materials, can be recycled and reused (Sittig, 1975; Skinner, 1975). There remain, however, substantial quantities of potentially harmful substances which are byproducts of industrial processes or which are concentrated in ordinary waste disposal procedures (Cheng, et al, 1975). These substances are neither recyclable nor readily detoxified. One group of such particularly troublesome hazardous wastes consists of materials such as cyanide metal plating solutions, that are not persistent under ordinary environmental conditions, but which are so toxic and act so swiftly that they cannot be allowed even brief contact with the biosphere. Another group, which may be even more difficult to deal with although they have less immediate impact, are extremely long-lived and require long-term isolation from the biosphere. These persistent hazardous wastes include toxic base metals concentrated in sewage sludge and dredge spoil, various "hard" pesticides including DDT, Kepone and Mirex, detergents of the ABS (alkyl benzene sulfonate) type, PCBs, high- and low-level radioactive wastes, and a variety of other substances (Glynis, et al, 1983; Metry, 1980-a; Pye, et al, 1983; Young, et al, 1975). The fundamental disposal problem which both groups share is the requirement that they be effectively isolated from the biosphere, in the latter case for periods of tens, hundreds or even thousands of years.

In the past, the practical disposal of hazardous wastes (as opposed to "temporary" storage of such materials) was dominated either by the concept of dispersal and dilution or by that of isolation and containment. In the first alternative, it was assumed that natural mixing processes in the environment were sufficient to dilute harmful materials rapidly and thoroughly enough to eliminate any serious ecological threat. Landspreading of sewage sludges, direct discharge of gases to the atmosphere, and dumping of chemical wastes into rivers and the sea were widely practiced; all relied upon the presumed efficiency of the natural dispersion effects. Recent experience has shown that today these natural processes are too often overwhelmed by the growing volume and increased toxicity of the waste burden thrust upon them. The repeated fish kills in the New York Bight during the late 1960s (CEQ, 1980; U.S. EPA, 1978), the pollution of Lake Superior by mill-tailings from the Reserve Mining Company at Silver Bay (Bartlett, 1980), and the disastrous results of the discharge of mercury into the sea at Minamata, Japan (Young, et al, 1975; Perkins, 1974), are but three examples of such environmental overload.

The alternative approach to hazardous waste disposal, isolation from human contact, too often took the form of concealment rather than true isolation. Even where effectively isolated from direct human contact, the waste was not isolated from the rest of the biosphere and many contaminants entered the food web where they jeopardized the general ecosystem directly. Their indirect effect upon man usually did not become apparent for months or years, making the source of the problem more difficult to determine, and corrective measures vastly more complex and costly. Ocean dumping, for example, offered effective concealment and protection against human intrusion, but did little to isolate the waste (U.S. EPA, 1979). The incidence of debilitating deformities, cancers and other diseases among fish in the New York Bight dumping area (NOAA, 1981; U.S. EPA, 1978), the development of antibiotic-resistant bacteria (Summers et al, 1972; Davies et al, 1972; Koditschek et al, 1974), and the bioaccumulation of toxic metals in the flesh of tuna, swordfish and other food fish (Perkins, 1974) afford clear evidence of the ineffectiveness of the isolation techniques practiced in the past. The disposal of hazardous waste in shallow burial sites on land also proved (too often) to be an exercise in concealment rather than containment. Leachate escaping from burial sites which were little more sophisticated than ordinary sanitary landfills was soon recognized as a major source of environmental contamination (Johnson, et al, 1981; McDougal, 1980; Subramanian, 1980). Public reaction to the disclosure of such contamination by chemical wastes in some cases approached panic (Holden, 1980; Kolata, 1980; Smith, 1982-a, -b, -c).

During the 1970's, growing public awareness and increasingly strict Federal regulation encouraged the development of alternative methods of disposing of all forms of hazardous wastes, including radioactive materials. Deep well injection of liquid wastes, which had been in limited use for many years, was examined as a disposal method of more general usefulness (Warner, et al, 1973; Rudd, 1972). The injection of grout mixed with finely ground solid waste, particularly radioactive waste, was also the subject of intensive study (Sun, 1982), and a major effort was made to develop artificial minerals, glasses and synthetic "rock" to fix especially hazardous waste (Ringwood, 1978; Roy, 1982). Finally, in the expectation that man-made containers would afford more reliable long-term confinement of hazardous materials, a variety of in-ground and above-ground structures were built. To date these artificial containers have proven to be only slightly more effective than natural disposal sites, provided the latter were selected with resonable care. Reinforced concrete vaults constructed by the Atomic Energy Commission at Eniwetok Atoll in the early 1950's to contain radioactive waste were broken open and emptied by chemical corrosion and wave action less than fifteen years later (Schmalz, personal observation, 1967). Vastly more sophisticated double-walled concrete and steel tanks erected at Hanford, Washington, for the "temporary" storage of high-level radioactive wastes have leaked repeatedly since 1958, with cumulative loss of nearly 500,000 gallons of highly radioactive and extremely toxic waste (Coates, 1981; Deutch, 1978). If the best economically practicable man-made containers can ensure isolation of hazardous materials for no longer than twenty to thirty years, as these observations suggest, then for the long-term isolation of waste (100 or more years) we must depend primarily upon the natural properties of the environment itself. The use of an artificial container will prolong the confinement of hazardous materials at the disposal site by a negligible amount, and will effect little improvement in site performance over that expected of simple burial in a geologically equivalent site.

BURIAL SITE REQUIREMENTS

Analysis of several unsuccessful hazardous waste burial sites supports the conclusion that failure of waste containers played an unexpected role in the events leading ultimately to condemnation of the site. The mechanical collapse and rupture of waste containers undoubtedly accelerated the release of waste materials, but the principal effect of container collapse was to weaken or destroy the integrity of the cell cap, allowing surface water more rapid access to the buried waste. In three of the most familiar low-level radioactive waste disposal sites (West Valley, New York; Sheffield, Illinois; Maxey Flats, Kentucky) and two hazardous chemical waste disposal sites (Stringfellow Quarry, California; Love Canal, New York) site failure resulted from unanticipated and inadequately controlled contaminant transport by ground- and surface-water (see among others, Kelleher, 1979; Foster, 1982; Dana, et al, 1982; McDougal, 1980; Grant, 1982). Typically, surface water entering the burial cells displaced liquid waste or leached contaminants from buried solid waste. The contaminated leachate then overflowed the burial cells, entering the biosphere directly as surface flow, or infiltrated the ground beneath the cells to contaminate the groundwater reservoir.

It follows from these observations, that the most fundamental requirement for a successful shallow land burial site centers upon effective control of the movement of surface and sub-surface water. In the simplest form, the site must be characterized by:

i) sufficient depth of soil (or regolith) to accommodate the burial cells,
ii) a regional water table at sufficient depth to ensure that the cells will be above the zone of saturation at all seasons,
iii) a surface water regimen which ensures (or can be adjusted to ensure) drainage away from rather than toward the burial cells,
iv) soil (or regolith) sufficiently homogeneous and impermeable to ensure slow, predictable groundwater movement away from the site, and
v) a sufficiently stable geological environment to ensure that conditions suitable for waste disposal at the time of burial will not be substantially altered during the anticipated useful life of the site.

Other site characteristics (slope, seismic activity, lithology, vegetation, climate, etc.) are significant factors in site performance primarily insofar as they affect one or more of the foregoing essential properties. Nearby landslides, for example, may not affect the site directly, but may seriously jeopardize site performance if they divert surface drainage onto the site, destroy the integrity of the cell covers or shorten groundwater flow paths to the surface, forming seeps of contaminated water. Highly deformed and fractured rocks at depth beneath the site, though apparently inconsequential, may so alter the rate and direction of groundwater flow as the make the site unacceptable. Additional examples could be cited, but it should be evident that a wide range of site characteristics and geological phenomena may critically affect performance through their impact on the flow of water across and beneath the site, and by altering its geological characteristics. It is appropriate, therefore, to consider the nature of the geological environment before continuing with the problem of site selection.

The Geological Environment
   The rocks which make up the earth's crust, though varied in both origin and composition, were formed over periods spanning thousands or millions of years. The processes which have subsequently deformed the rocks, and those which alter them when exposed at the earth's surface ("weathering" processes) are also very slow-acting. It follows that the geological characteristics of any potential hazardous waste disposal site have evolved over periods of thousands or millions, even tens of millions of years. They represent an approach toward equilibrium between the lithology, structure and tectonic development of the site on the one hand, and the climate, topography, vegetation and surface and sub-surface drainage of the site on the other. The mechanisms by which this adjustment takes place are exceedingly slow, and significant changes in site characteristics can usually be detected only after the passage of centuries or even millennia. Exceptions to this general principal are usually restricted to areas where rapid erosion is taking place today (coastal areas, river banks, oversteepened slopes, etc.) and to areas subject to flooding or active tectonism (earthquakes, faulting or volcanic eruption). Such areas of rapid change are often readily identifiable, and may be classed as geologically "immature", in contrast to regions of more advanced development where change, if detectable, is extremely slow.
   Obviously, geologically immature locations will be unsuitable for hazardous waste disposal. The exceedingly slow changes characteristic of geologically mature environments, however, promise greater site stability and more predictable long-term site performance than any man-made container or engineered structure. Further benefits of geologically mature sites may often include a deep soil and regolith profile, relatively uniform regolith mineralogy and chemistry, operationally desirable features such as low relief and gentle slopes, freedom from compaction subsidence, established patterns of surface and subsurface

water movement, and clear patterns of surface erosion and vegetation. To the extent that site preparation, operation and closure can be carried out without significantly altering the critical characteristics of the site (topography, vegetation, and drainage) a geologically mature location offers greatly enhanced long-term performance predictability. To realize the maximum benefit from a geologically mature site will require a detailed site-specific geological evaluation before starting operations, and the development of a joint geological and engineering plan for site operation and closure which must be rigorously adhered to during all operational phases.

Site Selection Procedures

The complex problem of site evaluation and selection has been the subject of extensive study and a voluminous literature. Preliminary regional screening may be carried out on the basis of existing Federal and state regulations, which have been summarized by Metry (1980-b), among others. Bain and Hensen (1981) have used this approach to delineate potential site locations in North Carolina. Site screening on the basis of the geological characteristics which ultimately will determine site performance appears to be a more direct approach to the problem. The method is summarized by Miller and Alexander (1981); Brown and LeGrand (1981) present a "hazard potential matrix" to guide the preliminary site screening process. More detailed examinations of the problem are presented by the Greak Lakes Basin Commission staff (Draft Report, 1981), by Lessing (1980), and by Schultz (1980).

The discussion which follows is based upon a geological selection scheme described in detail by Witzig, et al, (1983). The procedure presumes that waste containment will be achieved primarily through the geological characteristics of the site, rather than through the use of engineered containers or structures. It is also assumed that the procedure is intended to identify one or two optimum site locations in a previously unevaluated region of at least fifty thousand square miles (the area of an average state such as Pennsylvania). Finally, it is apparent that while final approval of a site will necessarily depend upon a detailed site-specific evaluation, the cost of such evaluations throughout so large an area would be prohibitive in both time and money. Some form of low-cost, rapid regional screening program is required as an initial step. The procedure presented consists of three levels of progressively more site-specific evaluation, designed to concentrate costly evaluative procedures as quickly as possible upon one to two potential sites. The fourth level of evaluation provides a means for weighing specific advantages of two sites considered comparable in terms of essential performance standards. These four levels of evaluation are presented as screening criteria:

EXCLUSIONARY CRITERIA - comprising those regional characteristics which so severely jeopardize site performance as to justify prima facie exclusion of the entire area or region from consideration;

SITE-SPECIFIC EXCLUSIONARY CRITERIA - including geological characteristics of individual (potential) sites which are likely to unacceptably compromise site performance;

INTERACTIVE SITE-SPECIFIC CRITERIA - including those interactive site characteristics which by their interaction ensure achievement of performance standards which could not be attained in the absence of such interaction; and

PERFORMANCE ENHANCEMENT CRITERIA - site-specific characteristics which are not essential to the achievement of site performance standards, but which may afford practical, socio-political or economic benefits.

Each of these groups of criteria is discussed briefly in the paragraphs which follow.

EXCLUSIONARY CRITERIA

Geologic processes which characterize volcanic terrains, seismically active regions, coastal zones and flood plains so seriously jeopardize site performance as to require their exclusion as potential waste disposal sites. A similar prohibition should apply to regions underlain by fractured or cavernous bedrock or by water-soluble rocks like limestone because of the rapid and often unpredictable movement of groundwater through such rocks. Finally, hazardous waste disposal sites must be located at sufficient distance from population centers and from public water supplies to afford reasonable protection in the event of any form of site failure.

SITE-SPECIFIC EXCLUSIONARY CHARACTERISTICS

Even minor earthquakes, exceptional storms or normal undercutting of slopes by stream erosion can trigger landslides which may endanger a hazardous waste site. Such hazards must be evaluated on a site-specific basis. Structural lineaments or fracture traces identifiable in aerial photographs give strong indication of fractured rock beneath and of rapid and unpredictable groundwater movement. Sites on or close to such linear features should be excluded. Steep topographic slopes, active or abandoned mines, exploratory or developmental oil or gas well drilling, nearby lake, stream, spring or exploitable aquifers, all may constitute site-specific exclusionary features of a potential site. Obviously, each potential site must be evaluated in the light of existing Federal, state and local regulations, and with due regard for the size of the available property and the anticipated useful life of the site.

INTERACTIVE SITE-SPECIFIC PERFORMANCE CHARACTERISTICS

Many geological characteristics of a potential site are related to other factors, such as site size, depth to water table, soil transmissivity or soil chemistry in such a way that predicted site performance depends upon the interrelation among two or more factors. For example, high soil transmissivity will allow adequate drainage of burial cells even if cell caps and surface water diversions permit unusually great influx of water. If soil transmissivity is high, however, the size of the site must be increased to ensure that contaminated leachate cannot escape from the controlled area before natural processes in the soil either "fix" the contaminants or detoxify them. Similar interrelationships exist between soil transmissivity and depth to the water table, and between surface water regimen and artificial diversionary structures, to identify but two.

SITE-SPECIFIC ENHANCEMENT CHARACTERISTICS

Many characteristics of a particular site may afford economic, operational or socio-political benefits which will influence the site selection process. Such features as transportation facilities, nearby sources of back-fill or capping materials, soil composition which may adsorb or neutralize one (or more) of the waste materials in the event of a leak or spill must be considered in making comparative evaluations of two otherwise equally suitable potential sites. Enhancement characteristics, however, are no more than that, and the primary site selection must be based on the achievement of performance standards as reflected in the foregoing criteria.

THE SITE EVALUATION PROCESS

The foregoing criteria are intended only to provide a guide in the site selection process. Applied with care, they should serve to eliminate regions in which there clearly exists little promise of locating a safe and effective hazardous waste disposal site. The ultimate site evaluation must be based on detailed site-specific investigation, for it is the actual performance of each individual site which provides the only meaningful measure of success.

THE DISPOSAL SITE AS A NATURAL RESOURCE

The foregoing discussion indicates that an effective and successful disposal site for hazardous waste must exhibit a complex but critical association of interrelated environmental characteristics. Although probably not rare, such sites are certainly relatively few in number, and both difficult and costly to identify. In the light of the urgent national need for disposal facilities, therefore, it is appropriate to consider them as geological resources analogous to ore deposits or oil fields. Indeed, they are no less vital to our modern way of life than automobiles and power plants. Their exploitation should be governed by the same concern for the environment and public health, and their operation may be expected to return the same kind of economic benefits to their operators and the communities where they are located. The search for suitable disposal sites may appropriately follow the procedures developed by mining and petroleum companies in their own exploration programs. The foregoing screening procedure was based upon just such a philosophy.

The parallel between waste disposal sites and mineral resources suggests one further inference which is worthy of consideration. It seems no more logical to assume that geologically suitable disposal sites will occur in comparable numbers in all political subdivisions of a region, a nation or a continent than to suppose that economically recoverable deposits of copper, coal, petroleum or iron will occur with a comparably convenient respect for political boundaries. The formation, accumulation and recovery of petroleum, for example, depend primarily on geologic factors. The petroleum company which disregards the geology and undertakes to produce petroleum from the granitic terrain of New Hampshire, solely because there is a local need for fuel, will not long survive. If it is unreasonable to expect that petroleum reserves will be distributed among the states in proportion to the need for fuel, is it any more reasonable to expect that suitable waste disposal sites will exist in every state with sufficient capacity to accommodate the waste each generates? A legal requirement that every state dispose of its own hazardous waste may be a political statement which does not adequately respect geological reality. Should a satisfactory waste disposal site, wherever it may be located, be considered a national resource and be made available to serve a public need at reasonable cost? Here in Pennsylvania, for example, we accept the environmental and health hazards of coal mining, because the product serves a national need and the economic benefits which we enjoy from the industry afford us reasonable compensation for the resulting damage. Exactly the same principle applies to waste disposal sites. Their operation is an essential public service without which our society cannot survive the century. But simply because disposal sites are public necessities does not imply that their operation need be more environmentally damaging or less economically rewarding than any of the infinitely more hazardous activities involved in the exploitation of conventional natural resources.

BIBLIOGRAPHY

Bain, G. L. and P. Hansen. 1981. Geologic and hydrologic implications of hazardous wastes management rules on siting in North Carolina. in: Geologic and Hydrologic Factors in the Disposal of Hazardous Wastes in the Southeastern United States (G.L. Stirewalt, Ed.), Bull. Assoc. Eng. Geologists, vol 18, pp. 267-275.

Bartlett, R. V. 1980. The Reserve Mining Controversy. Indiana University Press, Bloomington, Indiana. 293 pp.

Brown, H. S. and H. E. LeGrand. 1981. Perspectives on management and disposal of hazardous wastes. in: Geologic and Hydrologic Factors in the Disposal of Hazardous Wastes in the Southeastern United States, (G.L. Stirewalt, Ed.), Bull. Assoc. Eng. Geologists, vol. 18, pp. 231-235.

CEQ 1970. Ocean Dumping: A National Policy. Report of the Council on Environmental Quality. Washington, D.C. 1970.

Cheng, M. H., J. W. Patterson and R. A. Minear. 1975. Heavy metals uptake by activated sludge. Jour. Water Pollution Control Fed., vol. 47. p. 362-ff.

Claus, G., and G. J. Halasi-Kin. 1972. Environmental Pollution. in: Encyclopedia of Geochemistry and Environmental Sciences, (R.W. Fairbridge, Ed.) van Nostrand-Reinhold, New York, pp. 309-337.

Coates, D. R. Environmental Geology. John Wiley & Sons, New York, 1981. 701 pp & glossary, index.

Dana, R. H., Jr., S. A. Molello and R. H. Fickies. General Investigation of Radionuclide Retention in Migration Pathways at the West Valley, New York, Low Level Burial Site. U.S. Environmental Protection Agency, Washington, D.C. (NUREG/CR-0794) 1979. 99 pp.

Davies, J.E. and R. Rownd. 1972. Transmissible multiple drug resistance in Enterobacteriacae. SCIENCE. vol. 176, pp. 758-768.

Deutch, J. M. (Chm.) Nuclear Waste Management: Report to the President by the Interagency Review Group (Draft). (U.S.) Department of Commerce, Washington, D.C. 1978. 95 pp.

Foster, J. B. 1982. Lessons learned in a hydrogeological case at Sheffield, Illinois. in: Proceedings of the Symposium on Low-Level Waste Disposal, Volume 2. Arlington, VA., June, 1982. (M.G. Yalcintas, Ed.), Oak Ridge National Laboratory, Oak Ridge, TN. (NUREG/CP-0028 CONF-820674), pp. 237-246.

Glynis, M. Nau-Ritter, C. F. Wurster. 1983. Sorption of polychlorinated biphenyls (PCB) to clay particulates and effects of desorption on phytoplankton. Water Research, vol. 4, pp. 383-387.

Grant, J. L. 1982. Geotechnical measurements at the Maxey Flats, Kentucky, low-level radioactive waste disposal site - lessons learned. in: Proceedings of the Symposium on Low-Level Waste Disposal, Volume 2. Arlington, VA., June, 1982. (M.G. Yalcintas, Ed.), Oak Ridge National Laboratory, Oak Ridge, TN. (NUREG/CP-0028 CONF-820674), pp. 151-166.

Great Lakes Basin Comission Staff. 1981. Earth Science Considerations in Siting Secure Hazardous Waste Landfills; Workshop Results, Analysis and Recommendations. August, 1981. unpaginated.

Holden, C. 1980. Love Canal residents under stress. Science. vol. 208, pp. 1242-1244.

Hohenemser, C., R. W. Kates and P. Slovic. 1983. The Nature of Technological Hazard. SCIENCE. vol. 220. pp. 378-384.

Johnson, T. M., K. Cartwright, and R. M. Schuller. 1981. Monitoring of leachate migration in the unsaturated zone in the vicinity of sanitary landfills. Proc. First National Ground Water Quality Monitoring Symposium and Exposition: Natl. Water Well Assoc., United States EPA, Washington, D.C.

Kelleher, W. J. Water problems at the West Valley burial site. in: Management of Low Level Radioactive Waste (M.W. Carter, Ed.), Pergamon Press, NY. 1979. pp. 843-850.

Koditschek, L. K. and P. Guyre. 1974. Antimicrobial-resistant coliforms in New York Bight. Mar. Pollution Bull., vol. 5, pp 71-74.

Kolata, G. B. 1980. Love Canal: False alarm caused by botched study. Science. vol. 208, pp. 1239-1242.

Lessing, P. 1980. Guidelines for Geological Investigations of Hazardous-Waste Disposal Sites. Circular C-17, W. Virginia Geological and Economic Survey, Morgantown, W. Virginia. 19 pp.

McDougal, W. J. 1980. Containment and treatment of the Love Canal landfill leachate. Water Pollution Control Federation Journ., vol. 52, pp., 2914-2924.

Metry, A. A. The Handbook of Hazardous Waste Management. Technomic Pub. Co., Westport, Conn, 1980(a). 446 pp.

Metry A. A. 1980-b. Regulations affecting hazardous waste management. in: The Handbook of Hazardous Waste Management. (A.A. Metry, Ed.) Technomic Publishing Co., Westport, Conn., pp. 2-43.

Miller, D. G., Jr., and W. J. Alexander. 1981. Geologic aspects of waste disposal site evaluations. in: Geologic and Hydrologic Factors in the Disposal of Hazardous Wastes in the Southeastern United States (G.L. Stirewalt, Ed.), Bull. Assoc. Eng. Geologists, vol. 18, pp 245-251.

Nemerow, N. L. Industrial Waste Water Pollution: Origins, Characteristics and Treatment. Addison-Wesley Pub. Co. Reading, MA. 1978. 738 pp.

NOAA 1981. Report to Congress on Ocean Dumping Monitoring and Research, January through December 1979. National Oceanic and Atmospheric Administration, U.S. Dept. Commerce, Washington, D.C.

Parizek, R. R. Impact of highways on the hydrogeologic environment. in: Environmental Geomorphology (D.R. Coates, Ed.), Proceedings of the First Annual Geomorphology Symposia Series, Binghamton, NY. 1970. 262 pp.

Perkins, E. J. The Biology of Estuaries and Coastal Waters. Academic Press, New York. 1974.

Pye, V. I. and R. Patrick. 1983. Ground water contamination in the United States. Science. vol. 221, pp. 713-718.

Ringwood, A. E. Safe Disposal of Highlevel Nuclear Reactor Wastes: A New Strategy. Australian National University Press, Canberra, Aug. 1978. 64 pp.

Roy, R. 1982. Radioactive Waste Disposal. Volume I: The Waste Package. Pergamon Press, New York. 232 pp.

Rudd, Neilson. Subsurface Liquid Waste Disposal and its Feasibility in Pennsylvania. Environmental Geology Report #3. Commonwealth of Pennsylvania Department of Environmental Resources, Bureau of Topographic and Geologic Survey, Harrisburg, PA. 1972. 103 pp.

Shultz, D. (Ed.). 1980. Disposal of Hazardous Waste: Proc. Sixth Annual Research Symposium, March 17-20, 1980, Chicago, Illinois. EPA 600/9-80-010. 291 pp. NTIS, Springfield, Va.

Sittig, M. Resource Recovery and Recycling Handbook of Industrial Wastes. Noyes Data Corp., Park Ridge, NJ, 1975. 427 pp.

Sittig, M. Incineration of Industrial Hazardous Wastes and Sludges. Noyes Data Corp., Park Ridge, NJ. 1979. 347 pp.

Skinner, B. J. (Chm.) 1975. Resource Recovery from Municipal Solid Wastes. Sup. Report of the Committee on Mineral Resources and the Environment (COMRATE), National Academy of Sciences-NRC, Washington, D.C.

Smith, R. J. 1982-a. Love Canal study attracts criticism. SCIENCE. vol. 217, pp. 714-715.

Smith, R. J. 1982-b. The risks of living near Love Canal. SCIENCE. vol. 217, pp. 808-809, 811.

Smith, R. J. 1982-c. How safe is Niagara Falls? SCIENCE. vol. 217, p. 809.

Subramanian, V. 1980. Hydrogeologic processes and chemical reaction at a landfill: Discussion and reply. Ground Water, vol. 18, pp 504-505.

Summers, A. O. and S. Silver. 1972. Mercury resistance in a plasmid-bearing strain of Escherichia coli. Jour. Bacteriology. vol. 122, pp. 1228-ff.

Sun, R. J. 1982. Selection and Investigation of Sites for the Disposal of Radioactive Waste in Hydraulically Induced Subsurface Fractures. Prof. Paper 1215, U.S. Geological Survey, Washington, D.C. 87 pp.

U.S. EPA (1978) Environmental Impact Statement on the Ocean Dumping of Sewage Sludge in New York Bight. U.S. Environmental Protection Agency, Region II, Washington, D.C.

U.S. EPA (1979). Annual Report to the Congress on Administration of the Marine Protection, Research and Sanctuaries Act of 1972 as Amended (P.L. 92-532) and Implementing the International Ocean Dumping Convention, January through December, 1978. (U.S.) Environmental Protection Agency, Washington, D.C.

U.S. EPA (1980-a). Forecasts of the Quantity and Composition of Solid Waste. U.S. Environmental Protection Agency, Office of Research and Development, Washington, D.C. (EPA 600/5-80-001).

U.S. EPA (1980-b). Environmental Outlook, 1980. U.S. Environmental Protec Agency, Washington, D.C. (EPA 600/8-80-003).

U.S. EPA (1980-c). Report of the Interagency Ad Hoc Work Group for the Chemical Waste Incineration Ship Program. U.S. Environmental Protection Agency, Washington, D.C. 1980.

Warner, D. L. and D. H. Orcutt. 1973. Industrial Wastewater Injection Wells in the United States - Status of Use and Regulation. in: J. Braunstein (Ed.), Second International Symposium on Underground Waste Management and Artificial Recharge (AAPG, USGS, Int. Assoc. Hydrol. Sciences). George Banta Co., Menasha, Wisc., 2 vols., 931 pp.

Witzig, W. F., W. P. Dornsife and F. A. Clemente (Eds.). 1983. Low Level Radioactive Waste Disposal Siting: A Social and Technical Plan for Pennsylvania. Final Report for Subcontract No. C29-007909, EG&G, idaho, Inc., Institute for Research on Land and Water Resources, The Pennsylvania State University, University Park, PA. 4 volumes, LW8303-I through -IV.

Young, D. R., D. J. McDermott, T. C. Heesen & T. K. Jan. 1975. Pollutant inputs and distributions of Southern California. in: Marine Chemistry in the Coastal Environment. (T.M. Church, Ed.) American Chemical Society Symposium Series, #18. Am. Chem. Soc., Washington, D.C. 710 pp.

Hazardous and Toxic Wastes: Technology, Management and Health Effects. Edited by S.K. Majumdar and E. Willard Miller. © 1984, The Pennsylvania Academy of Science.

Chapter Ten

The Time Perspective in the Subsurface Management of Waste

Richard R. Parizek, Ph.D

Professor of Hydrogeology
College of Earth and Mineral Sciences
340 Deike Building
The Pennsylvania State Univ.
University Park, PA 16802

Surface-water, soil-water, and groundwater flow systems are dynamic and are driven by natural and man-induced stresses. When viewed in the short-term, these systems may appear to be unchanging or in a state of equilibrium. Rates of pollutant and water inputs into and their transport within these systems appear to be balanced by outputs. The frameworks of some flow systems are so large or so slow to respond that they appear to be unaffected by land use activities and waste disposal practices. However, evidence to the contrary is abundant. Water quality degradation has and is occurring at increasingly rapid rates both on a regional and local scale from pollutants associated with a host of land and material usages and waste disposal practices.

The dynamic nature of pollutant transport proceeses within air, surface-water, soil-water, and ground-water flow systems can best be appreciated when viewed in a long enough time perspective such that environmental changes are obvious or water quality changes imposed by man are measurable and can be separated from natural, long- and short-term background quality changes.

The time perspective for waste isolation within each of these systems is briefly reviewed here with emphasis on ground-water flow systems. A time perspective commensurate with the wastes at hand is necessary when attempting to isolate, store or dispose of wastes of all types. The control and management of diffuse and point source toxic and hazardous wastes will require the talents of individuals trained and experienced in various disciplines not previously routinely called upon to solve past point-source pollution problems. Non-point pollution problems are tied to the landscape and to surficial and subsurface processes that can have complex physical, geochemical, and biochemical interactions and time dimensions that must be understood when identifying the nature and sources of pollutants, and when selecting and designing remedial measures to control these pollutants (Parizek, 1980).

Liquid and solid wastes are inevitable by-products of advance technological societies. These wastes may result from energy development and production, commercial, agricultural and industrial development. Liquids, gases, sludges, and solids are inevitable by-products of secondary and higher levels of waste treatment. Ultimately, these wastes must be returned to the land in some form by land application systems, buried in landfills, injected into deep wells or returned as atmospheric fallout when incinerated. For land application systems, hazardous and toxic wastes must be stored in one form or another for the indefinite future, or their leachates attenuated by biochemical processes or chemically combined to produce stable or rather inert substances that will not contaminate or pollute surface or subsurface waters. If the conversion of wastes to more acceptable substances is not complete or wastes are not hydraulically isolated, pollutants may be remobilized in one form or another as leachates or solid particles. Leachates may be free to migrate within surface and subsurface flow systems or they may be carried with soil particles eroded from waste "disposal sites" (Parizek, 1980).

Original wastes or their secondary products (daughter products) of chemical weathering will have storage-time-constants, or retardation factors that may be measured in hours to thousands of years, depending upon the physical-chemical- and biological characteristics of the waste and of the land areas upon or in which they are applied. We have choices on how to maximize waste retention times through waste form selection, design of overpacks, site selection and leachate and gas management practices. Leachate, for example, can be isolated within predetermined portions of watersheds, within lakes or selected streams, or within segments of soil-water and ground-water flow systems. The residence times of leachates entrained within groundwater may be measured within days to thousands of years, depending upon the magnitude of flow system selected, and initial boundary conditions governing these flow systems. Subsurface flow systems may be local, intermediate, to regional in scale depending upon the hydrogeologic setting. The pathways of subsurface water flow and history of travel can be adequately predicted given sufficient knowledge of controlling parameters and existing theory that describes soil-water and ground-water flow. Still poorly defined, are the mass transport potential of leachates derived from a variety of hazardous and toxic wastes.

Subsurface flow systems can be unsteady. Initial and boundary conditions influencing subsurface water flow can be altered by natural or man-induced changes or by seemingly slow changes in geological processes and climate. Environmental changes and geological processes that respond so slowly that they are gauged against geologic time are being factored into high-level radioactive waste storage programs in the United States and elsewhere. However, time dimensions measured in hundreds to thousands of years have been largely ignored in most other previous waste disposal and management programs, even though some more toxic and hazardous wastes may linger in one form or another in the environment long after the design life of the facility that generated these wastes or that was selected to contain them has been exceeded.

For some waste types, the premature breakout of pollutants to the land surface may have more to do with the improper siting of waste disposal facilities than to the lack of subsurface flow systems ability to attentuate them. Wastes applied to ground-water discharge areas, for example, will not derive benefits from deep percolation within the unsaturated zone and long-term storage and transport within saturated soil and rock. Leachate may simply be flushed to land surface by upwelling groundwater for discharge into surface streams.

Surface-water and ground-water reservoirs are a geochemical and biochemical system with interconnected, water, sediment and chemical reservoirs. Sources of toxic substances, chemical sinks and pathways of material transport are the principal parts of the system to be inventoried for toxicants. Dissolved materials, biota and man are the main interacting components that may be impacted. Toxicants are transferred between these reservoirs and their component parts and from sources and sinks by complex and still poorly understood processes. While on pathways between sources and sinks, organic and inorganic substances may become altered or stored within sediments. This entire sedimentological, geochemical and biological system, or ecosystem responses to natural processes that may mask the influences of man in the short-term but in the long run, the accumulative effects of man's production, use and disposal of hazardous and toxic substances cannot but help to adversely impart surface and ground-water quality in increasingly more important and harmful ways.

Better definition of ground-water systems and their potential for transporting natural and man-made pollutants is in keeping with the Ecosystem approach to water quality management. An Ecosystem is defined as: The interacting components of air, land, water and living organisms, including man, within drainage basins of interest. Local and regional ground-water systems have to be more adequately considered if governments are to realize their common purpose "to restore and maintain the chemical, physical, and biological integrity of waters within our Ecosystem" (Parizek, 1983a).

INVENTORY OF MATERIAL USAGE

Pollution by toxic and hazardous substances from land-drainage for example, was regarded as an equal if not greater concern in the Great Lakes Basin ecosystem by the Pollution From Landuse Activities Reference Group (PLUARG, 1978). This group appointed by the International Joint Commission estimated that approximately 2,800 chemicals, including 2,200 organic compounds were being produced and used in the Great Lakes Basin by 1980. About 400 organic compounds were identified in the Great Lakes ecosystem including many of the compounds in the above inventory.

Residual levels of persistent compounds, specifically DDT, aldrin-dieldrin and chlordane continued to appear in Great Lakes biota, although their use in the basin had been banned or severely restricted prior to or during the PLUARG investigation.

PLUARG (1978) found unacceptable levels of industrial organic compounds, heavy metals and other trace elements in waters and sediments of the Great Lakes. Lakes Ontario and Erie sediments, particularly those adjacent to large urban areas, were found to be highly contaminated with PCB's. These compounds were thought to represent an environmental hazard because they are exceptionally stable and bioaccumulate readily through the food chain in fish and birds and have been detected in humans.

PLUARG (1978) inventoried land and materials usage within the Great Lakes Basin to better understand the potential sources of organic and inorganic contaminants impacting the Lakes. Similar studies of diffuse source pollutants have been conducted elsewhere in the U.S. where it was shown that ground-water and surface-water pollution has resulted from a variety of land uses and waste disposal practices.

An inventory of known or suspected solid and liquid waste disposal facilities known or suspected to contain hazardous wastes in the U.S. was updated by the U.S. Environmental Protection Agency (EPA) as part of its Comprehensive Environmental Response, Compensation and Liability Act (CERCLA or Superfund) responsibilities. This Act (Publication 96-510), passed by the U.S. Government in December 1980, resulted in the listing of more than 10,000 potential or known sites containing hazardous wastes. Of these, 419 were cited for immediate alteration by Superfund in 1982. This list has since been expanded to include 539 sites. Still others will be added to this cleanup program.

The disposal of hazardous or toxic liquid and solid wastes generated by the intense industrial activity throughout the United States should remain a matter of urgent and continuing concern. Despite the recent appreciation of the potential magnitude of the environmental and health problems associated with past disposal of these wastes, the potential for future pollution of lakes, streams, and ground-water remains. This results from past and even current materials usage and water disposal practices.

POTENTIAL CONTRIBUTING AREAS AND HYDROLOGICALLY ACTIVE AREAS

PLUARG (1978) showed that diffuse pollutants are not derived uniformly from whole watersheds or even sub-basins draining into the Great Lakes and that areas contributing pollutants to surface or gound water may reqpresent only a small portion of a drainage basin's area. This has since been found to be true for other watersheds. PLUARG developed criteria for the identification of potential contribution areas within these, the most hydrologically active areas, which are the zones most likely to produce water pollution from landuse activities. It was shown that the Great Lakes were being polluted from land drainage sources by phosphorus, sediments, some industrial organic compounds, some previously-used pesticides, and potentially, some heavy metals.

Hydrologically active areas were found to be a basic tool for estimating the level and location of potential pollutants derived from potential contributing areas and as a means for identifying where remedial programs might be adopted that were likely to maximize the reduction of pollutant loads for least costs. This same concept will be helpful to more adequately access the threat to water quality posed by hazardous and toxic waste disposal facilities.

Hydrologically active areas are land areas that contribute directly to ground and/or surface waters, even during minor precipitation and snow-melt events, because of their proximity to streams or aquifer recharge areas. The size of hydrologically active areas may vary basin by basin and with season of the year. They are a function of landuse and management, steepness of slope, infiltration rates and soil moisture content, position of water table, direction of ground-water flow, ground-frost conditions and other factors.

PATHWAYS INTO SURFACE AND SUBSURFACE FLOW SYSTEMS

Pollutants may be added to drainage basins through atmospheric deposition, dissolved in surface runoff and attached to sediment. Less data are available for accessing pollutant loadings to surface streams from ground-water sources, although it is recognized that ground water has been degraded and/or polluted by a variety of landuses at least locally in the United States. These same pathways are still available for transporting pollutants into aquifers that are derived from toxic and hazardous disposal sites and from other diffuse sources.

Atmospheric Sources:
Atmospheric loadings are aerial imputs that land directly on fluvial, bay or lake waters and within ground-water recharge areas. These sources may be derived from outside of the watershed under study and may even be transported across state and international borders as witnessed by the acid rain problem.

Atmospheric loadings may result from dryfall (dust) and as dissolved constituents in precipitation. Davis and Galloway (1981) have shown that small aerosols (particles $< 1\mu$) may have a residence time of 4.7 days hence, have the

(a)

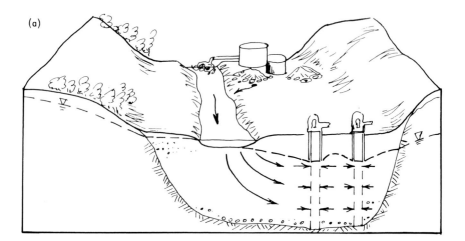

(b)

FIGURE 1. Diffuse source pollutant loadings within surface watersheds may become point source loadings to ground-water reservoirs. In (a) induced streambed infiltration results in ground-water pollution derived from surface water sources and in (b), surface runoff is recharged into shallow holes.

potential to be transported for significant distances from many upwind industrial areas. Global dissemination of such pollutants also appears to be possible.

Some data are available on the dryfall atmospheric deposition of metals (Lazarus et al., 1970; and Davis and Galloway, 1981). The concentration of metals in the atmosphere are proportional to the total amounts of metals released from fossil fuel combustion, manufacturing processes, other man-related and natural processes. Fallout on hydrologically active areas should find its way into surface water in a shorter period when compared to fallout in adjacent upland areas or in ground-water recharge areas. This may be as brief as a few hours or months in the spring of the year when most water in rivers and streams is derived from overland flow or precipitation occurs on frozen ground.

Riverine Sources:

Riverine sources are mass loadings that enter lakes or the sea at river mouths. Contaminants and nutrients may be either in solution or in suspension. These sources tend to show a strong association with the magnitude of runoff events. Riverine contaminants can impact ground-water quality where streams are perched above ground-water levels and rapid streambed infiltration results.

Influent streams are common where mountain streams cross alluvial fan deposits at the base of mountains, in desert basins, in many karst terranes and wherever extensive ground-water development has caused induced streambed infiltration to occur.

In karst terranes, surface watersheds may collect and integrate atmospheric loadings for direct recharge to swallow holes (sinkholes) where little or no additional water purification is possible. In this sense, diffuse source loadings to surface watersheds can result in point source loadings to ground-water reservoirs (Figure 1).

Pollutant loadings observed at river mouths may have been derived from a combination of various sources that are discharged into these rivers above the point where loadings are monitored. River-mouth loadings, therefore, contain fractions of both natural and man-related pollutants that enter rivers as atmospheric fallout, from sediment, overland flow or ground-water sources. Rivers serve as major pathways for the ultimate transport of these pollutants into lakes, or estuaries.

Riverine sources tend to contain heavy concentrations of airborne pollutants deposited in urbanized regions containing a high percentage of interconnected impervious surfaces and frequent storm water runoff events. Pb is still notably being released from the combustion of leaded gasoline, Zn from the abrasion of tires, and Cu and Cr from automobile brakeshoes, to mention a few sources for these contaminants. Rivers also receive significant amounts of effluent discharged from municipal and industrial treatment plants as well as runoff from intensively farmed regions of some watersheds, and sites of mining activity elsewhere.

Point sources include industrial and municipal loadings that discharge directly into lakes, or rivers at concentrated locations resulting from past and present waste treatment and disposal practices. A considerable inventory of toxic substances has entered lakes and streams in this manner. Some of these contaminants can impact ground-water quality.

Ground-water sources:

Ground-water sources are diffuse in nature and can contribute pollutants directly into lakes or tributaries. The ground-water pathway into lakes and streams involves far greater quantities of water, hence possibly greater contribution of pollutants than previously recognized. Pollutants within this pathway can show particularly long residence times due to slow rates of ground-water movement and various attenuation mechanisms that operate in the subsurface. These attentuation mechanisms may either slowup or eliminate pollutants entirely before water is discharged into surface-water systems (Parizek, 1983a).

Lacustrine and eustraine sources:

Processes of shoreline, river-bank and lake bottom erosion also contribute pollutants to lakes and streams. Sediments transported from these sources may contain geological material which are natural sources of organic and inorganic substances. However, lake bottom and shoreline erosion and sediment transport are mechanisms for stirring up, transporting and redepositing toxic substance previously deposited with sediment.

Metals, borne on suspended materials can be transported along two pathways within lakes and streams, a hydrodynamic route and a bioecological route. The hydrodynamic route is revealed by dispersion patterns of metals and organic substances in bottom sediments. Some metals may follow a bioecological path having been bio-magnified along the food chain (Parizek, 1983a).

The pollutant loadings into lakes and streams vary for each of these pathways. Contaminants attached to fine-grained sediments are more likely to enter surface-water bodies by atmospheric and riverine routes. In most cases, physical filtration processes are so efficient in the subsurface as to all but preclude significant transport of particulates within soil water and ground water except to shallow field drainage tiles, within channels developed in karst terranes, lava tubes or well-developed fracture systems in many other rock types. Dissolved substances may be transported along atmospheric, riverine and ground-water routes.

PLUARG (1978) investigated the diffuse sources of metals and organic compounds within the Great Lakes system, for example, but did not investigate the ground-water loadings of these substances to the Lakes in detail except within Pilot watersheds and near selected waste disposal facilities. It was realized that the pollution potential from hazardous waste disposal sites could pose significant water quality threats to the Great Lakes System in time because of the slow travel time of pollutants within ground-water flow systems. Pollutants entrained with sediments also may be placed into temporary storage during their step-wise migration into the Lakes, thereby delaying the delivery of pollutants from this source. Elsewhere, rather detailed studies of shallow aquifers such as the outwash deposits of Long Island have been conducted and widespread groundwater quality degradation and local pollution problems have been documented.

The time frames for pollutant transport between source areas and surface water and subsurface water bodies also vary for each route. Atmospheric fallout is all but instantaneous on lake surfaces and rivers but may be delayed to some extent where fallout occurs on hydrologically active areas before being flushed into surface water. Fallout elsewhere may be contained indefinitely on other portions of watersheds where sediment and chemical sinks are available to store particulates within uplands. Contaminated sediment may experience similar delays on route to surface water. Wetlands, kettle lakes, desert lake basins, and other closed surface depressions can collect and retain soils and associated contaminants for decades to thousands of years depending upon the topographic and fluvial setting. Transport of contaminated sediment in river channels and flood plains also can be periodic to rather rapid during peak storm events.

Atmospheric and riverine routes for pollutant transport into lakes and streams are best understood but not necessarily easiest to control. Contaminated sediments can be mobilized and transported by wind, gravity, and water. Rates of erosion have been measured or estimated for continental areas and shown to be highly variable depending upon climatic, vegetation, topographic and land use factors. Erosion rate data on average reveal that decades to thousands of years may be required before near surface pollutants entrained with sediment may be eroded and transported into surface water systems. Not all organic pollutants or toxic substances will survive for such long periods in their original toxic state. Metals may be an exception. Such substances can remain in temporary storage along with sediment until they are remobilized by surface water. However, changes in landuse can greatly accelerate rates of erosion and sediment transport. Processes of soil erosion in lands undergoing a transition from rural to urban, for example, typically are greatly accelerated as are lands adjacent to transportation corridors, or subjected to surface mining (Figure 2). Colluvial and alluvial fan deposits, stable since the retreat of the last continental glacier (12,000 to 15,000 yrs. B.P.) in Pennsylvania for example, have been subjected to sheet and gullie erosion within the vicinity of highways, urban areas, surface mines, etc., where more sediment may have been removed from slopes in the last decade, than has occurred since their original deposition more than 10,000 years ago.

Sediment deposited on pavements within urban areas from atmospheric fallout and other sources also is rapidly removed with stormwater.

Processes of pollutant transport along ground-water routes are the least well documented and understood. Renovation and attenuation processes are

FIGURE 2. Accelerate rates of erosion associated with surface mining of coal.

numerous. These alone or in combinations may help to contain toxic substances indefinitely in the subsurface or falsely appear to be containing them due to their slow rate of migration.

Pollutants transported along each pathway listed above will have its own travel time, attenuation mechanism, and control measure potential. Some of these attenuation mechanisms are listed in Table 1. The persistence of hazardous substances and attenuation mechanisms operating within soil-water and ground-water flow systems must be better understood in order to adequately access their long-term potential for lake and stream loadings via the ground-water route as well as the potential for local and regional ground-water contamination.

TOXIC SUBSTANCES

Toxic substances identified within lakes and streams include both metals and organic compounds which occur within the water column, lake bed sediments, suspended sediments and in aquatic organisms. Their sources and pathways to streams and lakes include atmospheric loadings, riverine and point sources, sediment sources within rivers, and resulting from shoreline and beach erosion and ground water discharged into lakes and rivers. Once added to the lakes and streams, contaminants may be released from bottom sediments to plants and aquatic organisms where they are free to be disseminated within surface water by

TABLE 1

Probable Effect of Various Processes on the Mobility of Constituents in
Subsurface Waters Contaminated by Waste Disposal (D. Langmuir, 1972)

Physical process
  Dilution—Favors reduction in solute concentrations. Less effective in unsaturated materials
   where caused by mixing with preexistent or infiltrating moisture in amounts generally below
   field capacity. Most effective in saturated materials of high porosity and permeability at high
   groundwater flow rates.
  Dispersion—Causes moisture movement at right angles to primary flow direction. Favors reduction in solute concentrations. Moisture flow mostly vertical in unsaturated soil and rock,
   and mostly horizontal in saturated soil and rock.
  Filtration—Favors reduction in amounts of substances associated with colloidal or larger-size
   particles. Most effective in clay-rich materials. Least effective in gravels or fractured or cavernous rock.
  Gas movement. Where it can occur, favors aerobic breakdown of organic substances, and increased rates of decomposition. Constituents mobile under oxidized conditions will then
   predominate. Restriction of gas movement by impermeable, unsaturated materials or by
   saturated materials can produce an anaerobic state, and reduced rates of organic decay. This
   will mobilize substances soluble under anaerobic conditions.
Geochemical processes
  Complexation and ionic strength—Complexes and ion pairs most often form by combination of
   ions, including one ore more multivalent ions and increase in amount with increased amounts
   of ions involved. Ionic strength is a measure of the total ionic species in a water. Both ionic
   strength and complexation increase the total amounts of species otherwise limited by processes
   such as oxidation, precipitation, or adsorption.
  Acid—base reactions—Most constituents increase in solubility and thus in mobility with decreas-
   ing pH. In organic-rich waters, the lower pHs (4—6) are associated with high values of car-
   bonic acid and often also of organic acids. These will be most abundant in moisture-saturated
   soils and rock.
  Oxidation-reduction—Many elements can exist in more than one oxidation state. Conditions will
   often be oxidized or only partially reduced in unsaturated soils, but will generally be reduced
   under saturated conditions when organic matter is present. Mobility depends on the element
   and pH involved: chromium is most mobile under oxidizing conditions, whereas iron and
   manganese are most mobile under reduced conditions.
  Precipitation—solution—The abundance of anions such as carbonate, phosphate, silicate,
   hydroxide, or sulfide may lead to precipitation especially of multivalent caions as insoluble
   compounds. The abundance of major multivalent cations such as calcium and magnesium
   may cause precipitation of some trace element oxyanions, including molybdate and vanadate.
   Dilution, or a change in oxygen content where precipitation has involved oxidation or reduction, may return such constituents to solution.
  Adsorption—desorption—Ion exchange can temporarily withhold cations and to a less extent
   anions, on the surfaces of clays or other colloidal-size materials. Amounts of adsorbed metal
   cations will increase with increasing pH. Molecular species may be weakly retained on
   colloidal-size materials by physical adsorption. Adsorbed species will tend to return to solution
   when more dilute moisture comes in contact with the colloidal material.
Biochemical processes
  Decay and respiration—Microorganisms can break down insoluble fats, carbohydrates, and
   proteins, and in so doing release their constituents as solutes or particulates to subsurface
   waters.
  Cell synthesis—N, C, S, P, and some minor elements are required for growth of organisms, and
   can thus be retarded in their movement away from a waste disposal site.

currents and the foodchain. Their presence even in small quantities, has significance because:

(1) they may be recycled within lake or stream bottom sediments, water and aquatic organisms for prolonged periods; (2) they can be biomagnified in the aquatic food chain way out of proportion to their original concentration, and (3) they may taint fish and drinking water making their health effects uncertain when present in game and commerical fish eaten by humans or when present in drinking water. These substances can be costly to remove from drinking water supplies. Because these substances can be both toxic and persistent in surface and subsurface waters, their initial presence in even small amounts and observed increase through time cannot be ignored. Their potential for future water contamination also must be understood when wastes are disposed of in tributary watersheds, when these chemicals are applied to the land for beneficial practices such as farming, or when they accumulate in response to atmospheric fallout resulting from other more distant activities.

Toxic substances are usually defined as chemicals or chemical compounds that can harm living plants and animals, including people, or that can impair physical and chemical processes. They may be grouped as inorganic and organic compounds. Inorganic substances include metals such as lead (Pb), mercury (Hg), arsenic (As), cadmium (Cd), chromium (Cr), copper (Cu), zinc (Zn), tin (Sn). These can be derived from natural source areas as well as from human activity. Manditory and recommended limits for inorganic substances in drinking water are included in Table 2.

Many of the organic substances found in ground water and aquatic systems result from human activiy. These include pesticides, phthalate esters, polynuclear aromatic hydrocarbons (PNA's), other chlorinated hydrocarbon compounds (PCB's, etc.) meta-organic compounds, alkyl-benzines, plasticisers, and other halogenated hydrocarbons.

PLUARG (1978) found that metals, like man-made organic substances, tend to be associated with fine-grained particulate matter such as detritus, silt and clay. This relationship to grain-size has also been recognized by other workers. Highest concentrations of metals were found in suspended organic and inorganic sediments that are the most difficult to control and retain in upland area given existing remedial programs. Metals tend to be found in suspended sediment plumes that enter lakes at river mouths and that are transported by nearshore and lake bottom currents. These contaminant enriched zones may occur in areas of high biological activity where organisms respire, reproduce and grow. Here, metals are available for uptake by microorganisms and aquatic organisms where they can be passed through the food chain of other aquatic organisms, birds, animals and man.

Bottom and suspended sediments within lakes and rivers both may contain higher concentrations of metals and organic substances than will be found in their associated water columns. Detection of long-term changes in water quality, therefore, must include the monitoring of water and sediment together with indicator organisms that have a narrow habit or range. Spottail shiners were found to be useful indicators for this purpose in the Great Lakes whereas, sea guls were found to forage for their food over large areas. Hence it was not possible to determine exactly where gulls were picking up contaminants. Suns et al. (1978) conclude that biological materials offer a more reliable base for environmental impact assessment than either sediment or water. No similar biological indicator has been found for providing early signs of pending ground-water contamination.

Davies and DeMoss (1982) indicate that because of polarity, some organics may be dissolved in water and exist below the detection limit of present day instrumentation. Monitoring water quality alone within lakes and rivers, therefore, is inadequate to define the magnitude of loadings of toxic organic and inorganic substances to surface waters.

In recent years the number of potentially toxic chemicals synthesized, and produced in the U.S. has increased. Although hazardous and toxic substance management programs have been put into place, it is inevitable that loadings of organic and inorganic substances to surface water and ground water will con-

TABLE 2

Manditory and Recommended Limits for Inorganic Substances in Drinking Water

Constituent	Recommended concentration limit* (mg/ℓ)
Inorganic	
Total dissolved solids	500
Chloride (Cl)	250
Sulfate (SO_4^{2-})	250
Nitrate (NO^3)	45†
Iron (Fe)	0.3
Manganese (Mn)	0.05
Copper (Cu)	1.0
Zinc (Zn)	5.0
Boron (B)	1.0
Hydrogen sulfide (H_2S)	0.05
	Maximum permissible concentration‡
Arsenic (As)	0.05
Barium (Ba)	1.0
Cadmium (Cd)	0.01
Chromium (Cr^{VI})	0.05
Selenium	0.01
Antimony (Sb)	0.01
Lead (Pb)	0.05
Mercury (Hg)	0.002
Silver (Ag)	0.05
Fluoride (F)	1.4-2.4§

TABLE 2 (Continued)

Manditory and Recommended Limits for Inorganic Subtances in Drinking Water

Organic	
Cyanide	0.05
Endrine	0.0002
Lindane	0.004
Methoxychlor	0.1
Toxaphene	0.005
2,4-D	0.1
2,4,5-TP silvex	0.01
Phenols	0.001
Carbon chloroform extract	0.2
Synthetic detergents	0.5
Radionuclides and	Maximum permissible activity
radioactivity	(pCi/ℓ)
Radium 226	5
Strontium 90	10
Plutonium	50,000
Gross beta activity	30
Gross alpha activity	3
Bacteriological	
Total coliform bacteria	1 per 100 mℓ

SOURCES: U.S. Environmental Protection Agency, 1975 and World Health Organization, European Standards, 1970.

*Recommended concentration limits for these constituents are mainly to provide acceptable esthetic and taste characteristics.

†Limit for NO$_3$ expressed as N is 10 mg/ℓ according to U.S. and Canadian standards: according to WHO European standards, it is 11.3 mg/ℓ as N and 50 mg/ℓ as NO$_3$.

tinue to increase as the intensity of landuses increase and as populations and industrial activities increase.

It is recognized that toxic substances accumulate in certain biota many thousandfold above ambient concentrations. The links between causes and effects within ecosystems are still poorly understood. Toxic chemicals may in time be responsible for the decline in essential biotic components in aquatic ecosystems hence, justify a high priority concern even when present in small amounts. Many substances are known or suspected to be carcinogenic to mammals, hence possibly also to man. These substances justify survaliance within aquatic ecosystems and further understanding of their long-term consequences.

Degradation of organic toxicants within uplands, lakes and streams and in ground water are still poorly defined processes. Most toxic substances of all classes entering aquatic systems may accumulate in the sediment; others degrade, and some have the potential to accumulate in biota or to be flushed from the system. Adsorbed chemicals contained in mud can be picked up by filter-feeders, metabolized by plankton and reach high concentrations within the food chain (Davies and DeMoss, 1982).

The regions within lakes and streams most likely to continue to receive enrichment by toxic substances may be expected where:

(a) the sources of supply are concentrated in hydrologically active areas within tributary watersheds;

(b) fine-grained sediment entrapment and accumulation is high; and

(c) where rates of sedimentation are moderate to fast;

(d) where atmospheric fallout is concentrated;

(e) where contamination of near shore sediments is common, below points where sewage effluents are discharged, near industrial facilities, mine tailings dumps, concentrated hazardous waste disposal facilities, and in intensive urban and agricultural runoff.

Aquifers within upland areas most likely to be impacted by these same toxic substances include:

(a) surficial sand and gravel with rapid to very rapidly drained soils;

(b) fractured bedrock lacking a soil cover;

(c) karst terranes with extensive development of sinkholes and shallow holes, and areas with thin to discontinuous soil;

(d) surficial aquifers adjacent to rivers and streams with a high induced streambed infiltration potential;

(e) immediately adjacent to solid and liquid waste storage and disposal sites;

(f) immediately adjacent to unprotected material storage facilities;

(g) in proximity to bulk storage tanks and transmission lines;

(h) near transportation corridors;

(i) where industrial and commercial activity is concentrated;

(j) where populations are concentrated;

(k) where intensive agricultural activity is concentrated;

(l) and where wastes were poorly managed in the past.

WASTE CHARACTERISTICS

Various schemes may be devised for classifying hazardous substances and risks associated with past waste disposal practices. Sites containing the most hazardous substances likely to impact existing populations are given high priority for cleanup by the U.S. Environmental Protection Agency. Toxicity and persistence of wastes are also extremely important. Table 3 lists the toxicity and persistence of some common chemical compounds that may be found at waste disposal sites. Table 4 shows the (persistence) biodegradability of some organic compounds in more detail.

These data are important in the design of hazardous waste disposal facilities, in the selection of waste types that may be mixed together at a common site and in attempting to assign priorities for the cleanup of abandoned or active hazardous waste disposal sites. These and other criteria were used in the selection of the initial list of Superfund Sites in the U.S.

TABLE 3

Persistence and Toxicity of Hazardous Substances. (From USEPA, 1982).

Class

I.   Easily biodegradable compounds.
II. Straight chain hydrocarbons.
III. Substituted and other ring compounds.
IV. Metals, polycyclic compounds and halogenated hydrocarbons.

Examples of Each Class

I.		II.		III.		IV.	
Acetaldehyde	3	Aniline	3	Chlorobenzene	2	Aldrin	3
Acetic Acid	3	Benzene	3	Cyclohexane	2	Carbon Tetrachloride	3
Acetone	2	Cresol-O	3	α-Trichloroethane	2	Chlordane	3
Ammonia,		Cresol-					
Anhydrous	3	M and P	3			Chloroform	3
Formaldehyde	3	Ethyl	2			Endrin	3
		Benzene					
Formic Acid	3	Isopropyl	3			Lindane	3
		Ether					
Hydrochloric Acid	3	Methane	1			PCB	3
Methly Ethly	2	Naphthalene	2			Trichlorobenzene	2
Ketone							
Methly	3	Petroleum,	3				
Parathoron in		Kerosene					
Xylene Solution		(Fuel oil No. 1)					
Nitric Acid	3	Phenol	3				
Parathion	3	Toluene	2				
Sulfuric	3	Xylene	2				
Acid							

SAX Toxicity Ratings

1. Slight Toxicity (Low)
2. Moderate Toxicity (Moderate)
3. Severe Toxicity (High)

Persistent compounds are likely to pose more serious pollution problems to ground- and surface water supplies because more processes are potentially available for transporting such wastes into streams and lakes either in solid or liquid form. Also, once introduced into water, they may remain hazardous for longer periods of time or available to biota. Biodegradable compounds on the other hand, are less likely to survive long-term storage in upland soils, in slowly moving soil water and ground water, and when being transported into the lakes and streams. They also have the potential of being further degrated once introduced into surface water.

To further characterize ground-water pollution potential, the solubilities of hazardous compounds should be known. Table 5, for example gives the water soluabilities of major herbicides which show a considerable range.

TABLE 4

Persistance (Biodegradability) of Some Organic Compounds.
(From J.R.B. Associates Inc., 1980).

Value—3 Highly Persistent Compounds

aldrin	heptachlor
benzopyrene	heptachlor epoxide
benzothiazo's	1,2,3,4,5,7,7-
	heptachloronorbornene
benzothiophene	hexachlorobenzene
benzyl butyl phythalate	hexachloro- 1,3-butadiene
bromochlorobenzene	hexachlorocyclohexane
bromoform butanal	hexachloroethane
bromophenyl phyntyl ether	methyl benzothiazole
chlordane	pentachlorobiphenyl
chlorohydroxy benzephanone	pentachlorophenol
bis-chloroisoprophyl ether	1,1,3.3-tetrachloroacetone
m-chloronitrobenzene	tetrachlorophenyl
DDE	thiomethylbenzothiazole
DDT	trichlorobenzene
dibromobenzene	trichlorobiphenyl
dibutyl phthalate	trichlorofluoromethane
1,4-dicnlorobenzene	2,4,6-trichlorophenol
dichlorodifluoroethane	tiphenyl phosphate
dieldrin	bromodichloromethane
diethyl phthalate	bromoform
di(2-ethylhexyl)phthalate	carbon tetrachloride
dihexyl phthalate	chloroform
di-isobutyl phthalate	chloromochloromethane
dimethyl phthalate	dibromodichloroethane
4,6-dintro-2-aminophenol	tetrachloroethane
dipropyl phythalate	1,1,2-trichloroethane
endrin	

Value—2 Persistent Compounds

acenapthylene	cis-2-ethyl 4-methyhl-1,3-dioxolane
atrazine	trans-2-ethyl-4-methyl-1,3-dioxolane
(diethyl) atrazine	quaiacol
barb tal	2-hydroxyadiponitrile
borneol	isophorone
bromobenzene	indene
camphor	isoborneol
chlorobenzene	isoprophenyl-r-isopropyl benzene
1,2 bis-chloroethoxy ethane	2-methoxy biphenyl
b-chioroethyl methyl ether	methyl biphenyl
chloromethyl ether	methyl chloride
chloromethyl ethyl ether	methylindene
3-chloropyridine	methylene chloride
di-t-butyl-p-benzoquinone	nitreanisole
dichlorouthyl ether	nitrobenzene
dihyrocarvone	1,1,2-trichloroethylene
dimethyl sulfoxide	trimethyl-troxo-hexahydroltriazine isomer
2,6-dinitrotoluene	

TABLE 4 (Continued)

Persistance (Biodegradability) of Some Organic Compounds.
(From J.R.B. Associates Inc., 1980).

Value—1 Somewhat Persistent Compounds

acetylene dichloride	1,2-dimethoxy benzene
behenic acid, methyl ester	1,3-dimethyl naphthalene
benzene	1,4-dimethyl phenol
benzene sulfonic acid	dioctyl adipate
butyl benzene	n-aecane
butyl bromide	ethyl benzene
e-caprolactam	2-ethyl-n-hexane
carbon-disufide	o-ethyltoluene
o-cresol	isodecane
decane	isoprophyl benzene
1,2-dichloroethane	
limonene	octane
methyl ester of lignoceric	octyl chloride
acid	pentane
methane	phenyl benzoate
2-methyl-5-ethyl-pyridine	phthalic anhydride
methyl naphthalene	propylbenzene
methyl palmitate	1-terpineol
methyl phenyl carbinol	toluene
methyl stearate	vinyl benzene
napthalene	xylene
nonane	

Value = 0 Nonpersistent Compounds

acetaldehye	methyl benzoate
acetic acid	3-methyl butanol
acetone	methyl ethyl ketone
acetophenone	2-methylpropanol
benzoic acid	octadecane
di-isobutyl carbinol	pentadecane
docosane	pentanol
eiccsane	propanol
ethanol	propylamine
ethylamine	tetradecane
hexadecane	n-tridecane
methanol	n-undecane

*JRB Associates, Inc., *Methodology for Rating the Hazard Potential for Waste Disposal Sites,*
May 5, 1980.

PCB's, for example, are reported to have low soluabilities in most natural water that restrict their theoretical starting concentrations at landfill sites in the absence of solvents. Higher PCB concentrations than can be released by solution processes alone nevertheless, have been observed in lake and stream bottom sediments. A significant portion of these wastes, therefore, must have been

Chemical Properties of Major Herbicides

CHEMICAL CLASSIFICATION

Common Name (Trade Name)	Manufac-[b] turer	Chemical[b] Name	Water Solubility[c] (ppm)	Vapor Pressure[a] (mm Hg x 10^{-6})	Molecular[c] Weight	Solvent Partitioning[d] K_{ow}
IONIC HERBICIDES						
I. Cationic						
Paraquat (Orthoparaquat)	Chevron	1,1-dimethyl-4,4-bipyridin-ium-dichloride	Low	Low	186 (257, Salt)	1
II. Basic						
Atrazine (Aatrex)	Ciba-Geigy	2-chloro-4-(ethylamino)-6(isopropylamino)-s-triazine	33	0.3	216	226
Simazine (Princep)	Ciba-Geigy	2-chloro-4-bis(ethylamino)-s-triazine	5	0.006 -0.036	202	88
III. Acidic						
2,4-D	Dow	(2,4-dichlorophenoxy) acetic acid	650	0.4	221	443
Dicamba	Veliscol	3,6-dichloro-O-anisic acid	4500	3570	221	—
NON-IONIC HERBICIDES						
IV. Substituted Anilines						
Trifluralin (Treflan)	Elanco	a-a-a-trifluro-2, 6-dinitro-N, N-dipropyl-p-toluidine	0.05 -24	114	335	1150
V. Phenylureas						
Linuron (Lorox)	Dupont	3-(3,4-dichlorophenyl-1-methoxy-1-methylurea	75	15	249	—
VI. Substituted Anilides						
Alachlor (Lasso)	Monsanto	2-chloro-2,6-diethyl-N-(methyloxymethyl) acetanilide	148 -242	22	270	434

[a]Vapor pressure at 20-25 °C (WSSA 1980, Mrak et al. 1974)
[b]WSSA (1980), Wauchope (1978)
[c]Water solubility at 23-27 °C (WSSA 1980, Stevenson and Confer 1978)
[d]Octanol-water partition coefficient, K_{ow} (Rao et al. 1981)

transported to the lakes and streams attached to sediment.

Soluabilities of chemical compounds in pure water may bear little resemblance to their soluabilities under field conditions where leachate may contain high concentrations of mixtures of organic and inorganic substances. The behavior of "chemical soups" within soil-water and ground-water systems also is difficult to access. The mobilities of toxicants under various field conditions are still poor understood and justify investigation if reliable mass transport predictions via soil water and ground water using various computer models are to be achieved.

It is also becoming clear that organic substances dissolved in ground water may alter the permeability of poorly permeable unconsolidated deposits with which they come in contact. Clay liners used to contain leachate within solid and liquid waste disposal facilities, for example, may undergo marked permeability increases when interacting with some dissolved organic substances. This can lead to the breach of engineered clay liners designed to contain leachate as well as more rapid transport of contaminants from hazardous waste disposal sites (Green, et al., 1983).

Site specific studies are required to determine the relative mobility of various pollutants under various geological, and geochemical settings. At the same time, more regional ground-water monitoring is needed to keep track of potential long-term changes in ground-water quality resulting from varous landuses in more densely populated areas.

Regulatory agencies must continue to investigate our ecosystem with increased emphasis on toxic substances and ground water. This is justified on a number of grounds:

(1) many fold accumulation of toxicants in lake and stream bottom sediments and biota, including sport and commercial fish than is found in ambient concentrations in water;

(2) carcinogenic nature of many of the organic compounds found in waste disposal sites and potentially, within surface water and ground water;

(3) increase in population growth, industrial and commercial activity within the U.S., and the widespread use of ground water to meet water supply needs today and in the future;

(4) past, present and projected future use and discharge of large amounts of potentially toxic substances;

(5) increases in the number of potentially toxic chemical being synthesized, produced and used within the U.S.;

(6) realization that control measured will never be completely adequate in handling hazardous and toxic substances resulting from projected future uses and past practices;

(7) and, realization of the significance of ground water to present and future generations of U.S. citizens.

Regulatory agencies must champion this effort. But they are not alone. The public at large strongly supports the need for protecting and improving water quality by a variety of means.

Detailed information on the subsurface behavior of synthetic organic compounds are especially scant. Many of these compounds have only been newly created and analytical instruments for their detection also have been only recently developed. Our knowledge basis, therefore, is scant for logical reasons.

Because toxicants may find their way into surface and ground waters from legal and illegal past, present, and future waste disposal practices, it is clear that laws and existing institutional frameworks alone are insufficient to protect these waters. Systematic chemical monitoring of surface and ground water, sediment and aquatic organisms is needed to provide early warning of increase loadings of toxicants, or the long-term buildup of toxicants within surface and subsurface aqueous systems. Thousands of compounds have the potential for being present in low concentrations. Monitoring programs need to account for this potentially wide compositional range of organic compounds having initially low concentrations. Such data are needed to detect anomalous concentrations of pollutants before they buildup to harmful levels and to establish baselines for future comparison. An effluent toxicity characterization program is needed to screen municipal and industrial effluents for toxic chemicals and to determine their degree of toxicity to individual species, to populations, to ecosystems and man.

The relative capacities of portions of surface and subsurface water systems to assimulate existing and future pollutants also must be understood in order to design meaningful remedial programs for waste management efforts. Toxicant buildups and transformations are possible through biological and sedimentological processes. From a national and international perspective, it is clear that many research problems still must be researched.

PHYSICAL FRAMEWORK

The potential mobility of hazardous substances within ground water relates to the hydrogeologic setting of waste application and disposal sites. Site specific studies are needed, for example, to determine the type of soils, surficial unconsolidated deposits and bedrock units present in a potential contributing area that receives hazardous or other wastes.

SURFICIAL DEPOSITS

The granular nature or texture of soils and overburden deposits together with surficial topography and ground-water flow systems have much to do with segments of watershed's containment potential for hazardous wastes. Permeable surficial deposits such as outwash sand and gravel (Table 6) will favor more rapid and direct recharge to shallow aquifers, less surface area to interact with pollutants, and more rapid flow to ground-water discharge areas located along

TABLE 6

Permeability of Geological Materials.
(From USEPA, 1982 as adapted from S. N. Davis, 1969 and Freeze and Cherry, 1979).

Material	Approximate range in hydraulic conductivity
Clay, compact till, shale; unfractured metamorphic and igneous rocks.	$< 10^{-7}$ cm/sec
Silt, loess, silty clays, silty loams, clay loams; less permeable limestone, dolomites, and sandstone; moderately permeable till.	$< 10^{-5} > 10^{-7}$ cm/sec
Fine sand and silty sand; sandy loams; loamy sands; moderately permeable limestone, dolomites, sandstone (no karst); moderately fractured igneous and metamorphic rocks, some coarse till.	$< 10^{-3} > 10^{-5}$ cm/sec
Gravel, sand; highly fractured igneous and metamorphic rocks, permeable basalt and lavas; karst limestone and dolomite	$> 10^{-3}$ cm/sec

rivers and lakes. Permeability contrasts with underlying, less permeable over-burden or bedrock may be great favoring mainly a lateral component of ground-water flow back to land surface. This can shorten the subsurface path length along which contaminants may travel between recharge and discharge areas as well as the residence time of pollutants within soil water and ground water. The attenuation or retardation potential of soil and rock is a function of residence time, path length, chemical and physical characteristics of aquifers and confining beds, chemistry of interstitial water and other factors.

Clay deposits such as lacustrine sediments are fine-textured in character. These deposits have large surface areas, and tend to contain significant portions of both clay-sized minerals and clay minerals which can interact with contaminants. The cation and anionic exchange capacities of clay minerals exceeds that of coarse-textured sand and gravel. Their fine-grain size also restricts their permeability and favor slow ground-water flow rates. Lacustrine sediments also tend to be associated with low-relief featureless plains resulting from slow deposition in lake basins. Lacustrine plains and their associated poorly permeable deposits favor the presence of shallow water tables, regions of rejected recharge, hence regions favoring overland flow, and regions of ground-water discharge. Typical-ly, they make up extensive hydrologically active areas and may border lakes and many major tributaries in glaciated areas. Lake-plain sediments frequently must be drained by tiles and drainage ditches to control ground-water levels and sur-face water in agricultural and urban regions. These conditions favor the transport of fine-textured sediment enriched with pollutants as well as degrada-tion or contamination of nearby surface water.

Till plains also contain fine-textured sediments deposited by the direction action of glaciers. The deposits in part, loaded by the weight of glacial ice, tend to be more dense and less permeable than lacustrine sediments. However, they also contain extensive networks of secondary openings or joints that enhance their permeability and reduce their easily accessible surface areas available for ion exchange reactions by channeling water flow and pollutants along joints.

Till may occur in gently undulating plains, as well as in hummochy terrane with numerous closed surface depressions. The latter result from drift deposition in terminal moraines, recession moraines and in regions of mass, late-stage ice stagnation. Hummochy terranes typically display closed surface depressions that act as sediment and contaminant sinks at various locations within glaciated regions. Water from kettle depressions may drain by ground-water and surface water evapotranspiration and/or by ground-water underflow. Such depressions may trap and contain sediment and pollutants for generations to thousands of years lacking the presence of integrated surface drainage networks.

Unconsolidated overburden deposits also may be deposited by the action of wind (loess, sand dunes, etc.), as stream alluvium, in deltas, as beach deposits and by other processes that control their permeability and chemical character. Uncemented alluvial deposits may exceed 1,000 to 5,000 feet is thickness in the Basin and Range province of the U.S. and are well developed on most flood plains.

BEDROCK STRATA

Bedrock units in the U.S. vary from igneous, metamorphic and metasediments of Precambrian to more recent age to sedimentary rocks ranging from pre-Cambrian to Recent age. Bedrock units may contain secondary openings such as rock fractures, joints, bedding plane partings, sheeting joints, faults and zones of fracture concentration depending upon their origin and structural history. Secondary openings are the only openings that afford water storage and transmission characteristics for most igneous and metamorphic rocks. Sedimentary rocks, as well as unconsolidated overburden deposits also contain intergranular or primary openings that were formed during original sedimentation. These openings can be cemented in part or nearly completely with secondary minerals following deposition and burial. The water storage and transmission properties of all sedimentary rocks may be enhanced by the imposition of secondary openings resulting from various geological processes.

Processes of physical and chemical weathering of bedrock units tend to enhance their water storage and transmission characteristics. For soluble limestone, dolomite, marble, gypsum and salt, weathering processes may greatly enlarge joints and rock fractures producing small to large voids that may serve as regionally interconnected conduits. Caves also may be present in some regions.

Soluble cements also may be selectively leached from sandstone and silt beds exposed to chemical weathering thereby adding to intergranular permeability and porosity of some sedimentary strata.

A variety of geological processes help to account for the development and distribution of hydraulic properties of bedrock and overburden deposits throughout the U.S. that influence amounts of ground water in storage, rates of ground-water flow and their pollutant attenuation potential. Repeated erosion by continental ice sheets approximately along a line north of the Ohio and Missouri Rivers has altered the bedrock topography and helped to remove the more permeable weathered top of bedrock. Glacial drift, in turn was deposited by these glaciers as till, outwash, lacustrine and windblown sediments. These deposits are variable in thickness, distribution and area extent and either help to isolate or connect bedrock aquifers with contaminants applied to the land surface.

Limestone, dolomite, gypsum and marble, for example, may contain well-developed networks of solution openings that favor the rapid subsurface movement of pollutants. In the absence of thick, poorly permeable residual or transported soils, such bedrock units would comprise hydrologically active areas for ground-water recharge. On the other hand, poorly permeable bedrock enhanced by interconnected networks of rock fractures may allow for the transport of pollutants at rapid rates where permeable overburden sediments directly overlie these bedrock units in recharge areas or where residual soils are thin to discontinuous. The extremes of several bedrock settings and their ability to retard pollutants is illustrated in Figure 3.

GROUNDWATER FLOW SYSTEMS

Ground-water flow systems operate in all saturated overburden and bedrock units. These flow systems develop in response to ground-water recharge and discharge governed by the thickness and distribution of aquifers and confining beds, topographic variability, position of barriers to ground-water flow and many other factors. Selected flowlines depicting the theoretical pathway along which ground water must flow between recharge and discharge areas are illustrated for several aquifer configurations (Figure 4).

The permeability of units is the same for Figure 4 A and B. Only the water table configuration has been varied when predicting resulting changes in the flow field. Note that in A, nearly the entire upland surface acts as a recharge zone. Path lengths and residents time would be maximized to the right of the river valley shown. Pollutants might be contained in transient storage for years to thousands of years allowing other retardation mechanisms also to operate to reduce or eliminate pollutants. For the case shown in B, five ground-water

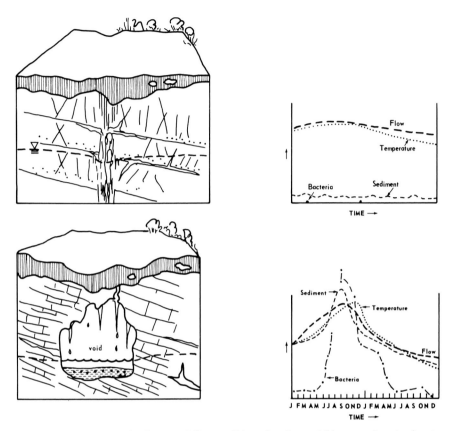

FIGURE 3. Range in bedrock permeability conditions favoring variable rates of contaminant transport within ground water. (See next page, from Parizek, 1983a).

discharge areas have been created where in A only one exists. Rates and directions of flow have been altered, and residents time have been shortened within the ground-water reservoir. Ground-water discharge areas shown now act as a hydrologically active areas to surface water systems precluding the deep penetration of pollutants within the underlying aquifer.

An appreciation of the rates of ground-water flow for various hydraulic conductivity and gradient values are given in Table 7. One can understand why pollutants, once added to the ground-water reservoir may take years to reappear at the land surface, even in the absence of retardation and attenuation processes. Detecting the presence of pollutants within the ground-water flow fields shown in Figure 4 A and B also requires the placement of monitoring installations in such positions as to insure they will eventually intersect pollutant plumes.

In Figure 5, density contrasts between water and pollutants result in differentiation between ground water and pollutants that must be understood when designing monitoring networks. Pollutants may float on ground water, sink

A

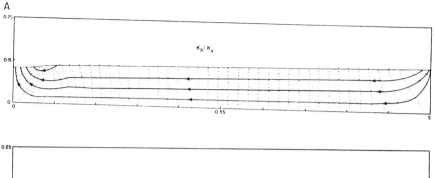

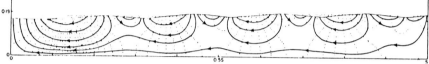

FIGURE 4. Theoretical Flow Lines for Several Aquifer Configurations. (From Freeze and Witherspoon, 1967).

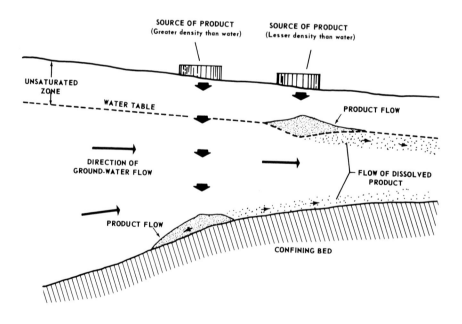

FIGURE 5. Effects of density on migration of contaminants. (USEPA, Basic Elements).

TABLE 7

Rates of Ground-water Flow for Various Permeability Values and an Assumed Hydraulic Gradient of 0.01. (Adapted from Cedergren, 1967)

Permeability (cm/sec)	Hydraulic Gradient(i)	Velocity (ft/day)
3.5×10^{-6}	0.01	4×10^{-4}
3.5×10^{-5}	0.01	4×10^{-3}
3.5×10^{-4}	0.01	0.04
3.5×10^{-3}	0.01	.40
0.035	0.01	4.0
0.35	0.01	40
3.5	0.01	400

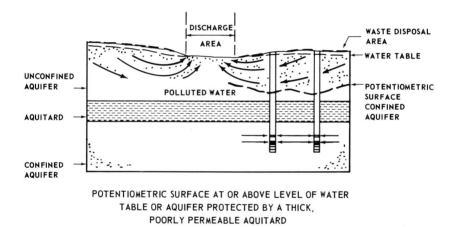

POTENTIOMETRIC SURFACE AT OR ABOVE LEVEL OF WATER
TABLE OR AQUIFER PROTECTED BY A THICK,
POORLY PERMEABLE AQUITARD

FIGURE 6. Comparison between confined and unconfined aquifers and their accessibility to pollutants. (Modified from Parizek, 1983b).

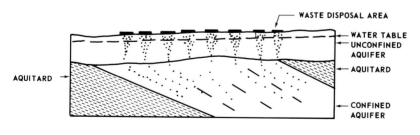

ABSENCE OF AQUITARD

FIGURE 7. Unconfined and confined aquifers separated by a confining bed effecting pollutant migration in groundwater recharge area. (Modified from Parizek, 1983b).

within less dense water, flow in opposite directions as ground water, or move in various directions, as multiphase flow.

Figure 6 shows how deep confined aquifers might be protected against pollutants introduced within shallow, unconfined aquifers, or how confined aquifers may become contaminated where they are exposed in recharge areas receiving pollutants (Figure 7). The presence of perched ground-water lenses (Figure 8A and B) also may influence the subsurface migration hence, early detection of shallow pollutants.

All of these factors affecting subsurface waterflow may lead to the mistaken impression that the ground-water route provide insignificant pollutant loads to rivers and lakes or that ground water has not been contaminated when in fact, toxic substances may be slowly building up in ground water within aquifers and confining beds and is moving to ground-water discharge areas.

It is important to realize that shallow unconfined aquifers commonly become polluted in densely populated industrialized and urbanized regions for a variety of reasons. These water supply sources, relied upon in early development of a region, tend to be abandoned in favor of deep sources of ground water when available, or water may be imported from distant lakes and rivers. In time, shallow wells are abandoned and water quality sampling points and quality data become scant. Stream quality data reveal the presence of pollutants within these industrial regions derived from a variety of sources. Rarely have scientists attempted to separate the amount of pollutants entering rivers and lakes from ground-water sources. Ground-water discharge areas are logical areas in which to look for advanced signs of more regional ground-water quality degradation.

GROUNDWATER QUALITY

Ground water is not pure to start with. Subsurface waters become mineralized as they infiltrate into the unsaturated soil zone and percolate to the water table. Ranges in quality are extreme. It may be as "pure" as rainfall (prior to man's impact on rainfall water quality) ranging all the way to brines more concentrated in total dissolved solids than sea water. Quality variations are a function of many factors including the age of water, minerals contacted in its history of flow between recharge and discharge areas, temperature, pressure, presence of gases, membrane effects, etc.

Table 8 shows the major inorganic species that occur in ground water. Ranges in concentrations are given only as a guide because it is not uncommon for these concentrations to be exceeded. The major ions in ground water that comprise nearly 90% of the total dissolved solids for potable to highly mineralized water include Na^+, Mg^{2+}, Ca^{2+}, Cl^-, HCO_{3-}, and SO^{3-}. These ions are released from the chemical weathering or solution of soluble minerals such as calcite, dolomite, and gypsum, that are present either as cements in sedimentary rocks or as

TABLE 8

Classification of Dissolved Inorganic Constituents in Ground water

Major constituents (greater than 5 mg/ℓ)	
Bicarbonate	Silicon
Calcium	Sodium
Chloride	Sulfate
Magnesium	Carbonic acid
Minor constituents (0.01-10.0 mg/ℓ)	
Boron	Nitrate
Carbonate	Potassium
Fluoride	Strontium
Iron	
Trace constituents (less than 0.1 mg/ℓ)	
Aluminum	Molybdenum
Antimony	Nickel
Arsenic	Niobium
Barium	Phosphate
Beryllium	Platinum
Bismuth	Radium
Bromide	Rubidium
Cadmium	Ruthenium
Cerium	Scandium
Cesium	Selenium
Chromium	Silver
Cobalt	Thallium
Copper	Thorium
Gallium	Tin
Germanium	Titanium
Gold	Tungsten
Indium	Uranium
Iodide	Vanadium
Lanthanum	Ytterbium
Lead	Yttrium
Lithium	Zinc
Manganese	Zirconium

SOURCE: Davis and De Wiest, 1966.

sedimentary rocks and from the slow chemical weathering of sedimentary, metamorphic and igneous rocks.

Minor and trace elements commonly are low in concentrations. They may be at or below drinking water standards in shallow aquifers with active ground-water circulation to in excess of recommended and maximum permissible concentrations in deep, confined aquifers, in metaliferous regions, and when mixed with shallow water resulting from deep drilling, uncased and unplugged abandoned wells.

On average, igneous and metamorphic rocks have low solubility. Their ground waters tend to have a very low concentration of total dissolved solids as do waters in some sandstone aquifers located in overlying sedimentary rocks.

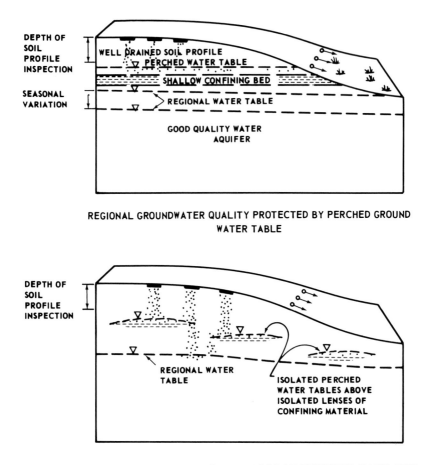

REGIONAL GROUNDWATER QUALITY PROTECTED BY PERCHED GROUND
WATER TABLE

REGIONAL GROUNDWATER QUALITY LOCALLY PROTECTED BY ISOLATED
LENSES OF CONFINING MATERIAL

FIGURE 8. Influence of perched ground-water lenses on pollutant migration to the regional water table. (Modified from Parizek, 1983b).

These waters are neutral to slightly acidic and have low buffer capacities. Gillham (1983) reports that these waters, lacking significant naturals controls on water chemistry, are particularly sensitive to activities that introduce contaminants into the hydrologic regime.

Carbonate rocks and calcareous glacial drift contain waters that are generally calcium and magnesium bicarbonate type. These waters are usually of good quality but contain high total dissolved solids particularly when more deeply buried in sedimentary basins or when located in regional ground-water discharge areas especially where ground water has traveled to great depths and distances within these strata.

Evaporite deposits such as gypsum and halite are present in various sedimentary basins. These highly soluble strata, are associated with shale bedrock containing poor quality water. They display high total dissolved solids, high concentrations of calcium sulfate, sodium chloride, and hydrogen sulfide. Evaporites also are activity being deposited in semi-arid to arid regions of the U.S. and account for abrupt changes in water quality within sedimentary basins.

Accessory minerals such as pyrite and marcasite and other sulfide minerals are present in some shale and coal beds as well as other sedimentary and igneous rocks. When oxidized, they may produce high sulfate, iron and total dissolved solids contents. They are a natural and man-induced source of acidity in coal and sulfide ore mining districts.

The base-flow of chemistry of streams resulting from ground-water discharge can be highly variable for natural reasons. Tributaries and lakes receiving relatively young ground water that has traveled relatively rapidly through its flow system tend to receive ground-water favorable in quality whereas basins receiving deep ground water tend to be more highly mineralized. These flow systems contribute chemical loadings to lakes and stream that were in dynamic equilibrium prior to man's development of the region. Ambient or background water quality data can still be determined by monitoring base flow water quality in tributaries least influenced by man and by monitoring water quality in well fields completed within various hydrogeologic settings still isolated from man's chemical influences.

Natural chemical loadings to lakes and streams are not uniform from drainage basin to drainage basin. Baseline ground-water quality data must be established in order to determine if increases in loadings in particular aquifers and tributaries are the result of man's activities. Baseflows to streams may range from 5 to 80 or more percent of total streamflow for various tributaries throughout a basin. Loadings of natural inorganic substances, therefore, should be significant.

FINAL STATEMENT

Definition and control of existing point and diffuse sources of pollutants resulting from past and current material usage and water disposal practices will require a knowledge of surficial and near-surface processes operating within watershed, their tributaries, master streams, and receiving lakes and estuaries. Similarly, the management and disposal of toxic and hazardous wastes also requires a clear understanding of the rates and directions of soil water and ground-water flow, attenuation mechanisms that may help to reduce or eliminate pollutants once introduced into soil or rock, their persistence, and toxicity. Rates of transfer of a host of toxic substances within the atmosphere, in surface water, soil water and ground water and their residence times within other segments of our ecosystem also must be better defined if we are to continue to demand a safe living environment and at the same time continue to enjoy the many products of a technological society to which we have grown accustom.

LITERATURE CITED

Cedergren, H. R., 1967, Seepage Drainage and Flow Nets, John Wiley and Sons, Inc., 489 pp.

Davies, T. T. and T. E. DeMoss, 1982, Chesapeake Bay program technical studies: A synthesis, U.S. Environmental Protection Agency, 634 pp.

Davis, A. O. and J. N. Galloway, 1981, Atmospheric lead and zinc deposition into lakes of the eastern United States; In Atmospheric Pollutants in Natural Waters. S. O. Eisenreich, ed., Ann Arbor Science Publishers, Inc., MI, pp. 401-421.

Davis, S. N., 1969, Porosity and permeability of natural materials in flow through porous media, R. J. M. DeWiest ed., Academic Press, N.Y.

Davis, S.N., and R.J.M. DeWiest, 1966, Hydrogeology. John Wiley and Sons, New York, 436 pp.

Freeze, R. A. and J. A. Cherry, 1979, *Groundwater,* Prentice-Hall, Inc., N.Y.

Freeze, R. A. and P. A. Witherspoon, 1967, Theoretical analysis of regional groundwater flow: Effect of water table configuration and subsurface permeability variation. Water Resources Res., Vol. 3, pp. 623-634.

Gillham, R., 1983, Great Lakes Contamination via Groundwater in Canada, Science Advisory Board Annual Report to International Joint Commission, Windsor, Ontario.

Green, W.J., G.F. Lee, R.A. Jones and T. Palit, 1983. Interaction of clay soils with water and organic solvents: Implications for disposal of hazardous wastes, Environ. Sci. Technol., Vol. 17, pp. 178-182.

International Reference Group on Great Lakes Pollution from Landuse Activities, (PLUARG) 1978, Environmental Management strategy for the Great Lakes System, International Joint Commission (Canada and the U.S.) Windsor, Ontario, 115 p.

J. R. B. Associates, Inc., 1980, Methodology for rating the hazard potential of waste disposal sites, May 5.

Langmuir, D., Controls on the Amounts of Pollutants in Subsurface Waters. Earth and Mineral Sciences, Vol. 42, No. 2, The Pennsylvania State University, 4 pp., 1972.

Lazarus, A. L., E. Lorange, and J. P. Lodge, Jr., 1970, Lead and other metal ions in United States Precipitation and Environmental Science and Technology, Vol. 4, pp. 55-58.

Mrak, E. M. ed., 1974, Herbicide report, Hazardous materials Advis. Comm. EPA-SAB-74-001, 196 pp.

Parizek, R. R., 1980. Non-point-source pollutants within the Great Lakes—A significant international effort, Part 5, pp. 435-472. Thresholds and man in *Thresholds in Geomorphology,* pp. 435-472, ed. D. R. Coates and J. D. Vitek, Allen and Urwin Inc., Winchester, Mass.

Parizek, R.R., 1983a, Overview of ground water quality and its significance to the Great Lakes Ecosystem. Report submitted to Science Advisory Board, International Joint Commission, Windsor, Ontario.

Parizek, R.R., 1983b, In Working Paper No. 2 Ground water characteristics for State of Delaware, prepared for Dept. of Natural Resources and Environmental Control. Dove Delaware, Tafman and Lee Assoc., Inc., Wilmington, Delaware, 122 pp.

Rao, P. S. C. et al., 1981, EPA Rept. (in press).

Stevenson, J. R. and N. M. Confer, 1978, Summary of available information of Chesapeake Bay submerged aquatic vegetation. U.S. Dept. Inter. FWS10BS-78166 NTIS, Springfield, VA, 333 pp.

Suns, K. et al., 1978, Organochlorine and heavy metals residues in the nearshore biota of the Canadian lower Great Lakes. Submitted to PLUARG TASK D (U.S. Section) activity 3.36 Windsor, Ontario, June 1978, 44 pp.

U.S. Environmental Protection Agency, 1980, A method for determining the compatibility of hazardous wastes, H. K. Hatayama, et al., EPA-60012-80-076.

U.S. Environmental Protection Agency, 1982, Part V, National Oil and Hazardous Substances Contingency Plan, July 16, 1982, Federal Register, Vol. 47, No. 137, pp. 31180-31243.

Wauchope, R. D., 1978, The pesticide content of surface water draining from agriculture fields. A review, J. Environ. Qual. Vol. 7, pp. 459-472.

Weed Science Society of America, 1980, Herbicide Handbook 35th ed., WSSA, Champaign, IL, 450 pp.

Hazardous and Toxic Wastes: Technology, Management and Health Effects. Edited by S.K. Majumdar and E. Willard Miller. © 1984, The Pennsylvania Academy of Science.

Chapter Eleven

Siting Hazardous Waste Management Facilities: Theory versus Reality

Richard F. Anderson, Ph.D.[1] and Michael R. Greenberg, Ph.D.[2]

[1]Department of Urban Affairs and Planning
Boston University
Metropolitan College
755 Commonwealth Avenue
Boston, Massachusetts 02215

[2]Professor of Environmental Planning and Geography
Co-Director Graduate Public Health Program
Department of Medicine and Dentistry
Rutgers-The State University
Piscataway, N.J. 08903

Siting new hazardous waste facilities has met stiff opposition from community residents because of the fear that past mistakes and their costly consequences will be replicated. Yet there is a need for new hazardous waste facilities. A technical solution is possible that can allow communities to maintain the integrity of environmental quality and protect public health and at the same time meet the needs of industry to manage hazardous wastes properly. With the Resource Conservation and Recovery Act in place to standardize and regulate hazardous waste generators, transporters and receivers, and with the emergence of siting criteria evaluation procedures it is clear that we can learn from past mistakes and avoid them. A case study is presented to compare the reality of past siting decisions to what would be a theoretically more appropriate siting decision protocol. After comparing the locational attributes surrounding 46 existing hazardous waste landfills to criteria designed to determine site-suitability it was found that past siting decisions were largely based on very narrow economic-based factors. The case study shows why most facilities were improperly sited, and how the siting procedure can be used to make more informed siting decisions. Hopefully, siting procedures like the one presented in this case study will help overcome public opposition to new facilities.

INTRODUCTION

Siting new hazardous waste management facilities in the United States has become an almost impossible task. Proposals to locate new facilities have met adamant public opposition from organized citizen groups formed in potential host communities. These groups argue that by allowing this noxious industry in the community it will introduce an unacceptable risk of danger to air resources, water supplies, and ultimately human health. The Love Canal incident is often referenced as the worst case scenario where property values deflated, residents were evacuated, and, damage to human health has been an issue of continuing concern. Communities are no longer willing to assume the responsibility of environmental externalities resulting from private sector decisions to dispose of wastes on cheap, and nearby land. The fear of threats to health and the environment have led some resident groups to practice civil disobedience to obstruct governmental decisions to land dispose hazardous wastes (1). The reality of the poor past record of hazardous waste management and damage incidents, denial of responsibility from the private sector, and the inability of government bureaucracy to promptly rectify such incidents are not compelling arguments to persuade local residents to allow new facilities into the community.

Proponents of proposals to construct new facilities argue that past disposal practices and uncontrolled land dumps are obsolete. The various requirements under the Resource Conservation and Recovery Act (RCRA) mandate a technical solution to the hazardous waste problem by designing the facilities to prevent migration of contaminants into the environment. Along with design standards, RCRA requires listing wastes that are hazardous and tracking them from their origin to destination. The legislative mandate will impose a new order on hazardous waste management that will theoretically minimize threats to health and the environment. Rather than dwelling on past mistakes, proponents of new facilities argue that society learns from past mistakes and that striking a balance between regulatory design and operations standards and a reasonable enforcement capacity should allow new facilities to be located and operated without danger to the community.

Methods designed to find the least disadvantageous sites are needed in order to avoid the serious siting errors of the past. The purpose of this article is to present one such method, a site screening method; and to show that it can be used to make better informed facility location decisions than were made in the past. Thus, the research hypothesis presented here is that given an appropriate set of hazardous waste facility siting criteria including demographic, environmental and economic factors, the vast majority of past facility siting decisions would probably be currently unacceptable.

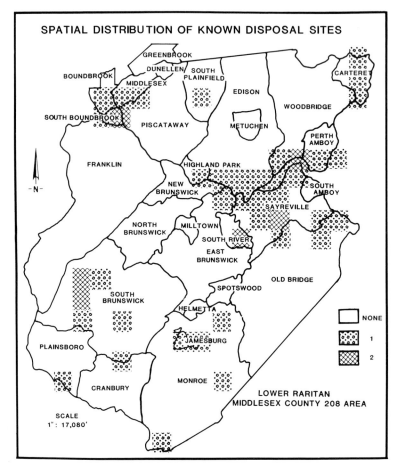

FIGURE 1. Spatial Distribution of Known Disposal Sites

MATERIALS AND METHODS

We use a case study approach to test the hypothesis in the Lower Raritan/Middlesex County 208 Water Management Area located in central New Jersey (Figure 1). In this study region we compare the characteristics of actual facility locations to a derived set of locations. In order to select sites that are better than existing sites it is necessary to have an understanding of the types of damage incidents associated with hazardous waste, and available technologies for managing these wastes. A set of facility siting criteria are derived from an understanding of the interrelationships between these factors: and these criteria serve as the centerpiece for comparing the past reality to the screening model approach.

The development of hazardous waste facility siting criteria

Past siting decisions did not consider safety as a major concern. The chief criterion was disposing of wastes as cheaply as possible. Nor is there a present consensus on what constitutes a "safe" place to manage hazardous wastes. Indeed, there is probably no completely safe palce: any place is likely to engender some problems. There are, however, some locations which are least disadvantageous, and it is the objective of the screening approach to identify them. The screening approach considers potential impacts to air and water resources and threats to human health concurrently with economic criteria.

Applying siting criteria in screening studies has been practiced by governmental agencies at different levels and through different arrangements for some time. For example, the Nuclear Regulatory Commission staff has had specific involvement with siting nuclear power plants. The Commission has developed an early site review guide that identifies the most important factors to be considered with regard to siting nuclear power facilities (2). The siting factors found in the guide include among other things: (1) geologic characteristics; (2) seismology; (3) hydrology; (4) meteorology; (5) accident analysis review (i.e., radiological consequences of accidents; geography; demography; other land uses; etc.); (6) radiological effluent assessment; (7) non-radiological impacts, (i.e., ecology and plant systems; water quality; etc.): (8) cost-benefit and socioeconomic reviews, (i.e., need for site; direct and indirect community/regional impacts; summary cost-benefit analysis); (9) other review areas, (i.e., details of land and water use; historic, cultural and natural features). The actual technical assessment of these factors are outlined in regulatory guides (3,4). The early site review procedure and regulatory guides offer a very good starting point for the development of hazardous waste facility siting criteria. The obvious distinction to be emphasized is that we do not ordinarily include radioactive wastes as constituents of hazardous wastes.

The EPA has more recently been involved with siting solid waste management facilities (5). Since passage of RCRA the EPA has been under mandate to aid the states in their development of hazardous waste management programs and provide information concerning adequate location of facilities, including consideration of regional, geographic, and climatic factors (6). The EPA proposed eight general site selection criteria, emphasizing areas that should be avoided. The eight criteria are; (1) active fault zones; (2) the regulatory floodway; (3) coastal high hazard area; (4) 500 year flood plain; (5) recharge zone of sole source aquifer; (6) wetlands; (7) endangered species and critical habitats; and, (8) a 200 foot facility setback (7). These general criteria are appicable across much of the continental United States, however, it is clear that the intent of EPA is to leave much discretion to the states since the proposed criteria are obviously incomplete.

The EPA promulgated interim final location standard regulations on January 12, 1981 that became effective in July of 1981 (8). Only two of the eight proposed location standards were retained. These are areas of seismic activity and the 100 year floodplain. The 100 year floodplain consideration is a considerable change from the proposed 500 year floodplain. The standard allows for a variance of this restriction if a developer can show that washout events in the 100 year floodplain can be avoided. The seismic consideration restricts facility location on or near faults experiencing displacement in Holocene time (roughly, in the last 11,000 years). The seismic consideration was modified in November 1981 by the EPA (9). The EPA altered the requirement for a geological investigation of seismic activity based on the assumption that differences in the geology of states east of the Rocky Mountains do not warrant seismic restrictions. The other proposed cirteria were either dropped or reserved for judgment by the EPA.

Siting criteria pertinent to evaluating location suitability for hazardous wate facilities are found in some state hazardous waste laws. For example, a review of hazardous waste legislation enacted by eight states in the Northeast specify siting criteria or location standards to be evaluated when siting new facilities. The eight states include Connecticut (10), Maine (11), Massachusetts (12), New Hampshire (13), New Jersey (14), New York (15,16), Rhode Island (17) and Vermont (18). Each state authorized different arrangements for siting. For example, some of the states (e.g., Massachusetts, New York, etc.) have empowered special Councils, Commissions or Board to oversee siting decisions according to an outlined procedure. Other states, like Vermont, seek to use existing laws and regulatory procedures to evaluate facility siting proposals. What is becoming increasingly clear is the fact that the states recognize the value of siting criteria as a useful aid in improving location decisions. Two exemplary studies show how criteria may be evaluated for a broad, multi-state region (19), or at the site-specific scale (20,21). Space does not permit a discussion of these otherwise relevant studies. The next section of this paper presents a test of a siting criteria evaluation procedure designed to show how the reality of past siting decisions can be much improved by considering factors which theoretically can identify adverse impacts and associated levels of risk.

Screening model approach
The siting model relies upon the identification of criteria appearing in the hazardous waste facility location studies, and the broader environmental impact literature. Twelve siting factor categories were selected to draw siting criteria from (Table 1). We evaluate these criteria for the study region to determine the least unsuitable areas for facility location. Next, we compare these least unsuitable areas to actual facility locations to determine how well or poorly they match. The 33 criteria listed in Table 1 are keyed to the selected region of study which is discussed along with the analytic method of comparison employed.

Many of the siting variables are easily understandable once they are identified. For example, variables numbered one through four represent the major aquifer groups in the study region. Floodlands consist of the floodway, fringe and prone

TABLE 1

Hazardous Waste Facility Siting Factors Important For The Study Region

Siting Variable	Factor Category
1. Englishtown Sands	1. Aquifers
2. Farrington Sands	
3. Mt. Laurel and Wenonah Sands	
4. Old Bridge Sands	
5. Floodway Area	2. Floodlands
6. Flood Fringe Area	
7. Flood Prone Area	
8. Newark Group	3. Unsuitable Geology
9. Diabase Sill	
10. Buried Extension of Diabase Sill	
11. Marshalltown Formation	4. Suitable Geology
12. Merchantville Clay	
13. Woodbury Clay	
14. Tidal Wetlands (1)	5. Wetlands
15. Tidal Wetlands (2)	
16. Fresh Water Wetlands	
17. Severe Septic Limitations	6. Permeability
18. Water Pumpage Centers	7. Water Pumpage Centers
19. Water Pumpage Center of 2 Million Gallons/Day or Greater	
20. Groundwater Contamination Area	8. Water Quality Problem Areas
21. Fresh Water 2 Streams	9. Surface Waters
22. Fresh Water 3 Streams	
23. Tidal Water 1 Streams	
24. Tidal Water 2 Streams	
25. Tidal Water 3 Streams	
26. Effluent Limited Segments	
27. Amphibian Habitats	10. Endangered Species
28. Reptile Habitats	
29. Total Population	11. Population Centers
30. Population, Ages 0-5 years	
31. Population, Ages 65 years and older	
32. Depth to Seasonal High Water Table	12. Water Table and Bedrock
33. Depth to Bedrock	

areas. Unsuitable geology consists of formations that are characteristically cracked, folded or fissured (i.e., the Newark Group), or formations that are extremely dense and thus not easily modified for engineered storage. Suitable geology, on the other hand, include clayey formations with characteristically low permeability rates which impede (though not halt) leachate and lessen potential groundwater contamination. Wetlands consist of tidal and freshwater wetlands that support diverse ecosystems and normally enhance potable water supplies in some groundwater recharge areas. Severe septic limitation areas represent areas where permeability, water table, and/or bedrock coditions preclude surface impoundments, and especially landfills.

The remaining groups of siting variables are self explanatory except for water quality problem areas and population centers. The groundwater contamination variable (number 20) consists of a small number of areas in the study region with documented contamination of ground water including chemical contamination and salt water intrusion. The population variable group includes, in addition to population size, the two population groups at the extremes of the age distribution pyramid. These groups—the young and elderly—represent persons at high risk of disease due to undeveloped or an aged immunological system.

The Study Region

The Lower Raritan/Middlesex County 208 River Basin in New Jersey was chosen as the study region. This region is comprised of 29 minor civil divisions, and covers an area of roughly 350 square miles. The study region is shown in Figure 1, and is shaded to point out various areas containing 46 known dumpsites, (22).

The study region currently hosts chemical landfills that are among the most serious hazardous waste problem sites in the United States. New Jersey ranks first in chemical manufacturing employment among the 50 states, and according to the U.S. EPA has 65 of the worst 419 abandoned dumpsites in the United States. Middlesex County, all of which is in the study region, has the largest concentration of hazardous waste generating chemical industries in New Jersey. These local industries as well as others, many from outside the state, uôlized local dumpsites for some time. One of the worst hazardous waste dumpsites in the United States is located in Middlesex County. The infamous Kin Buc Landfill accepted 75 percent of solid wastes and 50 percent of liquid wastes generated in New Jersey during the late 1970s (23) and has caused serious environmental concern (including the death of a worker) in the community since it was prone to fires and explosions and contributed large quantities of contaminated leachate daily to a nearby estuary. Should this facility and the others have been located where they were?

Method of Analysis

To answer the above question, least disadvantageous locations in the study region were sought and compared to the actual locations. The screening method used to identify the least disadvantageous locations is summarized below, and is reported at greater length elsewhere (24,25).

First, a grid system was overlaid on the study region to divide the land area into 272 cells of equal unit area (2 by 2 square kilometers). The cells serve the dual function of allowing comparison of areal locations within the region, and providing sufficient avreage parameters to facilitate a large scale, integrated hazardous waste treatment and disposal facility. In order to quantify tne variables the requirements of a large-scale integrated facility must be outlined. RCRA does not specify land area requirements for such facilities. However, the waste management industry suggests at least a 200 acre site would be needed to be consistent with RCRA regulations (i.e., a landfill with a 20 year expected operational lifetime, with segregated storage amenable for future materials recovery), (26). Add to the 200 acres additional space for treatment and recovery operations, and a reasonable buffer zone to separate the facility from other community activities (i.e., residential, institutional, etc.) and we are considering a regional facility requiring roughly 336 acres, or about one-third the area of a single 2 by 2 square kilometer cell in the study region (which is about 988 acres). Data for each siting variable was obtained from the natural resource inventory and special task studies under the 208 program conducted by the Middlesex County Planning Board. Additional data were gathered from maps and information provided by the New Jersey Department of Environmental Protection and the U.S. Department of Agriculture - Soil Conservation Series reports.

Despite the fact that governments have been developing siting laws and procedures, there is a substantial lack of scientific measurements that bear upon hazardous waste management facility siting criteria (27). A facility that is four miles away from an aquifer is relatively better suited to protect groundwater than a facility located directly above an aquifer. How much safer is a facility that is one mile away rather than one-half mile; or one-half mile rather than one-quarter mile? Such distances may in reality be equally suitable or unsuitable; and site-specific information may require resources well beyond what is reasonably available to find out. In the absence of scientific rules the researcher must fall back to relative distances between cells. Listed in Table 2 are the assumptions used in this effort. They are based on three types of measurement: (1) area occupied in a cell by a specific factor: (2) the linear distance of a cell centroid to a factor; (3) the numerical distribution of a factor. The first two measurements assume attenuation of impacts with distance, while the third measurement relies on judgment and natural breaks in the data. Each cell was assigned initial suitability ranks based on these measurements (see Table 2) for the siting variables in order to assess which existing dumpsites in the study region were improperly sited.

The 33 variables were reduced to 22 variables by aggregating some of the criteria. For example, the four aquifer variables were aggregated into one variable representing all aquifers by adding the percent land area for each cell containing any of the aquifers. This procedure was also used to aggregate floodlands, wetlands and geologic factors. While this procedure serves to simplify the analysis and presentation here it may be wise, for other purposes, to evaluate each variable separately. The final 22 variables to be analyzed are listed in Table 3.

TABLE 2

Criteria Ranking and Assumptions

Siting Factor Category	Type Measurement	Initial Suitability Rank	Siting Assumption
	Percent Land Area		
Aquifers	0.0-5.0	1	MOST SUITABLE
Floodlands			Probably no adverse impacts.
Unsuitable Geology			
Suitable Geology[a]	5.1-15.0	2	SOME CAUTION
Wetlands			Probably no adverse impacts but facility should avoid areas of criteria presence.
Permeability			
	15.1-66.0	3	CAUTION Treatment and disposal facilities potentially allowable with extreme caution and reasonable mitigation plans in engineering design.
	66.1-100.0	4	LEAST SUITABLE Facilities involving any type of surface impoundment should not be located.
	Linear Distance (in miles)		
Water Pumpage Centers	4.32 or greater	1	MOST SUITABLE
Water Quality Problem Areas[b]			
Surface Waters[b]	3.09-4.31	2	SOME CAUTION (same as above)
Endangered Species			
	1.86-3.08	3	CAUTION (same as above)
	0.0 -1.85	4	LEAST SUITABLE (same as above)
	Frequency Distribution and Natural Breaks in the Data		
Population Centers			
Total Population	< 252	1	MOST SUITABLE
	≥ 252 ≤ 895	2	SOME CAUTION
	> 895 ≤ 3463	3	CAUTION
	> 3463	4	LEAST SUITABLE

TABLE 2 (continued)

Criteria Ranking and Assumptions

Siting Factor Category	Type Measurement	Initial Suitability Rank	Siting Assumption
Population Age 0-5 (in percent)			
	≤ 7.1	1	MOST SUITABLE
	>7.1 ≤ 8.4	2	SOME CAUTION
	>8.4 ≤10.1	3	CAUTION
	>10.1	4	LEAST SUITABLE
Population Age 65 Plus (in percent)			
	≤ 4.2	1	MOST SUITABLE
	>4.2 ≤ 6.6	2	SOME CAUTION
	>6.6 ≤ 9.3	3	CAUTION
	> 9.3	4	LEAST SUITABLE
Water Table and Bedrock			
Depth to Seasonal High Water Table (in feet)			
	≤ 5.0 =	4	LEAST SUITABLE
	>5.0 ≤ 7.8 =	3	CAUTION
	>7.8 ≤10.0 =	2	SOME CAUTION
	>10.0 =	1	MOST SUITABLE
Depth to Bedrock (in feet)			
	>5.0 =	2	SOME CAUTION
	≥2.25 ≤5.0 =	3	CAUTION
	<2.25 =	4	LEAST SUITABLE

ªVariable measurements in units of area assume that a large scale - integrated facility will require roughly one-third of the land area in a cell. Therefore, a cell with over two-thirds of its land area situated in floodlands would be least suitable. Conversely, cells with under five percent of their land area designated as floodland would be most suitable. The suitability ranks of some caution and caution were chosen to represent different margins of safety. Some variables, such as suitable geology use the rank system in reverse since it is more advantageous to site a facility in suitable geologic formations such as thick clay.

ᵇSome variables are subject to debate, and rankings are not necessarily straight forward (see text). However, this analysis uses the ranks shown as a first approximation.

TABLE 3

List of Final 22 Siting Criteria

1. Aquifers	12. Tidal Water-1 Stream
2. Floodlands	13. Tidal Water-2 Streams
3. Unsuitable Geology	14. Tidal Water-3 Streams
4. Suitable Geology	15. Effluent Limited Segment
5. Wetlands	16. Population Centers
6. Severe Septic Limitations	17. Population 1-5 years
7. Water Pumpage Centers	18. Population 65 years +
8. Water Pumpage Centers 2 MGD +	19. Amphibian Habitats
9. Water Quality Problem Areas	20. Reptile Habitats
10. Fresh Water-2 Streams	21. Seasonal High Water
11. Fresh Water-3 Streams	22. Depth to Bedrock

Each of the 272 cells were represented by the 22 aggregate siting variables which were converted into initial suitability ranks as outlined in Table 2. Some of the siting variables have similar geographical patterns. To reduce the number of separate ciriteria maps that had to be made the 22 variables were factor analyzed.

Since the raw data had been converted to suitability ranks, Spearman Rank correlation was used to obtain the factor analytic solution. Resulting multivariable factors are used to describe the spatial clustering of demographic characteristics and environmental attributes. The final criteria evaluations provide a way of identifying different levels of site-suitability: places to avoid; cells which would require mitigative measures; and, places that are the least disadvantageous relative to all other cells in the study region.

Results from the factor analysis indicated that the 22 siting variables could be reduced to 9 composite statistical factors (Table 4). The first six factors listed in Table 4 are multiple-variable factors where the individual variables are listed in order of magnitude of statistical importance. The factor scores, which represent a standardized measure of association between individual cells and the statistical factors, were used to reassign ranks similar to the initial ranks (28) to the 272 cells. Factors seven through nine in Table 4 represent unique factors; or those siting variables that had little spatial association with other siting variables. Overall, each of the 272 cells was assigned nine ranks of suitability.

The theoretically least disadvantageous locations for siting an integrated hazardous waste management facility were determined by deriving a composite rating for each cell. The procedure used here was as follows. If a cell had nothing worse than most suitable or some caution ranking for the 9 final factors it was assigned a composite rating of one (or "most suitable"). Cells with only one caution rank for any of the 9 factors was assigned a compostie rating of 2 (or "some caution"). Cells with more than one caution ranking were assigned a composite rating of 3 (or "caution"). Cells with only one least suitable ranking were assigned a composite rating of 4 (or "extreme caution"). Finally cells with more than one least suitable ranking were assigned a composite rating of 5 (or "least

TABLE 4

*Multivariate and Unique Factors Derived From Factor
Analyzing the Siting Criteria*

Siting Factors	
1. Population 0-5 years Population 65 years or older Population Centers Water Pumpage Centers 2 MGD +	5. Wetlands Floodlands
2. Water Pumpage Centers Water Pumpage Center 2 MGD + Aquifers	6. Unsuitable Geology Seasonal High Water Table Fresh Water-2 Streams Suitable Geology Fresh Water-2 Streams Amphibian Habitats Severe Septic Limitations
3. Tidal Water-3 Streams Tidal Water-2 Streams Fresh Water-2 Streams Effluent Limited Segments	7. Water Quality Problem Areas
	8. Reptile Habitats
4. Tidal Water-1 Streams Effluent Limited Segments Population Centers	9. Depth to Bedrock

suitable"). The reasoning incorporated here was that cells should not be determined as strictly unsuitable because of only one least suitable initial ranking. Indeed, if only one poor ranking were found the land in the cell might be amenable to engineering controls designed to eliminate or mitigate potential adverse impacts. On the other hand, it was assumed that some cells would have more than one least suitable initial ranking; and that other cells would be identified which did not suffer from these siting constraints. The composite rating scheme is shown in Table 5.

An assumption of the composite rating method discussed above is that each of the nine rankings (composite and independent factors) is of equal importance. In reality some may be more important than others (e.g., water supply protection is more important than reptile habitat protection). Recognizing the limitation of this assumption, other screening analyses were made. First, water protection criteria were applied, and cells with poor results were eliminated. Then other criteria were similarly applied. The results were not substantially changed. Therefore the results derived from equal ratings for the nine rankings are presented below.

RESULTS

Results from the composite rating scheme indicate that there are 38 cells (or roughly 14 percent of the study region land area) that appear to be the least disadvantageous for siting new facilities. The remaining 234 cells had composite

TABLE 5

Composite Rating Scheme Based on Statistical Results and Initial Suitability Rankings

Composite Rating	Initial Rank and Statistical Results
1 Most Suitable	Cells with most suitable or some caution rankings
2 Some Caution	Cells with only one caution ranking
3 Caution	Cells with more than one caution ranking
4 Extreme Caution	Cells with only one least suitable ranking
5 Least Suitable	Cells with more than one least suitable ranking.

ratings of 2 through 5 signifying that they are less advantageous for siting new facilities. Of the 38 least disadvantageous cells only 4 were occupied by existing landfills. This means that 36 of the 40 cells hosting 46 landfills (6 cells hosted 2 landfills) were located in areas conducive to population exposure or adverse impacts on critical environmental resources, (Table 6, and compare Figures 1 and 2). Only 8.6 percent of existing landfills (4 of 46) were sited in cells with the least disadvantageous cells among the cells without landfills than among the cells with landfills, a random chance pattern would have been better from the public health and environmental protection perspective than the least economic cost approach that was used when the existing landfills were sited.

Most unfortunately, there is a positive statistical association between the location of landfills and areas unsuitable for their location. Initial Spearman's rank correlations indicate that existing landfills were sited with little regard for a variety of siting criteria. Some examples include correlations between the presence of landfills and: floodlands ($r_s = 0.23$, $P < 0.001$); wetlands ($r_s = 0.16$, $P < 0.0007$); water pumpage centers ($r_s = 0.35$, $P < 0.001$); population centers ($r_s = 0.31$, $P < 0.001$); tidal water 1 streams ($r_s = 0.26$, $P < 0.001$).

Next, a Chi-square test was applied to determine if the absence or presence of a landfill was related to the most or least disadvantageous character of the cells. Using the 2 by 2 contingency table (Table 6) the resulting Chi-square was not significant. Thus, past facility siting decisions exhibit no relationship to cell suitability and the presence of a facility in this study. There is little comfort, then

TABLE 6

Crosstabulation of Cells With/Without Landfills, and Cells Least/Most Disadvantageous

	Number of Cells With Landfills	Number of Cells Without Landfills	Total
Number of Least Disadvantageous Cells	4	34	38
ªNumber of Most Disadvantageous Cells	36	198	234
Total	40	232	272

a-includes cells with composite ratings of 2 through 5.

in discovering that managers probably did not go out of their way to pick a most disadvantageous area. However, the positive correlations between the factors and the presence of landfills, and the fact that ninety percent of the landfills in the study region are located in unsuitable areas are a sad commentary on the historical pattern of siting.

The poor past record of siting decisions is also shown in Figure 2. This map shows the 38 least disadvantageous cells, four of which host existing landfills. It also shows the remaining 234 cells that are most disadvantageous as facility locations. Clearly, the Kin-Buc landfill in Edison and nearly all of the landfills in the study region should not have been sited where they were.

DISCUSSION

Public opposition to proposals for new hazardous waste facilities is motivated by a convergence of factors including: fear of adverse impacts on public helath; degradation of natural resources; diminution of property values; and the perception of an overall decline in the quality of life in the community. There is a strong public concern that past mistakes will be repeated, and that government and industrialists are either ineffective or act irresponsible when a hazardous waste damage incident occurs. Judging from the empirical investigation presented here, most past decisions to site facilities in the Lower Raritan Middlesex County 208 Area were influenced by a desire to keep costs down, not by considerations of demographic patterns and environmentally sensitive resources. If environmental protection is the yardstick by which we measure any new proposal for siting a hazardous waste facility, then probably few proposals based solely on economic criteria will be successful in securing approval for permit and license applications.

The same yardstick can be used to identify the least objectionable areas for locating a facility. While national hazardous waste management capacity is close to levels of national demand, there are many states where the geographical shortage of capacity will lead to higher costs and perhaps elevated risks related to transportation of wastes (29). The regulatory appartus has been set in place under RCRA and the approved state hazardous waste management programs to track wastes, via annual reports by generators, manifest records from transporters and facility owner/operators. RCRA and its progeny (the state RCRA's) also require standards for facility design, operation, and contamination detection. These are the main ingredients necessary to avoid many of the past problems associated with these facilities. Yet, these concerns are very site-specific in character, and do not address a broader spectrum of locational considerations such as those evaluated in this research. New methods for siting facilities are becoming available which are designed to consider these demographic, natural resorce, and built environment factors.

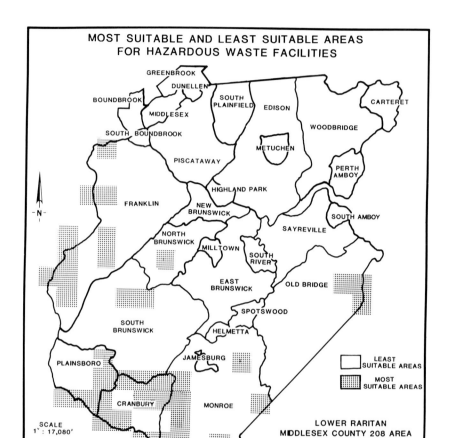

FIGURE 2. Spatial Distribution of Most Suitable and Least Suitable Cells for Hazardous Waste Facilities

The challenge for hazardous waste management in the 1980s is to better manage wastes, and one component of this objective will be to increase the number of new, technically well-planned and sited facilities. The development of various siting councils, boards, and advisory groups, that are drawing up what amounts to a protocol for making siting decisions signifies the intense scrutiny being brought to bear on land use decisions. Simultaneously, the level of accountability required of industry and government by the public with regard to facility proposals is quite severe. In order for these various waste management programs to become effective a rational-deductive approach must replace the highly-charged emotional confrontation between facility siting opponents and proponents.

REFERENCE

1. Over 277 arrests were made in Afton, North Carolina of citizens obstructing the land disposal of PCB laden roadside dirt, see The New York Times, October 3, 1982.
2. U.S. Nuclear Regulatory Commission, 1977, *Early Site Reviews for Nuclear Power Facilities-Procedures and Possible Technical Review Options*, (Springfield, Virginia: NTIS).
3. U.S. Nuclear Regulatory Commission, 1976, Regulatory Guide 4.2-Preparation Of Environmental Reports For Nuclear Power Stations, (Washingtion, D.C.: U.S. Nuclear Regulatory Commission).
4. U.S. Nuclear Regulatory Commission, 1978, Regulatory Guide 1.7-Control of Combustible Gas Concentrations in Containment Following A Loss-of-Coolant Accident, (Washington, D.C.: U.S. Nuclear Regulatory Commission).
5. U.S. Environmental Protection Agency, 1978, Siting Solid Waste Facilities-A Bibliography, (Washington, D.C.: U.S. Environmental Protection Agency), SW-722.
6. Public Law 94-580, 90 Stat. 2795; 42 U.S.C. 6901, Section 1008 (a) (3).
7. U.S. Federal Register, 40 CFR Part 250. 43-1, December 18, 1978, pp. 59000-59001.
8. U.S. Federal Register, 40 CFR Part 264, January 12, 1981, pp. 2810-2818.
9. U.S. Federal Register, 40 CFR Part 264, November 23, 1981, pp. 57284-57285.
10. Public Act No. 81-369, An Act Concerning Siting of Hazardous Waste Facilities, effective July 1, 1981.
11. Maine Revised Statutes Annotated, Title 38, Chapter 13, Waste Management Laws: Solid Waste, Hazardous Waste, Hazardous Matter, Hazardous Waste Fund, September 1981.
12. Massachusetts General Laws Chapter 21C, Massachusetts Hazardous Management Act, 1979.
13. New Hampshire House Bill 468, Chapter 567: An Act Relative to Hazardous Waste Facility Review and Waste Management, Chapter 147-A.
14. N.J.S.A. 13: 1 E-49 et seq., Major Hazardous Waste Facilities Siting Act, adopted January 13, 1981.
15. N.Y. Environmental Conservation Law, Article 27, Title 11, Industrial Siting. Hazardous Waste Facilites, December, 1981.
16. ECL, Section 27-1103, Part 361.7, Siting of Industrial Hazardous Waste Facilities.
17. See State of Rhode Island and Providence Plantations Department of Environmental Management-Division of Air and Hazardous Materials-Hazardous Waste Management Facilities Operating Permit Rules and Regulations, Providence, Rhode Island.

18. Vermont Act 250, Title Ten, Part 5 Land Use and Development, Chapter 151, 1969.
19. Delaware River Basin Commission and New Jersey Department of Environmental Protection, 1979, Technical Criteria For Identification And Screening Of Sites For Hazardous Waste Facilities, prepared by Environmental Resources Management, Inc. (West Trenton, New Jersey: Delaware River Basin Commission).
20. Clark-McGlennon Associates, 1980a., *Criteria for Evaluating Sites for Hazardous Waste Management*, (Boston: New England Regional Commission).
21. Clark-McGlennon Associates, 1980b., *A Decision Guide for Siting Acceptable Hazardous Waste Facilities in New England*, (Boston: New England Regional Commission).
22. Subcommittee on Oversight and Investigations, 1979, *Waste Disposal Site Survey*, Committee on Interstate and Foreign Commerce, House of Representatives, 96th Congress.
23. Neamotolla, Mounir, 1977, Middlesex County: *A Program for the Safe and Economical Disposal of Hazardous Wastes*, Program Manager, Middlesex County Department of Solid Waste Management, New Brunswick, New Jersey.
24. Anderson, Richard F., and Greenberg, Michael R., 1981, *A Macro Screening Process For Siting Hazardous Waste Management Facilities: A Case Study On The Lower Raritan/Middlesex County 208 Area*, report submitted to the New Jersey Department of Environmental Protection.
25. Anderson, Richard F., Greenberg, Michael R., 1982, "Hazardous Waste Facility Siting: A Role for Planners", *Journal of the American Planning Association*, (48:2), pp. 204-218.
26. Lurcott, Jack, 1979, Rollins Environmental Service, Director of Corporate Development, 1 Rollins Plaza, Wilmington, Delaware, personal interview.
27. Anderson, Richard F., and Wilson, Margaret, 1981, "Assessing Land Use Suitability For Hazardous Waste Mangement Facilities", *Northeast Regional Science Review*, Vol. 11, (Amherst, Massachusetts: University of Massachusetts), pp. 90-106.
28. The shadings for each cell were detemined vis-a-vis the standardized factor scores resulting from the rotated factor solution. If a score was slightly above one half of a standard deviation (or about 0.65) then the cell was categorized as problematic via the factor. Other efforts using this technique may experiment with this technique as the categorization scheme may be interpreted differently by different researchers.
29. Booz-Allen and Hanilton, Inc., et al., 1980, *Hazardous Waste Generation And Commerical Hazardous Waste Management Capacity-An Assessment*, (Washington, D.C.: U.S. EPA), SW-894.

PART 3

Transportation, Emergency Response and Preparations

There is a growing awareness that spills of hazardous and toxic wastes can create a dangerous environmental situation. The proper management of toxic waste spills is an increasing concern to the citizens of the nation. Consequently, regulations and procedures are being formulated to handle hazardous wastes in emergency situations.

Part Three begins with a discussion of the transportation and disposal of refinery wastes. Petroleum refineries produce from 50 to over 100 waste products. Of these, the Environmental Protection Agency, under its Resource Conservation and Recovery Act, lists five classes of wastes from refineries that are hazardous. These include dissoluble air flotation float, slop oil emulsion, heat exchange sludge, separate sludge and leaded tank bottom solid, oil and water mixtures. The ultimate goal in management of these refinery wastes is their disposal at the least cost while maintaining an acceptable environmental quality.

As the variety of hazardous and toxic wastes has increased and the potential danger grows, attempts to control their disposition are developing at the local, state and federal levels. Of the statutes, rules, regulations, ordinances and laws that have evolved, the Environmental Protection Agency has assumed a major role in establishing procedures to control emergency environmental spills. The EPA control measures are implemented through its National Contingency Plan.

This Federal plan, in order to respond to hazardous waste spills, consists of four phases. These are the discovery and notification of authorities of the spill, preliminary assessment and initiation of action, containment of the spill and the initiation of countermeasures to control the situation, and finally, documentation and cost recovery in order to once again have a satisfactory environment. The Federal plan is a major working document for dealing with emergency hazardous and toxic spills. Many other plans at the state and local level duplicate these procedures.

The final chapter in this section considers the needs of the local community to respond to toxic waste spills caused by transportation accidents, deliberate or accidental releases of toxic materials, and explosions or fires releasing toxic substances. It must be recognized that a plan cannot be developed at the time of the spill, but must be implemented prior to the emergency.

Hazardous and Toxic Wastes: Technology, Management and Health Effects. Edited by S.K. Majumdar and E. Willard Miller. © 1984, The Pennsylvania Academy of Science.

Chapter Twelve

Disposal and Transportation of Refinery Wastes

Theodore M. Grabowski[1] and Arthur J. Raymond[2]

[1]Environmental Engineer
Sun Refining and Marketing Company
P.O. Box 426
Marcus Hook, PA 19061

[2]Environmental Engineering
Sun Refining and Marketing Company
P.O. Box 426
Marcus Hook, PA 19061

Like many industrial facilities, petroleum refineries are faced with the complex problem of effectively managing the wastes they generate. Waste management activities have come under increasing scrutiny from both the general public as well as government regulators. With the current body of specific regulations promulgated under the Federal Resource Conservation and Recovery Act (RCRA) and various state laws, it is clear proper waste management is becoming a major endeavor. The fact that agency regulations and public perceptions are constantly changing serves to further complicate an already complicated task.

Unlike many industrial facilities, petroleum refineries generate a variety of waste materials encompassing many physical forms, chemical characteristics, and generating rates. This variability presents its own set of operating problems, since one waste management scheme seldom matches the needs of so many different wastes. As a result, the "typical" petroleum refinery is forced to implement several management options to accommodate the many wastes generated. Faced with these realities, refinery managers must be concerned with the design, implementation and monitoring of an effective waste management program.

Any systematic program aimed at proper refinery waste management must address several, often conflicting, goals. It must accommodate a variety of concerns ranging from economical disposal fees through effective liability management. Since it may be many years before today's wastes become a problem, it is difficult to monitor the program's effectiveness in "real time". What seems like an effective disposal option today can become a remedial action site tomorrow. Table 1 summarizes the major objectives of a good waste management program.

TABLE 1

Waste Management Program Major Objectives

Provide permitted, environmentally acceptable disposal options for every generated waste. Maximize the number of available options. Minimize accumulation and storage times.

Minimize waste treatment, transportation and disposal costs whenever consistent with risk assessments and environmental compatibility.

Comply with existing agency regulations, anticipate future regulatory and process changes.

Identify and control possible employee and public exposure routes.

Accommodate all generated wastes. Monitor chemicals and commodity purchases to "flag" incoming materials.

Identify, assess and manage waste related risks. Potential tools include:

- Waste elimination or reduction.
- Recycle or reuse.
- Treatment or destruction of "residual" wastes.
- "Delisting" hazardous wastes.
- On site management.
- Good vendor contracts, control and monitoring.
- Comprehensive control, from "cradle to grave."

REFINERY WASTE INVENTORY

The first step in designing an effective waste management program for an individual facility is the preparation of a comprehensive waste inventory. For some facilities, this inventory may be simple and consist of a few "major" process wastes and several small "intermittent" wastes. In the typical petroleum refinery, however, this inventory consists of fifty to one hundred individual wastes. For the purposes of organization, the refinery inventory can be divided into four distinct classes of waste: "listed" hazardous wastes, "characteristic" hazardous wastes, nonhazardous wastes and special wastes.

Listed Hazardous Wastes: The regulations promulgated by the United States Environmental Protection Agency (EPA) under RCRA specifically list five wastes from the petroleum refining industry as hazardous. Although refineries generate a number of wastes considered hazardous due to their characteristics, it is these "listed" wastes that demand specific attention. The RCRA regulations stipulate that listed wastes remain hazardous independent of their disposition. With this provision, simple treatment does not render "listed" wastes nonhazardous. Table 2 summarizes the "listed" wastes found in many petroleum refineries.

Dissolved Air Floatation (DAF) float is the waste stream produced by intermediate wastewater treatment processes in the petroleum refining industry. The material is generated when dissolved air is used to float or strip solids and oil from wastewater prior to final treatment. DAF float is an aqueous mixture con-

TABLE 2

Refinery Waste Inventory "Listed" Hazardous Wastes

Name	Description	EPA ID #	Physical Form	Recycle/Reuse Potential
DAF Float	Secondary wastewater treatment sludge	K048	Aqueous solution	Thicken sludge, recover oil
Slop Oil Emulsion Solids	Solids from slop oil storage and treatment tanks and separators	K049	Oily sludge	Chemically treat emulsion to recover oil
Bundle Cleaning Sludge	Solids from heat exchanger cleaning	K050	Heavy sludge	Little potential
API Separator Sludge	Sludge accumulated in API separators	K051	Heavy sludge	Decant oil and water
Leaded Tank Bottoms	Sludge accumulated in leaded product storage tanks	K052	Scale laden heavy sludge	Decant oil and water
Benzene, Toluene, Xylene	Off-spec or contaminated products	U019, U220, U239	Clear liquid	Recycle through slop
Carbon Disulfide	Container rinsate	P022	Clear heavy liquid	Add rinsate to charge
Spent Solvents	Solvents used in maintenance and housekeeping activities	F001, F002, F003, F004, F005	Oily liquids	Off site reclaimer
Cresols, Phenols	Off-spec or contaminated lube oil mfg. solvents	U052, U188	Amber liquids	Introduce to process recovery system
Methanol, Alcohols	Octane enhancers	U154	Clear liquid	Remove water, burn as fuel
Furfural	Lube oil/wax extraction solvent	U125	Amber liquid	Introduce to process recovery system
Methyl Ethyl Ketone	Lube oil/wax extraction solvent	U159	Clear liquid	Introduce to process recovery system
Tetra Ethyl Lead	Octane enhancer	P110	Colorless liquid	Little potential, contact supplier

taining several percent solids and oil. Although the total amount of solids generated is small, their presence in many volumes of water complicate and restrict the available options. Management steps can include thickening prior to filtration or solidification.

Slop oil emulsion solids may be generated in tanks and separators used to recover slop oil. Slop may be accumulated in large tanks where oil is removed for reprocessing and water is decanted for subsequent treatment. The remaining emulsion ("rag layer") may be further treated with chemicals and heat to recovery additional product. In any event, oily solids are eventually accumulated for disposal. The consistency of this waste can vary from a mixture of only a few percent solids up to a heavy sludge.

Heat Exchanger Bundle Cleaning sludge is generated during the cleaning of heat exchangers to regain heat transfer efficiency and is normally conducted on site during plant shutdowns. The RCRA regulations (see 40 CFR 261.3(a)(2)(iv)(C)) provide that bundle cleaning sludges may be introduced into permitted wastewater treatment systems for disposal.

API separator sludge is the accumulated material that settles in oil-water separators used in the primary treatment of refinery wastewaters. This sludge may be periodically removed by vacuum trucks or may be continuously cleaned with sludge pumps. As it appears in the separator itself, it consists of water, up to 40% solids (usually silt, sand, rust or other solids introduced into the refinery sewer system), and 10 to 25% oil. It can be successfully managed by a variety of methods including filtration, solidification, incineration and land treatment.

Leaded tank bottoms are the solids, oil and water mixture which accumulates in leaded product storage tanks - principally gasoline. This material contains organic lead which has precipitated from the product. Effective management is complicated by the presence of rust, scale and other corrosion products. The use of tetra-ethyl lead (TEL) as an octane booster is gradually being reduced and should result in smaller quantities of this waste being generated. Disposal can occur using specially designed "landfarms" or solidified for subsequent landfilling.

In addition to these five industry specific listed hazardous wastes, petroleum refineries can generate a number of other listed chemicals. Examples include:

Benzene, Toluene, and Xylene (BTX): Off-specification products or storage tank bottoms in refineries with BTX production capabilities.

Carbon Disulfide: Small quantities or residue may be removed from containers of carbon disulfide used to presulfide catalysts.

Spent Solvents: Some quantities of spent solvents such as trichloroethane and methylene chloride may be generated from cleaning and housekeeping activities. In addition, certain solvents such as methyl ethyl ketone may be used as extraction agents in the refinery process.

Cresols/Phenols: Off-spec or contaminated phenols may be generated in refineries using phenol as an extraction agent in lube manufacturing.

Methanol/Other Alcohols: Alcohols and alcohol mixtures are increasingly being evaluated for use as octane enhancers in gasoline. Cleaning of associated equipment will generate small quantities of these products.

Characteristic Hazardous Wastes: Petroleum refineries purchase, process and produce a wide variety of materials that if wasted, could required management as hazardous waste. In many cases, these wastes can easily be recycled (such as waste gasoline) or treated (such as spent acid.) However, the presence of such materials on site does require close monitoring. Equally important is the fact that spills of these materials can easily generate a hazardous waste in the form of clean-up residue. This residue is normally difficult to manage since the recycle/reuse potential is complicated by its contamination and physical form. Table 3 summarizes several common refinery wastes that are considered hazardous due to their characteristics.

Nonhazardous Wastes: In addition to a number of hazardous waste streams, petroleum refineries generate a variety of other solid, nonhazardous wastes. Although such materials are only nominally regulated under RCRA, they require proper management and disposal nonetheless. It should be noted that many states choose to regulate the disposition of oily wastes in excess of those considered hazardous under RCRA. For this reason, individual refinery waste management programs must be designed with consideration of state and local rules as well as federal.

Like every industrial and commercial facility, refineries generate day-to-day office trash, cafeteria waste, and packaging material. These wastes are commonly compacted on site and disposed at local sanitary landfills. As the number of commercial landfills decrease (due to sitting problems), interest in on site or regional trash incinerators should grow. Several have already been proposed with consideration given to cogeneration of steam. The viability of these projects should increase as disposal costs (tipping fees and transportation) rise.

Petroleum refineries, as a result of construction and repair activities, generate a significant amount of overburden, excavation soil and demolition debris. Much of this material is either reused on site as fill or carted off in dump trucks to demolition or sanitary landfills. Although much of this material is suitable for use as fill, risk management considerations discourage its "sale" to third parties. The disposition of these wastes may be complicated by the presence of small quantities of oil or oil products as a result of small leaks, spills or drips. Several states hve adopted finite limits for oil and grease in waste slated for disposal at sanitary landfills. When this soil or debris exceeds these limits, sanitary landfill disposal is prohibited. In some cases, it is necessary to dispose of such material in "Class I" "hazardous" waste facilities at costs several times those of sanitary landfills. Given this disparity in disposal costs, a clear incentive exists to manage refinery operations to minimize the number and extent of oil spills and leaks.

TABLE 3

Refinery Waste Inventory "Characteristics* Hazardous Waste

Hazardous Class	Description	EPA ID #	Physical Form	Recycle/Reuse Potential
Ignitable	Waste gasoline, aviation fuel, some grades of kerosene, etc.	D001	Liquids	Decant water, recover free oil
Corrosive	Spent sulfuric, hydrochloric acids; spent sodium hydroxide	D002	Liquids	Neutralize and discharge, send to supplier for recovery
Barium	Certain off-spec lube additives	D005	Liquids, viscous liquids	Accumulate on site, reuse slowly or recycle as slop
Chromium	Certain cooling tower bottoms	D007	Solids, heavy sludges	Dewater through filter press or solidify, dispose of solids
Lead	Certain slop oil tanks or separators servicing gasoline loading racks	D008	Sludges, heavy sludges	Decant to remove free oil, filter or solidify to remove water, dispose of solids

*40CFR261 defines four hazard characteristics: ignitability, corrosivity, extraction procedure toxicity and reactivity.

A number of operating units in petroleum refineries use catalysts to promote product formation or reaction. As a result, refineries can generate several difficult types of spent, non-reusable catalysts. In most cases, these materials are nonhazardous under RCRA. In addition, they are usually free of hydrocarbon contamination since entrained oil is "flashed", steamed or otherwise purged prior to removal from the unit. Certain spent catalysts may have value due to potential metal recovery. These materials are normally accumulated on site and sent to commercial reclamation firms or the original supplier. Catalysts unsuitable for recovery are usually disposed of in sanitary landfills. The largest concern with such material is usually minimizing dusting during transportation. To do so, covered, sealed bulk containers are used.

Several processes employed at petroleum refineries use a final "polishing" step prior to product shipment. This step can include product filtration through clay or other filter media to remove color bodies and impurities. The filter media is usually regenerated on line to reduce filtering costs. At times it becomes necessary to waste a fraction of the filter media to retain product quality. The spent material is a dry, soil-like solid that may contain from one to several percent oil. Although the total oil content may be high enough to warrant close scrutiny, this oil is usually intimately tied up in the porous clay. In addition, the retained oil is usually high-boiling, asphalt-like material. As a result, spent clay exhibits little propensity for leaching and is suitable for disposal at sanitary landfills.

Special Wastes: In addition to waste materials unique to the petroleum industry, refineries also generate a number of closely regulated wastes common to many industrial plants. For the most part, three individual types are common: asbestos, PCBs and potential pathological agents.

In refineries, asbestos is generated during the demolition, reinsulation or repair of existing equipment. Due in large part to the age of these facilities, the insulation encountered is predominately asbestos or asbestos containing. Federal regulations under the Clean Air Act and Occupational Safety and Health Act require certain management techniques for the removal and disposal of asbestos. These include:

- Notification to a state or federal agency prior to removal and disposal. In addition, the disposal site and other agencies must be notified prior to shipment.
- Work rules that minimize the generation of air borne asbestos. This includes wetting the material prior to removal and placement in sealed containers or bags.
- Labeling of all containers or bags. Also, the work area must be roped-off to minimize unauthorized access to the area.
- Disposal only at permitted sanitary landfills. The asbestos must be covered immediately upon placement in the fill, hence the requirement to notify the landfill operator prior to shipment.

• Safety rules that require worker protection such as use of respirators, gloves, glasses and protective clothing.

Due to the extreme need for explosion and fireproof equipment in petroleum refineries, many existing electrical transformers were filled with fluids containing poly chlorinated biphenyls (PCBs). Regulations promulgated under the Toxic Substances Control Act (TSCA) have banned the further manufacture and distribution of PCBs, but permit their continued use in a "totally enclosed manner." These rules stipulate the disposal options to be used for PCB or PCB containing materials. In summary, these regulations require:

• Proper labeling of existing equipment.
• Periodic inspections.
• Annual preparation of an inventory.
• Proper storage of items removed from service.
• Disposal of all PCB fluids by incineration and PCB solids by landfilling in specially permitted facilities. Certain exceptions to these disposal rules exist, yet many companies have chosen to maximize the incineration of PCB material as a risk management tool.

Petroleum refineries employ a large number of persons necessitating the presence of many support services. Typically included are reasonably complete medical departments. These facilities are used to conduct employee physicals, new-hire screenings and emergency response. In some cases, these facilities generate wastes that are considered potentially pathological by agency rules. The proper management of wastes such as bandages, body fluid samples, cultures and other contaminated materials, is normally controlled by State Health Department rules. Disposal, in some cases must occur at permitted incinerators. In any event, proper packaging is normally called for.

TRANSPORTATION OF REFINERY WASTES

The transportation of waste from its point of generation to the ultimate disposal site carries with it significant risk and liability. In many cases, the cost of transportation equals or exceeds disposal costs. If only for these reasons, significant attention should be given to this area in any waste management program.

Significant safety and cost considerations dictate that travel distances should be minimized whenever possible. Although economics have historically kept travel distances low, the reduction of available disposal sites have resulted in longer trips. Shipments of 500 to 1500 miles are now common for certain highly regulated wastes. This trend will continue for some time due to the inability to site appropriate disposal facilities.

Given this need to traverse longer distances, the subject of payload has taken on more prominence. At one time most waste shipments occurred in relatively small five to ten ton payloads vehicles. With transportation costs and distances

FIGURE 1. 1982 R Model Mack Tractor, 1982 Accurate Tilt Frame Trailer and an Accurate 30 yd³ secure container. (Photo courtesy of Industrial Waste Removal, Inc.)

rising, the use of twenty ton "dump trailers" is now common. As costs and distances continue to escalate, barge and rail shipments of waste will bear closer study.

In many cases, the ultimate solution is the siting of appropriate waste treatment and disposal facilities on the generator's property. Although permitting is seldom easy, significant cost and liability benefits accrue to those that succeed. In addition to these benefits, refineries with on site facilities are also able to employ company owned transportation means in lieu of third party contractors. This presents its own economic and liability advantages since control of waste is vested in the generator from cradle to grave.

The transportation contractors, vehicles and methods used for refinery wastes must be reviewed and monitored for compliance with agency rules and company safety standards. Contracts formerly used for the purchase of transportation and disposal services are commonly inadequate in today's highly regulated environment. Even with effective contracts, vendor employees should show evidence of training. Many companies conduct annual vendor audits to achieve this goal. These efforts may include vehicle inspections, record reviews and site visits. It should be noted that many state regulations require special permits for transporters of hazardous waste. In some cases, specific insurance or bonding requirements apply.

For the most part, the transportation methods, equipment and contractors used to move waste are dictated by the material's physical form and characteristics. For transportation purposes, refinery wastes can be divided into four general categories: containerized wastes, liquids and liquid sludges, heavy sludges and bulk solids.

Containerized Wastes: Containerized wastes include those materials managed using drums, pails, fiber packs and cans. It may include containers as large as

FIGURE 2. 1980 Presvac® Vacuum Trailer with a 1973 GMC Tractor (Photo courtesy South Jersey Pollution Control, Inc.)

PCB transformers. Recently, a major disposal firm has developed large reusable cartons for use in transporting small quantities of bulk materialsl. These cartons are "sold" to the generator for use throughout his location and periodically transported to the disposal firm where the contents are removed and carton returned.

Containerized wastes are typically transported using flatbed or box trucks. These vehicles enjoy significant payload advantage. This advantage is tempered by the need to handle individual containers as well as the loss of these containers upon disposal. Small containers are effectively used to manage wastes that are generated over a long period of time. Examples include waste laboratory reagents, small PCB capacitors and cleaning solvents. In some cases, the use of containers permits the segregation or recoverable wastes.

Liquids and Liquid Sludges: Liquid wastes generated by petroleum refineries include spent acids and bases as well as sludges with low solids contents such as DAF float or slop oil emulsion solids. If the waste is generated in high volumes, ideal transportation is by pipeline. Obviously, this method only applies to refineries with on site treatment or reuse capabilities or those located very near commercial facilities.

Many liquid wastes are moved using bulk tank trucks. These may be similar to bulk product vehicles or may be specifically designed for waste service such as

FIGURE 3. 1981 Vactor® Model 2045 High Suction Vacuum Truck. (Photo courtesy SNOW Environmental Services, Inc.)

vacuum trucks. Such vehicles perform well with sludges of up to approximately forty percent solids and some are available with "tilt and dump" capability. Although several sizes are available, payload considerations usually dictate the use of 5000 gallon vehicles.

Some commercial treatment and disposal facilities actively promote the use of barges for liquid waste movements. This option is complicated by the need for loading and unloading facilities which may be difficult to site. At least one firm is actively attempting to locate a waste terminal for use by their incineration ship. Potentially promising for some refinery services is the use of bulk rail transport. Although more expensive than barges, it enjoys less siting hurdles.

Heavy Sludges: Unlike many industrial facilities, petroleum refineries generate a significant amount of heavy, solids-laden sludges. These can include materials such as API separator sludge and certain infrequently cleaned product tanks. In addition, certain materials are inherently more difficult to manage due to their high wax or heavy oil contents. To remove such heavy materials from the generating locations (tanks, separators), relatively specialized equipment must be used. High volume, high suction vacuum equipment (originally developed for sewer cleaning) has found wide applicability in this service.

An apparent drawback with this equipment is its relatively small payload - approximately 3000 to 3500 gallons. In some applications, high suction equipment is used only to remove the sludge from its vessel; after which it is transferred to large tank trucks for subsequent transportation. Although some interest has been shown in using barge and rail for movement of heavy sludges, their physical form and tendency to deposit solids may prove difficult barriers to overcome. The cleaning charges for "butterworthing" barges or rinsing rail cars must be considered in the overall transportation cost.

solids: Wastes that exist in solid form have historically been transported in relatively small bulk containers. These range from portable five cubic yard "lugger" buckets for intraplant movements to larger twenty or thirty cubic yard "roll-offs" for interplant shipments. Each method uses a portable container that is located at the generating site, filled and eventually carried to the ultimate disposal site. This method enjoys the advantage of portability but suffers from a maximum payload of approximately fifteen tons. For large volume, low density materials such as office trash or catalyst fines, this payload restriction is less important. In these cases, it is important to provide covered transportation means to minimize dusting.

The transportation of dense solid wastes to distant landfills is currently best accomplished using sealed, twenty ton bulk dump trailers. Although portability is sacrificed (demurrage charges prohibit "spotting" the trailer at the generator's site), increased payloads do offer cost savings. Several firms even offer covered dump trailers for transportation of dust prone material. With this method, a premium is now placed on proper scheduling to reduce or eliminate demurrage charges which can range from $25 to $75/hour.

Depending on the location of the ultimate disposal site, transportation firms may be able to arrange for a "backhaul" of compatible material. This potential may reduce transportation fees by spreading the cost over two or more customers. It remains to be seen whether this arrangement will prove useful in light of the risk of contaminating the backhaul with waste residue. In any event, it is important for generators to realize that transporters may already engage in backhauling activities without their knowledge.

At least one commercial disposal firm is actively investigating the use of rail cars such as gondolas for the bulk transport of solid wastes. The rail industry already possesses significant expertise in solids handling. Some of this expertise and equipment may be amenable for use in waste handling. Potential barriers exist in siting loading and unloading stations, construction of needed spurs and acceptance by waste generators. If the trend towards longer transportation distances continues, it is safe to assume this method will garner additional attention.

DISPOSAL OF REFINERY WASTES

The ultimate goal of any waste management program is to provide for the least costly, most environmentally acceptable disposal mechanism for every generated

TABLE 4

Potential Waste Management Techniques for Petroleum Refining Industry

Method	Description	Examples	Effects
Waste Elimination/Reduction	Conscious changes are made in the manufacturing process to eliminate or greatly reduce the formation of waste.	Residue in boiler water treatment drums is rinsed with clean water and used as make up in boiler water feed.	"Empty" drum problem eliminated. Reduced chemical costs.
		Process water withdrawn from surface waters is prefiltered prior to use.	Reduces the sludge formed in API separators and cooling towers. May help reduce scaling in process units.
Waste Recycle	Waste streams, or a fraction thereof, are recycled through the process to beneficially reclaim valuable products. Note: Certain recycle/reuse equipment may be regulated under RCRA.	Decant storage tanks are used to accumulate slop oils. Water is gravity separated and oil is reprocessed.	Increased throughput. Reduction in refinery losses.
		Spent solvents in extraction processes are recycled through distillation or stripping towers.	Reduced raw material costs. Increased operating efficiency.
		Spent solvents used in cleaning operations are accumulated in drums for off site reclamation by the material supplier.	Reduced solvent costs.
Waste Reuse	Substances formerly considered wastes are used as raw materials in compatible applications.	Spent sulfuric acid is accumulated for eventual shipment to the supplier for regeneration.	Reduced raw material costs.
		Spent lime from water treatment plant is used to neutralize spent hydrofluoric acid from alkylation units.	The basic nature of spent lime is used to the fullest extent possible in neutralizing spent acid.
		Spent catalyst is used to solidify sludges that contain free water prior to disposal.	The amount of free liquid in waste is reduced resulting in smaller "leaching" potential.

TABLE 4 (Continued)

Potential Waste Management Techniques for Petroleum Refining Industry

Method	Description	Examples	Effects
Waste Destruction	Waste materials are consciously destroyed in controlled processes.	Incineration or thermal oxidation of medium to high BTU wastes such as slop oils in units such as fluidized bed incinerators and waste boilers.	Disposal liability is effectively eliminated. Tremendous volume/weight reductions are realized. Fuel value of hydrocarbon may be realized with heat recovery equipment. Low BTU wastes can be co-fed for disposal.
Chemical Treatment	Wastes are subjected to specifically designed treatment processes rendering them nonhazardous.	Spent acids or bases neutralized to yield a neutral solution which is discharged to a water treatment plant.	Disposal liability is eliminated. Safety hazards greatly reduced.
		Wastes with "leachable" metals are solidified and/or stabilized with lime or other proprietary chemicals.	"Leachability" of constituents is greatly reduced by raising the pH of the material. As "leachability" is reduced so is disposal liability.
Physical Treatment	Wastes are subjected to specifically designed physical treatment processes that reduce their volume, weight or hazard potential. Certain process may be used in conjunction with recycle/reuse provisions.	Low-solids sludges are introduced into decant-settling vessels to remove free water and "thicken" remaining solids.	Total volume of waste is reduced resulting in smaller disposal costs and equipment needs.
		High pressure filter presses are used to strip solids from oily wastes. May be used in conjunction with lime treatment.	Total volume is greatly reduced. Resulting waste is more suitable for landfilling. If used in conjunction with lime, resulting waste may be nonhazardous due to pH adjustments.
Land Treatment	Oily sludges are biodegraded on specially constructed land application areas.	Application of API separator sludge.	Water content reduced by evaporation, oil content degraded by natural biological action. Residual material may prove nonhazardous.

TABLE 4 (Continued)
Potential Waste Management Techniques for Petroleum Refining Industry

Method	Description	Examples	Effects
Landfilling	Residual wastes from prior treatment operations are interned in secure, lined landfills.	Landfilling of incinerator ash and filter cake.	Landfilling of most wastes result in significant long term liability. The success of this option depends on construction and operation of the facility as well as the "hazardousness" of the waste. Landfilling should be viewed as the option of last resort.

waste. Although an almost infinite number of techniques exist to meet these objectives, they can be loosely grouped and ranked in a hierarchy according to preferred implementation. A summary of the major techniques used for refinery wastes appears as Table 4.

CONCLUSION

Petroleum refineries generate a large number of different waste streams each requiring proper management and disposal. This task is best accomplished under the auspices of an organized, systematic program. Program objectives must include risk management and cost effectiveness as well as compliance with all applicable federal, state, and local regulations. Transportation of wastes from their point of generation to their ultimate disposal site deserves specific attention under this program. Distances traversed should be minimized while payloads carried are maximized. In this regard, barge and rail transportation will receive closer scrutiny. One facet of the ultimate solution to waste disposal and transportation concerns will be the siting of appropriate treatment facilities on the generator's property. Risk is effectively managed in this scenario since wastes are now subject to company control from cradle to grave.

For most industrial facilities, the largest area of waste related exposure lies with the final disposal method used. Significant efforts have already been expanded in the petroleum refining industry to reduce the amount of waste generated while increasing oil recovery. Landfarms are widely employed in the industry to capitalize on nature's ability to effectively treat oil wastes. The industry continues to examine destructive technologies such as thermal oxidation, incineration and biodegradation to help reduce the need for landfills and eliminate the risk of future remedial measures. These destructive methods will become more attractive as the risks associated with landfills are better defined.

Hazardous and Toxic Wastes: Technology, Management and Health Effects. Edited by S.K. Majumdar and E. Willard Miller. © 1984, The Pennsylvania Academy of Science.

Chapter Thirteen

Emergency Response To Environmental Spills: The U.S. Environmental Protection Agency's Role

Charles H. McPherson, Jr., B.S. and Al J. Smith, P.E.
Emergency and Remedial Response Branch
U.S. Environmental Protection Agency, Region IV
345 Courtland Street
Atlanta, Georgia 30365

The natural evolution of public concern about the environmental and public damage potential of oil and hazardous substance spills correlates closely with the development of the chemical and fossil fuel industries. There are perhaps 50,000 to 60,000 different chemical compounds in existence in the various market systems in this country today, and the list is growing. The use of petrochemicals has experienced tremendous growth and the fossil fuel problem is known to even the disinterested. The demand for these products in the consumer market is the genesis of our problem. The compounding factor is, of course, the effect that use of these products may have on the environment, and/or public health and welfare. Couple these considerations with dozens (conceivably more) of local, state and federal statutes, rules, regulations and ordinances articulating various and, at times, overlapping interests and one can begin to imagine the scope of the problem generated when a spill-type emergency happens. How this problem is effectively dealt with is what will be discussed here.

SECTION I - HISTORY AND LAW/REGULATORY FORMAT

In 1970, Public Law 92-500, the Federal Water Pollution Control Act, was put into effect and required persons with knowledge of a spill of oil or hazardous substances onto or threatening a navigable water of the United States to immediately notify the appropriate Environmental Protection Agency (EPA) Regional Office. This legislation provided the EPA and United States Coast

Guard (USCG) with the necessary authority to regulate spills of oil and/or hazardous substances through the use of spill prevention regulations and/or enforcement/cleanup activities. Later, in 1978, the Clean Water Act (CWA) was amended to carry out Public Law 92-500 and a National Response Center (NRC) was also established to accept spill notification reports over a toll-free number (1-800-424-8802). This amendment provided the hazardous substance spill regulations and listed chemicals with their reportable quantities. The most recent legislation to deal with spills or releases of hazardous substances, as they are referred to now, is the Comprehensive Emergency Response, Compensation and Liability Act of 1980 (CERCLA or Superfund). This Act, Public Law 96-510, together with the CWA, requires the immediate notification of any release of a hazardous substance under any conditions as defined in the law and the immediate notification of a spill of oil, if that spill is in or threatening waters of the United States. Both Acts address the spill or release of chemicals but only the CWA addresses the spill of oil/petroleum-based products. To delineate the two Acts: CERCLA is directed toward response to releases of hazardous substances from spills and/or waste sites; the CWA addresses response to oil spills and establishes reportable quantities for hazardous substances. The reportable quantity (RO) for various chemical substances is still under study by EPA. However, in the interim all substances covered under the following Acts have RQ's of one pound:

(a) Any hazardous waste under Section 3001 of the Solid Waste Disposal Act;

(b) Any pollutant listed under Section 307(a) of the Federal Water Pollution Control Act;

(c) Any hazardous air pollutant listed under section 12 of the Clean Water Act; and

(d) Any imminently hazardous chemical substances or mixtures with respect to which the EPA Administrator has taken action pursuant to Section 7 of the Toxic Substances Control Act.

In the event there is a spill of oil onto or threatening a navigable water of the United States, the CWA basically provides for the following:

(a) A civil fine up to $250,000 for the spill.

(b) A criminal sanction (up to $10,000 and/or up to one year in jail) for failure to notify EPA/USCG about the spill.

(c) Access to a USCG-managed fund for clean up of the spill when the responsible party is unknown or will not clean up the spill.

A release of chemical(s) from a common emergency (train/truck accident, warehouse fire, etc.) or from a waste site could require the use of CERCLA monies if the responsible party cannot or will not assume the clean up responsibilities. CERCLA basically provides the following:

(a) A civil fine of up to three times the cost of cleanup (if the responsible party does not provide the cleanup).

(b) A criminal sanction (up to $10,000 and/or up to one year in jail) for failure to notify EPA/USCG about the release.

(c) Access to an EPA-managed fund for cleanup of the release when the responsible party is unknown or will not clean up the release.

The CWA and CERCLA furnish the mechanism for EPA or the USCG to respond to releases of chemicals or oil spills and if necessary mitigate the problem using federal funds. The working concept for applying these Acts comes through the National Contingency Plan (NCP) and the Regional Contingency Plan (RCP). The NCP establishes the framework for the federal work forces that may be on-scene at one of these incidents. The framework that the NCP utilizes for response to oil or hazardous substances spills or releases is divided into phases.

Subpart E - Operational Response Phases for Oil Removal

Phase I - Discovery and Notification
During Phase I, the report of a spill is made to the NRC by the spillor or anyone who has knowledge of the spill (i.e., fire department, private citizen). This information is then relayed to the appropriate individuals at the federal and/or state and/or local level.

Phase II - Preliminary Assessment and Initiation of Action

Phase II begins when either a first responder arrives on-scene or adequate information can be obtained to evaluate the need for response efforts. After the needs have been determined, some type of response activity will take place, either through activation will take place, either through activation of a cleanup contractor, federal/state/local response groups notified and responding, or just notification and monitoring by various agencies until the incident is cleaned up.

Phase III - Containment and Countermeasures

Phase III consists of the actual containment, countereasure, cleanup and disposal operations that will take place at the scene of the incident. These operations could take place via the responsible party or through the Federal government's cleanup fund.

Phase IV - Documentation and Cost Recovery

This final phase occurs both during and after the cleanup. Documentation takes place during the entire course of the removal so that adequate records will be available for cost recovery and other critical reports. Cost recovery will take place if a federally-funded removal is initiated or if fines are to be levied as a result of the spill.

Subpart F of the NCP deals with hazardous substance response and basically follows Subpart E, Operational Responses for Oil Removal. There are, however, different approaches to solving the long-term problems that occur when there is a hazardous substance release and these different approaches are explained in the NCP in Phases IV, V and VI. Phase III of Subpart F deals with emergency response or, as it is referred to in the NCP, immediate removal. Phase III, Immediate Removal, has established the guidelines for actions that may be taken at the scene of an emergency incident. This phase is the basis for whatever action may be taken at an emergency incident by EPA or the USCG and these actions will be discussed in more detail later.

The groups that are tasked with carrying out the NCP, the National Response Team (NRT) and the Regional Response Team (RRT), are established under Subpart C of the NCP. The U.S. Environmental Protection Agency chairs the NRT with the United States Coast Guard acting as vice chairman. Other members consist of representatives from the following federal agencies and departments:

Department of Agriculture
Department of Commerce
Department of Defense
Department of Energy
Federal Emergency Management Agency
Department of Health and Human Services
Department of the Interior
Department of Justice
Department of Labor
Department of State

The NRT sets national policy, goals, and direction while giving guidance on training, response equipment, and research and development ideas to the Regional Response Teams (RRTs) - both inland anc coastal.

As mentioned earlier, the Regional Response Team is also established under Subpart C of the NCP and also consists of membership from the above listed federal departments and agencies. State and local agencies are members or potential members of the RRT. The major function of the RRT is to provide assistance/direction to the OSCs in the area of expertise of each participating member. This RRT concept is crucial in the decision-making process of major and many minor incidents and will be discussed in more detail in "On-Scene Response" (Section II).

The working document of the RRT is the Regional Contingency Plan (RCP). This RCP utilizes the format of the NCP and also contains other important information concerning regional resources. Named within the RCP are the Federal On-Scene Coordinators (FOSC), RRT members and alternates, RRT Co-Chairman, additional resources such as cleanup contractors, industry representatives/experts, and the EPA/Coast Guard boundaries delineating inland and coastal zones. This plan along with the concepts, guidance and laws described is the tool which is utilized by the On-Scene Coordinator when responding to an environmental emergency.

Basically, we have discussed the various Acts which give the EPA or Coast Guard the authority to respond and mitigate whatever problems may arise from the resulting spills. We have also touched upon the framework for management of an environmental emergency utilizing the NCP, RCP and the working members of both the NRT and RRT. In this next section we will further discuss the actual responsibilities that are taken by the Federal On-Scene Coordinator and discuss his/her resources and responsibilities as they relate to EPA's role at an emergency incident. We will also briefly discuss the relationship with other response personnel.

SECTION II - ON-SCENE RESPONSE

It is 2:30 a.m., Saturday morning. The county fire department has just received a telephone call about smoke coming from a warehouse in the industrial park area. Upon arriving at the scene, the firefighters observe a light yellow/gray smoke and the smell of chemicals. Efforts to locate the source of the smoke are abandoned due to the fact that the smoke is causing extremely irritating reactions (even beneath turnout gear) on the firemen.

Some immediate questions may be asked:

Do we approach the building under a water mist and try to extinguish the source? Do we attack the structure with copious amounts of water? Do we evacuate nearby residences, and if so where do we evacuate to and how far should the evacuation zone be? Do we block off paths where water from the firefighting effort might leave the scene? Do we allow the structure to burn and protect nearby buildings/property?

This initial response is referred to as Stage 1 (these stages are not to be confused with the phases discussed earlier in relationship to the NCP) and generally occupies the first four to six hours of an incident. This stage refers to the behavioral characteristics of the incident as it proceeds from time zero to final removal effort. The initial response is the most difficult and often the most important, because the actions taken in this stage will affect the tasks and duties that will have to be performed later. As we have already illustrated, many important questions have to be answered in a matter of minutes and often have to be done with little or no information about what *may be* burning or leaking from a fixed facility or transportation incident. The local community usually has its own contingency plan and the phone numbers for sources of useful information to answer many of the numerous questions that will arise. The National Response Center, Chemtrec, various states' 24-hour emergency numbers and others should be well known to all involved in emergency response.

At this point, we want to briefly point out the need for local contingency planning. Local OSCs should know what type of products are stored in their jurisdiction and learn how to utilize a manifest or warehouse inventory. They should obtain information from a responsible company official about the location of products (where stored) at the facility, especially flammables, water reactive compounds, acids, pesticides, etc. Local OSCs should know their evacuation routes and alternate traffic routes for schools, public buildings, nursing homes, and other densely populated buildings, in the event a transportation incident blocks a normal evacuation route. Also important to know are the physical resources that may be available to the local response group such as heavy equipment, piping, sand, clay, etc. Finally, the local responders should know who is in charge of coordinating all of the local response efforts for this will be a bonus in terms of time and efficiency in utilizing all their resources during an event.

The problems that occur in Stage 1 of an incident often do require the use of several resources to obtain an accurate evaluation for coping with these initial concerns. One point where EPA can provide useful information and aid to initial responders is to provide them with a basis for determining if an evacuation is appropriate (i.e., chemical constituents) and, if so, how far or how large an area should be evacuated. Technical information on the chemicals may be provided so a decision on firefighting methodology can be made. Often individuals on-scene may not be able to make contact with the responsible party. EPA and Chemtrec often can provide a quick relay to make sure the responsible party is aware of the incident. In addition, a commitment on the removal operation is asked of the responsible party by EPA. This is done in the early stages so that mobilization of a cleanup contractor can begin and also so that if the responsible party is not going to accept responsibility the local, state or federal government can initiate some type of action to get a cleanup contractor on-scene. Under Superfund (CERCLA) and the CWA, the Federal government, EPA, and the USCG, have the authority to hire a contractor to clean up the spill if the responsible party cannot or will not conduct the removal operation. This authority allows the Federal government to pay for the cost of cleanup, disposal, investigation, epidemiological studies, assessing natural resource damage, and restoration costs. Some states also have funds for cleanup and this could be used in lieu of federal monies. Stage 1 generally covers through evacuation and typically ends with the arrival of the secondary responders.

Stage 2 of the incident generally begins four to six hours after the initial response and, as stated, is characterized by the arrival of the secondary responders, namely state and federal officials. Remember the primary objective of these secondary responders is to aid the first response group in quickly mitigating the possible health/environmental threats that may be present. General characteristics of the response that have taken place to this point are: completion of the evacuation; the fires/explosions are under control or more predictable; and other reactions (chemical, etc.) are beginning to stabilize. As the state officials move on-scene, many additional resources become available to the local community. The state fire marshall's office may be there to investigate the cause of the incident and help prevent this type of event from reoccurring. The state patrol works with other local law enforcement officials in crowd control and the security of the evacuation perimeter. Environmental agencies will be on-scene to assist in monitoring, technical assistance, possibly coordinating overall state response efforts, and other related duties. The state civil defense agency will aid local civil defense efforts and could also act as the state's coordinating official (SOSC). Other government officials that may be present are local government officials and state legislators. These individuals may serve as a very useful resource in relaying information of specific concern to the local population. This type of assistance by public figures is invaluable and aids in keeping the public trust be relaying information to affected parties from someone they recognize.

All of these officials on-scene, and there could be over a hundred "officials", are there under some type of jurisdiction. All of these individuals should have a working knowledge of the Regional Contingency Plan and/or appropriate State Contingency Plan so that everyone may work within the RRT. This "working together" concept is vital if the emergency is to be handled as efficiently, both environmentally and economically, as possible. RRT meetings may be scheduled at regular intervals at one of these incidents or held whenever needed. These impromptu meetings are crucial because of the everchanging conditions on-scene at one of these accidents, and this again points out the need for personnel on-scene at the accident that can make or give recommendations or opinions concerning their area of expertise. The Regional Response Team concept gains its support from the participating members and their areas of expertise, knowledge and personal experience in past emergency episodes. The working membership of the RRT is often tasked with the responsibility to determine the answers to such critical questions as: expanding or reducing an evacuation zone; opening/closing water supplies; deciding how to handle potential long-term fires associated with the accident; and so forth.

As stated previously, one of the problems associated with incidents of a large magnitude is that possibly hundreds of representatives could be at the site. Trying to assemble all of the officials concerned into a working group, the RRT, should be aided by EPA/USCG and their local and state counterparts. Through preplanning, simulated exercises and training, most of these people with a duty on-scene have been associated with one another, as well as during actual emergencies in the past. If there is some confusion on-scene, which often times can result, then the EPA/USCG and their state counterparts must organize everyone into a functional group.

During Stage 2 many decisions that will affect the long-term mitigation efforts will also be made. If there is a fire, will it be allowed to burn out? Should the fire be extinguished, and if the fire is put out, how will this be accomplished? Should the spill be neutralized or treated through a type of on-scene treatment system? Will soil have to removed? Will groundwater/drinking water be affected? These and many more questions will have to be addressed.

The Regional Response Team should be utilized by the On-Scene Coordinator to help evaluate the decision-making process and develop the best technical approach to solving the seemingly endless problems on-scene. Many of the problems have overlapping impact/jurisdiction that could only be addressed using the RRT concept.

As Stage 2 phases out many of the emergency response personnel are no longer on-scene. Fires have been extinguished, people have been returned to their homes and businesses, and only cleanup activities remain. However, these removal activities are a vital portion of the response effort.

Stage 3 involves the long cleanup of the accident site and addresses numerous problems and situations. As stated before, the number of individuals on-scene representing varous jurisdictional concerns has also decreased. The RRT, however, is still a functional entity and is utilized as a forum for addressing questions regarding health and cleanup objectives and methodology. EPA, local and state health/environmental departments, and the U.S. Department of Health and Human Services (HHS) all work closely together with other RRT members to resolve the many challenges and concerns that arise. EPA and the state environmental offices will have experts on-scene (normally the OSCs) who can assist in removal methodology. The cleanup portion of Stage 3 can often take several days to several months to complete dependent upon all the conditions of the site and the products involved. Various sampling techniques for determining the extent of contamination need to be evaluated, as well as different removal or treatment methods also need to be reviewed. However, these methods are dependent upon the type and quantity of the released material and the environmental makeup of the area where the product was spilled, i.e., sandy soil, clays, loams, water bodies, etc. EPA's OSCs can provide removal/treatment information to whomever is responsible for the removal activities. Most state agencies also provide individuals with removal/treatment methodology expertise on-scene as well as provide information about disposals and transporting requirements.

HHS, state, and local health offices are becoming more and more involved in these types of incidents in relationship to the safety of the workers who may be on-site conducting the cleanup. Advice from these health officials is valuable when the OSC is trying to determine what level of protective clothing and safety should be exercised on-site. These health officials can also address many of the concerns local residents may have about how this incident and the products involved could or could not impact on them. This type of question is often extremely difficult to answer but having qualified medical personnel with this type of experience on-scene will make the task easier.

The roles and lines of communications between the responsible party, EPA, and state and local environmental offices must remain open even though a formal RRT setting may not now be occurring. These lines of communication are extremely important, because of the many decisions that must be made regarding final cleanup levels and the costs that are associated with long-term cleanup techniques. Different approaches need to be discussed and analyzed quickly so that action can be taken swiftly to minimize not only the environmental damage but to control the removal costs.

As Stage 3 concludes, the visible contamination has been removed and properly disposed of. Long-term plans may have been made concerning monitoring the site to determine if any further work or sampling may need to be done, but basically the event has ended. The RRT members may still have some communication, but if any work is left to be done, it will not immediately impact the population or the environment.

CONCLUSIONS

With the passage of Superfund and the promulgation of the National Contingency Plan the Federal government has a working plan for dealing with emergency response activities. Many states have also developed their own contingency plans for response to emergency incidents by establishing response teams and providing an On-Scene Coordinator to direct the state's activities.

The recommendation to everyone in industry, county/city government, state government, nursing homes, hospitals, and any other entity that could be involved in an emergency incident involving oil or chemicals . . . establish a contingency plan. Have all your resources (team members, functions, physical resources) outlined and know who is in charge or can approve use of these resources. Most important, the OSC must be available *on-scene* during the entire course of the incident. Manaement from the office is extremely difficult when decisions needs to be made quickly "in the field" with information that changes minute-by-minute.

REFERENCES

1. Smith, A. J. 1979. The Dynamics of Regional Response Team Decision-making. A paper presented at the "1979 U.S. Fish and Wildlife Pollution Response Workshop", St. Petersburg, Florida.
2. Smith, A. J. 1981. *Managing Hazardous Substances Accidents,* McGraw-Hill, Inc., New York, N.Y.
3. Environmental Protection Agency, 1982. *National Oil and Hazardous Substances Contingency Plan,* 40 CFR Part 300 (SWH-FRL 2163-4), July 16, 1982.
4. Environmental Protection Agency, 1980. *Comprehensive Emergency Response, Compensation and Liability Act* (Superfund).

5. Bennett, G. F.; Feates, F. S.; Wilder, I. 1982. *Hazardous Materials Spills Handbook,* McGraw-Hill, Inc., New York, N.Y.

ACRONYM LIST

NRT	-	National Response Team
RRT	-	Regional Response Team
FWPCA	-	Federal Water Pollution Control Act
CWA	-	Clean Water Act
CERCLA	-	Comprehensive Emergency Response, Compensation and Liability Act (Superfund)
OSC	-	On-Scene Coordinator
NCP	-	National Contingency Plan
RCP	-	Regional Contingency Plan
NRC	-	National Response Center
RRC	-	Regional Response Center
RQ	-	Reportable Quantity

Hazardous and Toxic Wastes: Technology, Management and Health Effects. Edited by S.K. Majumdar and E. Willard Miller. © 1984, The Pennsylvania Academy of Science.

Chapter Fourteen

Emergency Response to Toxic Material Spills: The Role of the Local Community

David M. Lesak
Chief Technical Consultant
Hazard Management Associates
North Krocks Rd.
Box 3004
Wescosville, PA 18106

Scarcely a day goes by that somewhere in Pennsylvania, there is not some discharge of a toxic material into the environment. The large majority of these releases are of minor proportion and produce little, if any, effect. However, with ever increasing regularity, and in ever increasing numbers, these releases are producing a sizeable impact. The materials involved range from hazardous materials to hazardous, toxic wastes. They occur as the result of transportation accidents, accidental or deliberate releases and as the result of other types of incidents such as fires and/or explosions. The single common feature that the majority of these releases have is that the local community has to provide some type of response to the particular incident. In fact, the local community and its agencies have the primary responsibility of handling these incidents. This is not to minimize the responsibilities of state and federal agencies or other outside agencies that are involved, but rather is to emphasize the importance of local preparedness and response.

The local government and its agencies occupy a niche within the overall response that can not be filled by any other agency or organization. They have a vested interest in the progression and outcome of any incident. They are the ones who will feel the most impact and pressure that will result from such an incident. They also have a responsibility to provide for and protect the best interests of their community. They have a familiarity with local conditions and considerations that no one else can match. They have the responsibility to be prepared to the best of their ability to handle any eventuality that may develop. They must evaluate, coordinate, and prepare in order to fulfill the public trust that they have been granted.

PREPAREDNESS - THE ROLE OF THE LOCAL COMMUNITY

In order for a local community to properly handle its role in any incident, it must prepare. If preincident planning is not undertaken in a realistic organized fashion, the out-come will be of dubious value. The planning process must involve local government, its agencies, concerned citizens, business and civic organizations, county, state and federal agencies as well as business and industries involved with the manufacturing, transportation and use of hazardous materials or wastes. There must be a concerted effort to produce an organization with resources that will function as a well oiled machine and that can be mobilized on a moments notice. The organization will require the cooperation of not only governmental agencies, but also relief agencies, industry and a whole myriad of related organizations. All members of the organization must be well trained and thoroughly familiarized with the operations of the system. Response agencies such as the fire and police departments and emergency medical services must be fully trained in all aspects of on scene operations, as well.

At this point the primary concern is exactly what steps must be taken in order for the community to prepare itself for a hazardous materials incident which would include toxic waste releases. The first and foremost consideration is the attainment of a commitment from local government officials and agencies to develop a preplan. It must be remembered that the development of this plan and its management can act as a comprehensive disaster plan. The authority for the development, by local government, of a disaster plan is provided by Pennsylvania Law, under Act 1978-323 PL1332 enacted in November 26, 1978. PL 1332 requires local government to assign a local level Emergency Manager (C.D. Director) and establish a Disaster Plan. After the commitment has been made, there is a multitude of things that must be done.

RISK ANALYSIS

The process of developing a risk analysis is a very basic and usually a rather simple straight forward operation. It involves finding out what the exact situation is in the local community with respect to the types, quantities, locations and hazards of materials that can be found at any given time. The first step is to locate fixed places, that is businesses, industries, commercial establishments, that use, manufacture, or generate hazardous materials and/or wastes. Such facilities would include locations such as hospitals, colleges and universities, laboratories as well as manufacturing and industrial plants.

After identification of these locations has been accomplished, information relative to the exact materials involved must be gathered for later review. Usually, if approached in a professional, courteous fashion, there will be little or no problem obtaining the information that is necessary. Special care must be taken not to cause the impression that a "witch hunt" is under way. It is also important that the individual seeking the information is familiar with hazardous materials and wastes, so a positive impression about the planning process will be instilled. Some of the information that is gathered may also fall into the category of proprietary information. In this case every effort must be made to accommodate the particular facility in keeping the information gathered confidential.

However, if unreasonable or strong objections are encountered, it has been found necessary to introduce disclosure legislation in some communities. As an example, Philadelphia went this route. Also, if a facility handles hazardous wastes in any way, such as generating, handling, storing, reprocessing, refining and disposing, the facility is required to develop and maintain a contingency plan for its operations. In Title 40 of the Code of Federal Regulation, it further states that copies of the contingency plan must be supplied to local government and its public safety agencies, for their *APPROVAL*. Depending on the specific interpretation, it can be required that a listing of materials be provided as well. The justification for this requirement could be the necessity to know the materials involved in order to determine if the contingency plan is sufficient to handle the potential problems that specific materials could produce. Also, the waste handler must notify local hospital facilities of the specific materials handled and their possible effect. These contingency plans can be a source of vital information.

The next step to be taken is an evaluation of the transportation modes of the materials encountered in the facility. This entails surveys of highway, rail lines, airports, flight paths, water ways, and pipelines. In most areas the highways or rail lines will be the primary carriers. In order to properly evaluate the number of shipments and the materials involved, spot checks are usually necessary. This involves stationing personnel in locations where they can note the number of placarded vehicles that pass by and noting the carrier. After this information is gathered, carriers may be contacted to gain further information. For pipeline information it may require further research. The Federal Department of Transportation may be contacted in order to ascertain the availability of pipe line maps. Also, markers are required at specific intervals along the pipeline. These markers usually contain information as to who owns the pipeline and the right of way. These companies can be contacted to gather additional information.

Other information that can be very helpful in the analysis process is fire, police and other emergency response agency reports relating to previous incidents. These reports can provide significant insight into the present situation. Traffic accident reports may also be of benefit in pin pointing highway locations where accidents are a regular occurrence. All reports should be evaluated in order to determine the locations and severity of previous incidents, the types and quantity of materials involved, and the actual and potential impact that can and did result.

After the raw data has been collected, it must then be evaluated in order to establish both the possibility and the probability of incidents occurring. It must constantly be remembered that just because an incident has not *YET* occurred, does not mean that one will not occur in the future. In other words, the common impression that such incidents only occur in *other* jurisdictions and not here must be avoided. Such thinking is rather common among the general public and must be considered unrealistic. It must be remembered that the greater the quantity and number of shipments and materials, the greater the likelihood of an incident.

The next consideration that must be examined is the potential impact a major incident could produce. In the process of preparing a plan, it is important to consider the "worst case" situation. The reason for this is simply that if a community is prepared to deal with a "worst case" situation, it will be prepared to handle any eventuality. Also, incidents that do not develop to such a magnitude will be handled with much less difficulty and confusion.

Several important questions must be asked when attempting to establish a worst case scenario, in order to properly identify the potential impact. Some of these questions are as follows:

1. Are the materials flammable and if so, what is the magnitude of the potential fire development?
2. Is there a potential for a three dimensional liquid fire? This specifically addresses the problem of a flammable liquid spill, that when ignited, will spread out in down hill directions, into sewer systems, water ways and so on, causing a rapid and massive fire spread.
3. Is such a flammable liquid fire likely to occur in populated areas?
4. Is there the potential of resulting explosions that will further intensify the problem? These explosions may take the form of container over-pressurization or BLEVE (BLEVE refers to a Boiling Liquid Expanding Vapor Explosion many times involving liquified gas products such as liquified petroleum gas).
5. If BLEVE is a potential, how can the area within a 2,500 foot radius be evacuated in a timely, orderly fashion?
6. Is there the potential for the generation of vapor clouds? In one incident involving anhydrous ammonia a vapor cloud one-half mile wide and 9 miles long was generated within half an hour after the incident started. The product was in a 33,000 gallon rail tank car.
7. Again, if the potential for vapor cloud production exists, how can the affected area be rapidly be evacuated?
8. If mass evacuations are necessary, how will the evacuees be managed?
9. If spills of liquid product occur, how can containment be accomplished in a timely fashion? Is heavy equipment, manpower and materials readily available?
10. If spills occur and do enter water ways, will they effect public water supplies?
11. What is the potential environmental impact if spills do occur?

12. If a fire occurs and produces highly contaminated smoke, what is the probable direction in which it will travel? Also, what are some of the probable impacts of such smoke generation?

The list of such questions can go on and on and may be dictated by the specific materials that may be encountered. It is of the utmost importance that such questions be answered in an honest straightforward manner. If this is not done, there is little value in asking these questions in the first place.

THE MANAGEMENT ORGANIZATION

As a next step, it is important to establish a management organization. That is, to establish an organization of individuals that will have specific responsibilities and functions. These particular individuals will make up the nucleus of the Emergency Operations Center, or the EOC. On the local level, this group will consist of local governmental officials, including elected officials and more than likely, department heads. It is also possible that community volunteers may be required to fill positions such as the emergency medical coordinator, the police agency coordinator, the local level emergency manager, the public information officer, the liaison officer and so on.

The best way to understand how the management organization is to work will be to list the various positions and their corresponding responsibilities.

1. *Highest Local Level Elected Official*
    This individual has overall responsibility for operation during the emergency and its follow up. This is not to say that this individual will actively command on scene units and operations but rather oversee these operations. This individual will also provide for the overall coordination of all agencies involved and also provide intergovernmental communications and declarations.

2. *Other Local Elected Officials* - (Supervisors, Councils etc.)
    These individuals will aid the chief official in fulfilling his responsibilities and assist in providing emergency legislation as required. They will also help establish additional human service needs with the assistance of other agencies in the EOC.

3. *Public Health Coordinator*
    The primary responsibility of this position is to establish and coordinate all emergency medical service functions including ambulance services, triage, hospital designations, establishment of a morgue and so on. This individual will also coordinate all matters concerning public health in the community and evacuation centers, as well as aiding the emergency manager in the evacuation of all health care facilities as needed.

4. *Public Works Coordinator*
   Coordinate and assist response agencies with regard to providing equipment, manpower, materials. Such equipment would include heavy equipment (graders, front end loaders, dump trucks, etc.); materials such as crushed stone, sand, roadway barricades, etc.; manpower to provide the mentioned materials and equipment as well as supplemental personnel to aid utilities and other agencies as needed.

5. *Emergency Manager and Assistant*
   Coordinate the operation of agencies involved with incident management as an aid to the highest elected official. Also, develop, maintain, and update resource lists as supplied by his or her office and other management areas. A primary responsibility is to coordinate evacuation, transportation and care of residents and to establish, maintain and oversee operations of evacuation centers.

6. *Fire Coordinator*
   Relay information as to the status of all rescue, fire suppression and prevention activities as well as all hazard abatement activities of the fire department and other agencies, to other officials at the EOC. Also establish and coordinate all fire mutual aid activities including:
   -ensuring optimum capabilities of on scene operations in the way of manpower, equipment and apparatus.
   -ensure optimum fire protection for the remainder of the community.

7. *Police Coordinator*
   Coordinate all crowd and traffic control, evacuation and security activities associated with the incident. This individual must also keep all other members of the EOC appraised of the status of operations relative to police operations.

8. *Public Information Officer*
   Act as media liaison for public service information, bulletins, status reports and media news releases. This individual should also work to establish standard procedure for the release of such items as well as provide locations for media people to meet.

9. *Communications Officer*
   Establish and maintain an effective communications system for all agencies and organizations involved. This is to include the establishment of a *RELIABLE* two way radio communications network for all response agencies. If multiple frequencies are a problem (in all but some very large municipalities this will be very true), one alternative is the use of programmable scanners for the cross monitoring of communications. There must also be contingencies established with local phone companies to insure the ability to install 20 to 50 additional temporary phone services for the EOC as well as the media.

10. *Engineering Department or Firm*
Coordinate and aid Public Works, utilities and response agencies in the maintenance of roadways and utilities. Aid in the determination of possible impacts and problems associated with the incident. Also in the restoration of public utilities and roadways.

Another phase of an overall management system is the coordination and management of emergency response agencies. In most locations, the police agencies usually have some type of system developed. It is important to review such pre-existing systems to ascertain their practicality and usefulness relative to this type of incident. If the existing system is found to be lacking, it must be changed or modified to properly handle an incident of such magnitude.

Fire department operations are a totally different operation. In most areas, fire protection is provided by volunteer departments, which in most cases, are not prepared to deal with such massive operations. Coordination and delegation of authority by function are paramount to the successful handling or a major incident.

An excellent system for managing such large operations has been developed in California. The system is known as the Incident Command System and was designed specifically to handle such incidents. This system can be used as an excellent model to develop a management system in any locality. It has several advantages that make it a very attractive system to develop and employ. First, the system will not only work for major incidents involving hazardous materials and wastes, but can also be used in everyday fire fighting activities. This is a very important point, because using it in everyday operations allows fire fighters to become very proficient in its use. Secondly, after a training period, all departments that are proficient in its use will function very well together on the scene of any incident. For example, if 8 fire departments are proficient in its use, they will act not as 8 separate departments on the scene, but rather as one department. The advantages are, thus, quite evident.

Finally, emergency medical organizations must also be well coordinated and managed, especially if large numbers of casualties are present or expected. Again, if a system is presently in operation, it must be reviewed to ascertain its usefulness. If none exists, the Public Health Coordinator should take the lead role in establishing such a system. For all of the emergency response agency management systems, the Coordinator must develop a management system that will assure the agencies, that they will be able to accomplish the tasks assigned.

Overall, the establishment and effectiveness of each separate portion of the total management system will require commitment and training by all of those involved. The final evaluation however, should not be left to chance or in other word, the occurrence of an incident. Effective testing is required to evaluate the strengths and weaknesses of each individual portion of the system as well as the overall performance of the total organization. In order to accomplish an accurate evaluation it is essential to run tests of all phases of operations through the use of individual agency drills and eventually major disaster drills. The specifics of drills, evaluations and critiques will be covered later.

Master Resource List

The master resource list in its essence is simply a listing of resources including manpower, materials, equipment, services and organizations that are available to the community and which can aid in handling all aspects of an emergency. This listing must be in black and white, readily accessible and prepared in advance of its need. It should be broken down by the types of resources needed and should include names of agencies, organizations, businesses, individuals or any other pertinent listing as well as addresses and phone numbers (especially emergency phone numbers for 24 hour notifications) to contact emergency personnel. Prior contact with responsible individuals in a given organization is a must. A group or organization must be informed as to what will be needed and expected. This also offers an opportunity to develop a rapport and knowledge of people you may be working with under very trying circumstances.

In the process of establishing the resource list, it is important to ask several questions in order to properly prepare the list. The first question is, what is available through local governmental agencies? This involves examination of the capabilities and resources available from the local government, its equipment and materials as well as its personnel. It is also important to know the time factors involved in such a mobilization.

Next the local police agencies must be examined as to the resources they possess. If no local police agency exists, examine whoever provides police protection. Most likely this will be the State Police. Such questions must be asked as: How many men will be available and how quickly? Do they possess any specialized equipment such as mobile command post or communications capabilities? Are any of their personnel specially trained in areas that may be pertinent? If there is not a sufficiently large manpower pool, where can additional manpower be acquired? Will fire police or mutual aid fire fighters be required to supplement manpower needs? How will these individuals be notified, deployed and managed?

For the fire department, the same types of questions must be asked. What is its manpower status? If it is not sufficient to handle a major incident, do mutual aid agreements exist that can help alleviate the short fall? Does the department possess a Hazardous Materials Response Team? If not, is there one available through mutual aid agreement? (The availability of a Response Team is a major benefit in any such incident and will be discussed later.) Does the department have individuals who have received intensive specialized training in handling chemical incidents? What is the existing management system capable of handling? Is the present situation within the fire department going to dictate a very defensive handling of all but the most minor incidents? Is the department properly trained in general response techniques for chemical responses? Has the depart-

ment provided for a local individual to provide some on scene technical information rapidly? (This is a difficult person to find in some cases. A high school
chemistry teacher, a pharmacist, an industrial chemist and so on could prove
quite valuable.) Is the department aware that they will more than likely be faced
with making well informed decisions with very little outside help for an average
of one and a half to three hours until outside help arrives? How can manpower
requirements be met if the incident continues for an extended period of time?
(Some incidents are capable of going on for a week or more, with many averaging from 8 to 16 hours.) Is the department aware that it is the primary response
agency involved with the direct handling of the incident?

With respect to emergency medical needs, are EMS personnel trained in
handling chemical contamination victims? Is decontamination equipment carried on the units and the personnel trained in the importance and procedures used
in decontamination? (NO victim that has been contaminated should be placed in
a unit or examined until decontamination has been completed. If this requirement is not strictly followed there is an ever present danger of contaminating personnel, equipment and units with possibly fatal results.) Have standard
operating procedures been established for triage of victims? How will hospital
designations be handled and which hospital can handle this type of victim? If a
local poison control center does not exist, who can be contacted for the appropriate information? If a poison control center is available, how can communications be established with the field operations?

Now consider the resources that are available through the local community.
Are there any experts in the local area, are they available and how can they be
contacted? Who is to be contacted to arrange for buses to aid in evacuation? Can
buses be obtained from the school district or is there a private or public operation
available? Where will evacuees be told to report? How can sanitary food and
water needs for evacuees be met? How can rapid notification of evacuees be accomplished? Where are additional sources of heavy equipment and materials
such as stone or sand and how will they be notified? How can civic organizations
aid in the operations at evacuation centers or at the EOC? What other organization can provide some type of assistance in an emergency?

Concerning outside governmental agencies and private interests, what other
agencies have authority in such incidents? Who is to be contacted and how are
the contacts to be made? What assistance can these agencies provide? Who will
these agencies have responding and what do they expect from local officials?
What clean-up companies are available and which ones operate in a safe, ex-
peditious and proper fashion? Are there any other specialized companies or
organizations available to aid in any aspect of operations?

The preceding questions are just a few of the primary concerns that must be
addressed in order to prepare a useful, effective resource list. Any other considerations that seem to be important and necessary should be examined and

possibly added to the list as seen fit. It is also important that the list be periodically updated to insure its value is not compromised. There is nothing more frustrating and compromising than to need to contact some individual or organization in an emergency situation and find the phone number has been disconnected or the individual has moved.

Response Teams

As previously mentioned, a Hazardous Materials Response Team comprised of well trained, well equipped members is the best way in which such an incident can be handled. However, such response teams at present are rather few in number. If one is available to a community from its own fire department or through mutual aid agreements, the job will be greatly simplified. If a response team is not available, some very difficult considerations must be given to the overall response capabilities of the community.

A decision must be undertaken with full understanding of the consequences. If a team is to be started, it will require a core of dedicated individuals who are willing to embark upon an obligation that is dangerous and will require a commitment to extensive planning and training. Monetary aspects must also be considered. Some of the specialized equipment that is necessary to function *safely* and effectively is extremely expensive and has definite limitation for duration of use and applicability to all situations. It must be realized that the deployment of such a unit will more than likely generate additional responses outside of the local community. Ideally the response team should be established with a regional approach. That is to say, more than one department should be involved in providing personnel and possibly equipment.

If a response team is not deemed necessary, all involved in that decision must be aware of the impact that will result. There will be incidents where emergency response personnel will not be able to do anything other than evacuate the area and withdraw until the proper equipment becomes available. This is not to imply that a response team can always handle every situation, for this is indeed not true. However, without the proper personal protective equipment, especially when dealing with toxic materials, any action to do something about a spill or leak may be a *FATAL* move for those involved. Again the ramifications of a pro or con decision must be fully examined and understood by all.

If, on the other hand, a response team is deemed to be necessary, much additional planning and preparation must go into the process. In this case, haste can lead to wasteful expenditures for the purchase of equipment, training or apparatus. It is advisable to select key personnel and provide them with rather intensive schooling prior to the initiation of a team. There are quite a few schools available, with two of the best being the National Fire Academy and one in Jacksonville Florida. Other schools that are available are at Texas A&M, Denver as well as some industry sponsored training. By having these key personnel attend schools prior to the initiation of a team, many of the pitfalls encountered by other organizations, may be eliminated.

TESTING AND EVALUATION OF THE PLAN

After the proceding steps have been completed, many officials are under the mistaken impression that the planning process has been completed. True, the majority of the initial work has been completed. However, an integral portion of the planning process must still be completed. This portion is the evaluation of just how well the plan will work if it is in fact needed. In most cases, there will be a certain degree of "fine tuning" or adjustments that must be done in order to make the plan as functional as possible. The way in which the evaluation and fine tuning is accomplished is through the process of test drills.

Test drilling is probably one of the most misunderstood processes of all emergency procedures. There is a great tendency to overload everyone involved in the operations, and provide arbitrary circumstances with which all are faced. More simply stated, the first test many times involves a "worst case" situation that totally overwhelms all involved. A common scenario would include a school bus loaded with children involved in a traffic accident with a tractor trailer tanker carrying 9000 gallons of a flammable, toxic chemical. The accident happens to occur at a rail road crossing and a train also carrying hazardous substances is also involved. Arbitrarily, a wind direction is selected and is almost inevitably opposite of conditions at the time of the drill. The only thing that a test like this suceeds in doing is instill bad practices and makes the incident commander consider submitting his retirement papers.

All test drills should seek to provide those being evaluated with real life problems and considerations. In the early stages, special efforts must be made not to overwhelm the capabilities of those involved. Remember, this testing process, to be effective, must be a learning process as well. Improper and unsafe practices must not be tolerated, because the practices used during "training" will be followed during an actual incident. A striking example of this is the way police officers once trained on the pistol range. They were always told to save the brass shell casing when reloading. In order to do this, the empty shells were dumped into the officers hand and then placed in his pocket. The problem with this type of training became apparent, when officers were found severely or mortally wounded after gun fights with empty revolvers and empty shell casing in a hand inserted in a pocket. The training did not reflect the way things should occur in the field. TRAINING AND DRILLING MUST SIMULATE THE PROCEDURE TO BE FOLLOWED DURING AN ACTUAL INCIDENT.

All response agencies should first be trained and tested on their own operations before interagency drills are attempted. The drills should also start with a very managable scenario. The type of scenario used will in large part depend on the capabilities of each given agency when the process is started. In the case of the fire department, this may simply be an incident involving a leaking 55 gallon drum of a flammable liquid. As the capabilities of each agency of faction of the management system improve, the scenarios should increase in difficulty and complexity. If shortcomings are noted in the plan as it stands, appropriate changes should be made to eliminate the problem. Eventually, a major incident scenario should be developed and tested. Again, keep it realistic!

Another major problem associated with drills is the regularity with which they should be conducted. This aspect is a particularly nebulus area. To say the least, annual and semi-annual drills are insufficient. During the development stages, drills must be conducted on a very regular basis, most likely monthly. As all those involved in operations become more proficient at their functions, this time frame may be extended. It is also important that individual sectors of the management system conduct inhouse drills at regular intervals.

Possibly a good way to highlight the regularity with which drills should be conducted can be gained from others experience. A major metropolitan area wanted to test their effectiveness in handling major emergency medical operations. (This was a payed department) so a drill was initiated. After the initial drill many short comings were observed. Another drill was scheduled for the next month, and then one for the following month. The final outcome was that it took a total of 18 monthly drills until all involved felt that the proper degree of proficiency had been gained. At that point, it was decided to conduct drills every three months so the level of proficiency could be maintained. The degree of proficiency that is desired is a local decision that must include all aspects of the operations.

CRITIQUING

Critiques are a major factor in developing, maintaining, and evaluating the effectiveness of the plan as well as the overall operations of all agencies involved. Critiques should become part of the standard operations of portions of the management system and its agencies. Without the use of critiques, a major portion of the learning process and the evaluation process will be missed. The critiques will also point out short comings and improper practices that can then be corrected.

Critiques should be conducted after every drill or training session, as well as after actual incidents. It is important that all who have been involved in the operations also be involved in the critique. Every effort must be made to take an HONEST look at how the total operation went. If the critique turns into a pat on the back session, no value will be gained. Problems that were noted must be discussed in an attempt to prevent them from occurring at another time. The criticisms must be presented and ACCEPTED in the light of CONSTRUCTIVE CRITICISM. Personality clashes and power struggles must not be tolerated as they will only be counterproductive and compromise the integrity of the entire plan. However, if one agency or portion of the plan is found to be lacking or not functioning the way it should be, every effort must be made to correct the situation. If this requires a change in personnel, then that situation must be met and handled. If this type of situation is allowed to continue, the plan will be in serious trouble.

Depending on the size of the operation the critique may require several different meetings to accomplish its mission. It may first require discussions by members of the response agencies with the results of these discussions being relayed to a meeting of management heads. In turn recommendations and problems gleaned from the management head meeting must be returned to the response agency personnel. The bottom line in effective critiquing is the establishment of realistic open communication of problems and shortcomings throughout the entire Management Organization.

CONCLUSION

It is the responsibility of all local governmental agencies and organizations to undertake the process of preparing for the eventuality of the release of toxic materials within their community. The best way in which they can prepare is the development of a comprehensive Disaster Plan.

The planning process must be approached with a dedicated and realistic manner. It must be realized that just because an incident involving toxic or hazardous substances has not yet occurred that does not mean that one will not occur at some future time. If the planning process has not been undertaken and an incident should occur, it will tax to the limits the capabilities and resources of the local community. The need for an effective management system is paramount to the successful completion of an incident with such wide ranging impacts and ramifications. It is the obligation of each locality to prepare the best way it can to provide for the protection of its citizens and planning is really the only way to fulfill that obligation.

PART 4

Management, Regulations, and Economic Considerations

In the past hazardous and toxic wastes were frequently disposed of with little regard to the pollution of the environment. These practices have created environmental problems that have been the basis for the establishment of regulations and management programs to control the disposition of toxic substances. To illustrate, Pennsylvania's Solid Waste Management Act provides the basis for an effective management program by providing regulations for solid and hazardous waste collection, transportation, processing, treatment, and disposal.

The initial six chapters of this section consider present-day regulations and management of hazardous wastes in a wide variety of settings. The first chapter discusses the clean-up of the hazardous waste site created by the Drake Chemical Company of Lock Haven, Pennsylvania during the twenty-year-period 1962-1981.

The next three chapters are devoted to hazardous waste control programs that have been established in Pennsylvania, Kentucky and Western Europe with examples of national legislation from Denmark, West Germany and France. In the United States the basic Federal legislation to control hazardous waste sites, the Resource Conservation and Recovery Act, was enacted in 1976. From this Act the nation's regulatory program has been achieved. Common to most waste management programs in the nation are such aspects as the permitting of facilities, enforcement of controls by monitoring, and the development of hazardous waste facilities plans.

Two chapters are devoted to the handling of hazardous wastes within an industrial facility. Probably the single most important factor that distinguishes management of hazardous materials in an industrial facility from management of wastes in other settings is the ability to know what toxic materials are present, their properties, and the means of responding when problems occur.

The final three chapters of Part Four treat economic considerations of waste problems. It is sometimes not recognized that there are economic benefits of environmental regulations. To measure these benefits, models can be prepared to illustrate what is called the damage function approach. The damage function approach relies on four basic modelling components: facility failure, transport and fate, exposure, and effects.

The second economic paper presents the methodology for estimating the economic costs incurred by a society in the disposal of hazardous wastes. This paper focuses on disposal sites that contain some of the same elements as coal wastes as a basis for estimating damage costs.

The final paper presents viewpoints on requirement liabilities of waste generators. The primary theme of this paper is that the responsibility of waste management lies with the generator of the waste.

Hazardous and Toxic Wastes: Technology, Management and Health Effects. Edited by S.K. Majumdar and E. Willard Miller. © 1984, The Pennsylvania Academy of Science.

Chapter Fifteen

Case History: A Superfund Cleanup in Central Pennsylvania

Kenneth Caputo, Jr., B.S. and Richard L. Bittle, B.S.
Department of Environmental Resources
Bureau of Solid Waste Management
200 Pine Street
Williamsport, PA 17701

The Drake Chemical Company operated a chemical intermediate manufacturing plant in Central Pennsylvania from 1962 until petitioning for reorganization under the bankruptcy laws on August 11, 1981. During the years of Drake's operation, the company had been cited for violations under the Pennsylvania Clean Streams Law, the Pennsylvania Solid Waste Management Act, along with the federal Occupational Safety and Health Act.

On February 28, 1982, in a coordinated effort between the U.S. Environmental Protection Agency and the Pennsylvania Department of Environmental Resources, a Superfund immediate removal action was initiated at the Drake Chemical Site to remove approximately 1,700 drums containing process wastes, to clean tanks and reactor vessels containing chemicals, and to contain the site to abate the imminent threat to the public and to the environment.

The purpose of this report is to provide an overview of the Superfund program, to describe the events and activities leading up to and including a specific Superfund immediate removal project, and to discuss current efforts to study, design, and implement remedial action at the Drake Chemical site.

SUPERFUND BACKGROUND

There are three major federal laws which deal with uncontrolled hazardous waste sites and the response to spills of hazardous substances onto land, into the air, or in non-navigable waters. These laws include the Clean Water Act, the

Resource Conservation and Recovery Act, and the Comprehensive Environmental Response, Compensation, and Liability Act (Superfund).

The Clean Water Act provides for emergency response when oil or designated hazardous substances are discharged into navigable waterways, but does not permit the government to act when hazardous substances are released elsewhere in the environment, threatening to contaminate groundwater or to emit dangerous fumes.

The Resource Conservation and Recovery Act (RCRA), passed in 1976, contains the hazardous waste authorities, which include the development of criteria for determining which wastes are hazardous, the institution of a manifest system to track these wastes from point of generation to point of disposal, and organization of a permit system, based on standards, for hazardous waste treatment, storage, and disposal facilities. RCRA is designed to prevent the creation of uncontrolled hazardous waste sites, but it does not permit the government to respond directly to the problems caused by improper hazardous waste disposal sites already in existence.

The Comprehensive Environmental Response, Compensation, and Liability Act of 1980 (CERCLA) (Public Law 96-510) commonly referred to as "Superfund", was passed by Congress and signed into law on December 11, 1980, to "provide for liability compensation, cleanup, and emergency response for hazardous substances released into the environment and the cleanup of inactive hazardous waste disposal sites." The U.S. Environmental Protection Agency, by Executive Order, is responsible for managing the Superfund program.

Costs of Superfund cleanup actions are covered by a $1.6 billion fund, 86% of which is financed by taxes on the manufacture or import of certain chemicals and petroleum, the remainder coming from general revenues.

CERCLA provides for the establishment of Imposition of Taxes of Petroleum and Certain Chemicals (Subtitle A, Section 211), a Hazardous Substance Response Trust Fund (Subtitle B, Section 221), and the Post-Closure Liability Trust Fund (Subtitle C, Section 232).

The Tax on Petroleum and Certain Chemicals provides for the collection of taxes for crude oil received at a U.S. refinery and petroleum products entering the U.S. for consumption (79¢ per barrel) and the imposition of a tax on 42 taxable chemicals sold by the manufacturer, producer, or importer.

The Hazardous Substance Response Trust Fund pays for cleanup when a responsible party cannot be found or cannot finance the cleanup.

The Post-Closure Fund is used for response actions at hazardous waste disposal facilities which have received a RCRA permit, have complied with Subtitle C of the Solid Waste Disposal Act, and have been closed in accordance with conditions placed upon the RCRA permit. All liability from the disposal facility is transferred to the Post-Closure Fund 90 days after the owner or operator of the facility notified the EPA Administrator that the facility has been closed in accordance with the permit.[1]

Responses to incidents involving hazardous wastes are tailored to the specific needs of the particular site or spill. Depending on the nature of the incident, a Superfund response can take the following form:

1. *Immediate Removal* - This type of action is taken when a quick response is required to mitigate harm to the public or to the environment. Incidents which require Immediate Removal Actions include threats of fire or explosion, groundwater contamination, or other similarly acute situations. Generally, Immediate Removal Actions are limited to $1 million and must be completed within 6 months.

2. *Planned Removals* - When a prompt, however not immediate removal is necessary to reduce imminent and substantial dangers to public health or the environment that would otherwise occur if response were delayed. Planned Removal Actions are also limited to $1 million and must be accomplished within 6 months.

3. *Remedial Actions* - Actions which are intended to provide a permanent resolution to the hazardous waste incident. Remedial Actions are generally more expensive and require a longer time to implement. Remedial Actions are taken only at those sites recognized by the National Priorities List of 418 sites. Remedial Actions are not subject to the 6 months/$1 million provision.

The state in which the site is located must agree to maintain the site after the remedial action or planned removal is completed. In Remedial Actions for which the Federal Government has the lead responsibility, the U.S. Army Corps of

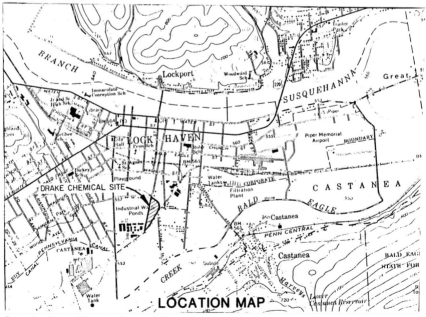

Location Map of Drake Chemical Co.

Engineers manages the design and construction stages for EPA. Private contractors perform the work under Federal and/or State government supervision.[2]

SITE HISTORY

Drake Chemical Company is located in Lock Haven, Pennsylvania (population 11,200) and is within an area that is residential and commercial, including an apartment complex inhabited mostly by senior citizens and a large shopping center located within ¼ mile. Schools, churches, and Lock Haven University are located within 2 miles of the site. The site is situated between the West Branch of the Susquehanna ¾ mile to the north and Bald Eagle Creek ½ mile to the south (see location map). Drake is contigious to the American Color & Chemical Company production facilities to the west.

The eight-acre site contains production facilities, an effluent treatment building, warehouses, and office buildings. Also located on site are two lined wastewater treatment lagoons along with an unlined lagoon in which process wastewater and sludge were placed for on-site storage. Through subsurface leakage and stormwater overflows, the constituents of Drake's process wastes have been carried into an unnamed tributary across a floodplain, through the Castanea Township Park, then into the Bald Eagle Creek which feeds the West Branch of the Susquehanna River.

The soil around Drake is classified as a mottled, silty-clay that extends to a depth of 15 to 40 feet, where it merges with the limestone bedrock formation. The limestone bedrock formation extends to approximately 110 feet.[3]

The groundwater table generally follows the surface gradient at an average depth of 5 to 15 feet.[3]

Drake specialized in making batch quantities of various organic compounds using the following synthesis schemes: aromatic substitution reactions, displacement reactions, amine functional reactions, carboxylate functionalization reactions, and oxidation/reduction reactions. The specialty intermediate chemicals were used in the production of dyes, pharmaceuticals, cosmetics, textiles, pesticides, and pesticide percursors. Drake manufactured over 150 different chemicals during its existence at the Lock Haven facility. Prior to ceased production in July 1981, 3,000 pounds of sludge per day was generated from the pretreatment of process wastes.

During the years of Drake's operation, the company had been cited for numerous violations of state and federal health and environmental regulations.

In 1973, a citation was issued by the Pennsylvania Department of Environmental Resources for violation of the Clean Streams Act in association with heavy flooding in June 1972. In November 1973 and again in January 1977, Drake was cited for non-compliance.

On April 4, 1979 the Bureau of Water Quality Management of the Pennsylvania Department of Environmental Resources issued a Consent Order and Agreement addressing the handling and disposal of solid waste, the disposal of pretreatment sludge and wastewater in the unlined impoundment, stream discharge, the preparation of a plan for phasing out unlined impoundments, and penalty provisions for non-compliance of the Clean Streams Law.

On August 18, 1980, pursuant to the newly promulgated hazardous waste regulations, Drake notified the EPA and the Pennsylvania Department of Environmental Resources that they were a generator of hazardous waste. This triggered the Bureau of Solid Waste Management to begin conducting hazardous waste inspections at the site. The first hazardous waste inspection was conducted in January 1981 and several violations were noted. In March 1981, in an effort to correct the violations, Drake discovered that the price of cleanup and the disposal of the hazardous wastes which had accumulated on-site was more than the company could afford. In July 1981, the company ceased production and in August 1981 Drake filed for reorganization under the bankruptcy laws.

In May 1981, the National Institute for Occupational Safety and Health (NIOSH) visited the Drake facility to evaluate reports of health complaints, bleeding gums, respiratory problems, skin conditions, and of unsafe working conditions.

During the NIOSH inspection, the medical officer had found that the processes at the plant were not highly automated. The "charging" of reactor vessels was often done by hand, and workers often poured the contents of bags or barrels directly into reactor vessels, putting the workers in close contact with chemical dust and vapors rising from the vessels. The only protective equipment observed were short rubber gloves worn during charging, cleaning, and most other work activities. The equipment within the plant appeared to be in poor condition: leaking pipes, dripping tanks, spills, and corrosion in areas where synthesis takes place. Splashes and spills of chemicals resulting in skin contact were not infrequent occurrences.

It was concluded by NIOSH that because of the combination of a deteriorating physical plant, inadequate engineering controls, inadequate use of personal protective equipment, and poor work practices created a work environment that contained many opportunities for exposure and injury.[7]

On January 25, 1982, the Bureau of Solid Waste Management issued an Administrative Order against Drake Chemical Company addressing the correction of the numerous hazardous waste violations that existed at the site and was again cited for non-compliance of the 1979 Consent Order and Agreement.

On February 10, 1982, Drake was referred to the U.S. Environmental Protection Agency by the Pennsylvania Department of Environmental Resources as a candidate for emergency cleanup under the Federal Superfund program.

Emergency Response

On February 23 and 24, 1982, a team composed of personnel from the U.S.

FIGURE 1. Outside storage of drums at the Drake Site.

Environmental Protection Agency, Region III (USEPA); Pennsylvania Department of Environmental Resources, Region IV (PADER); the USEPA Environmental Response Team of Edison, New Jersey; and the USEPA Technical Assistance Team entered the Drake Site and observed the following:

- Approximately 1700 drums filled or partially filled with unknown process waste and sludge. Many of the drums were rusted through, allowing the contents to spill onto the ground. Drums of cyanide reactant were also stored outside in a haphazard manner. (Figure 1).
- An unlined pond, located at the rear of the property, seeping through an adjacent railroad bed, created a leachate stream which flowed through a township recreational area and eventually to the Bald Eagle Creek, a major tributary of the Susquehanna River. Earlier sampling had shown the pond to contain certain priority pollutants.
- Unlined impoundments in which the sludges from the pretreatment process of wastewater were deposited along with demolition debris. Drums are presently emerging at the surface of these unlined impoundments. (Figure 2).
- Hazards with the potential to be passed through various media including air, groundwater, surface water runoff, and soil.

On February 25, the On-Scene-Coordinator (OSC) for USEPA transmitted a 10-point document (Fund Authorization Report) through EPA Region III to EPA Headquarters, Washington, D.C. The purpose of the report was to obtain necessary Superfund allocations for the initiation of an emergency clean-up of the Drake Chemical Company.

On February 26, verbal permission was given to the USEPA OSC to provide security, proceed with the construction of a fence to contain the site, drum removal, process building cleanup and the study portion of the unlined lagoon. At this point, $400,000 of the requested funds had been approved.

On March 2, EPA Region III, in cooperation with PA DER commenced emergency action in order to abate the imminent threat to public health and the environment.

Prior to working with the flammable or explosive materials on-site, a complete protective action plan was implemented for the workers at the site and the citizens of the Lock Haven area. Safety issues are a primary concern as they relate to the sampling, storage, transportation and ultimate disposal of hazardous wastes.

A Site Safety Plan was developed on March 1 to establish three zones of contamination containment at the site. Zone 1, the area contained by the fence, was defined as the Exclusion Area in which the workers were required to wear disposable outfits along with a full-faced mask with cannisters (Level C protection). Within the buildings, Level B respiratory protection (self-contained breathing apparatus) was required. Within Zone 2, the Contamination Reduction Area, was the area where decontamination of personnel and equipment took place. Zone 3 was designated as the Clean Zone into which no contamination was to enter.

An emergency evacuation plan for incidents at the site was developed under

FIGURE 2. Unlined impoundments with drums emerging at the surface.

the authority of and in accordance with the provisions of the Pennsylvania Emergency Management Services Act of 1978. An incident was defined as an event or condition at the Drake Site which could result in an impact on public health and safety. The plan was specifically outlined for the protection within a one-mile radius of the Drake Site in the event of potential or actual incidents of the site.[6]

Four protection level actions, ranging from less to more serious had been identified. The four protection actions were:

1. *On-site emergency* with no off-site action required.
2. *Protection in place* by taking precautionary measures to minimize exposure to the outside environment.
3. *Protection in place with selected evacuation* as a precautionary measure for individuals known to be seriously ill at home and/or those suffering from various debilitating health related disorders.
4. *General evacuation* triggered by "worse case situation" at the site and executed as a precautionary measure to minimize the threat to health and safety.

These safety plans were implemented immediately and were in effect until the immediate removal phase was over.

The clean-up phase at Drake began in a work assessment stage which included identifying chemicals, making contractual arrangements and setting up of work trailers.

During the first few days, legislators, local officials, the news media, and the citizens of Lock Haven were briefed and kept informed of the progress of the activities.

During the first week of the clean-up, the prime contractor, OH Materials of Findlay, Ohio, conducted an inventory of wastes at the site and labeled the numerous tanks and process lines. Various fencing contractors were contacted and bids were accepted for the construction of the fence. A study conducted by the PA DER located and mapped all underground utilities leading to the property.

By March 8, construction of the fence around the site was underway and a master inventory of known and suspected chemicals at the site was compiled. The work to be done at the Drake Site was prioritized by the PA DER and the US EPA. The first priority was to empty and to dispose of materials in the outside storage tanks. Second was to clean and secure the tanks located within the process buildings. Third, the removal of over 1700 drums containing process wastes were to be removed and disposed of at a site permitted to accept hazardous wastes. Concurrent with the above activities, an extent of contamination assessment was begun on the leachate stream.

Extent of Contamination

On March 8-9, 1982, the Environmental Response Team conducted an Extent

TABLE 1

Major Chemical Groups Identified in Leachate Stream

1) Trichlorophenyl Acetic Acid (all isomers) (TCPAA)
2) 1,2-Dichlorobenzene and 1,4-Dichlorobenzene (DCB)
3) Dichloroaniline (all isomers) (DCA)
4) Nitrobenzene
5) Phenol
6) Nitrotoluene (all isomers)
7) Naphthol
8) 2-Nitrophenol
9) 2,4-Dichlorophenol (DCP)
10) 2,6-Dinitro-o-Cresol
11) Chloromethyl Aniline (all isomers)
12) Methyl Nitroaniline (all isomers)
13) Diethylene Glycol

of Contamination Study of the area at the request of US EPA. The purpose of the study was to determine the degree and extent of hazard associated with toxic releases to the soil and surface waters from the Drake property.[4]

The surface design consisted of establishing a series of thirteen sampling stations along the 1,620 feet of the leachate stream which exits the Drake property, passes through a township park, and flows into the Bald Eagle Creek. Control sediment and water samples were taken from the Bald Eagle Creek upstream of the tributary.

Fifty-seven soil and sediment samples were collected within the leachate drainage area to define the extent of contamination. To determine whether contaminated sediments were migrating into the creek, a control baseline was established by sampling Bald Eagle Creek upstream and downstream of the tributary.

Soil and stream bottom samples were screened for 13 chemical substances of special interest in this particular investigation (See Table 1). One compound, Trichlorophenylacetic Acid (TCPAA), a herbicide commonly known as Fenac, was ubiquitous and was used as a representative of the waste.

Results of the sampling clearly show that most contaminants were distributed in the centerline of the leachate stream and in ponding areas where sediment is likely to accumulate. The TCPAA was primarily confined to the stream bed whereas the dichlorobenzene (DCB) displayed a wide degree of distribution and considerable deposition on the floodplain. These two pollutants were used to describe the extent of contamination at the site.

Results from the sampling indicate that the other contaminants tend to behave like one or the other of the TCPAA or DCB, which was sufficient to characterize most of the transport mechanisms in operation at the site.

As a result of the Extent of Contamination Report, it was recommended to township officials that the park be closed. The Environmental Response Team had recommended, as a mitigative option, the removal of 7,500 cubic yards of soil and sediment from the stream along with the removal and disposal of the contents from the unlined lagoon.

Problems Encountered

During a dilution stabilization operation of oleum from a tank on March 15, a small amount of material escaped causing an acid mist cloud of vapor to leave the site. The Pennsylvania Emergency Management Agency (PEMA) was notified and the Lock Haven Fire Department (LHFD) was summoned to the scene for standby. Approximately one and one-half miles from the site, five highway workers reported that they suffered respiratory irritation after being exposed to the cloud. The contractor performed repairs on the oleum tank piping to stop the release of oleum.

On March 16, in recovering a #955 loader which had sunk into soft ground, a 2-inch diameter natural gas line was accidently ruptured. The PA DER requested the LHFD and local gas company assistance through the PEMA "hotline". The fire chief reported gas fumes in the nearby shopping center and proceeded to evacuate the store. The gas was shut off by the gas company and normal operations resumed.

On March 23, oleum violently vented through the 14-inch tank dome hole, erupting as high as 100 feet into the air. PA DER notified PEMA and a Condition #2 was declared, whereas residents were alerted to stay indoors as a precaution. Once again, the LHFC was called to assist with water spray curtains downwind of the tank. The acid mist had already exited the site. It had taken the paint off of hundreds of cars parked downwind of the tank. As a result, the insurance company representing the prime contractor inspected the affected vehicles and provided repainting for each. Pictures taken of the inside of the oleum tank the next day confirmed that the tank was emptied.

Cleanup Continues

Because of the need for additional funding, EPA Headquarters approved a $750,000 ceiling for the Drake project. As of April 7, 1,717 drums of wastes had been sampled to determine waste compatability for disposal purposes. At this point, it was determined that the wastes could be classified into 6 categories:

Hexachlorocyclopentadiene
Trichlorophenylacetic Acid
Base-neutral compounds containing cyanide
Strong acids
Chlorinated hydrocarbons
Non-chlorinated hydrocarbons

Three 5-gallon containers of benzoyl peroxide, an extremely reactive compound which is shock sensitive and explosive, were found along with several small cylinders of phosgene oxide and chlorine. These cylinders were in marginal condition and required off-site treatment and disposal.

On April 11, the EPA Headquarters granted a ceiling increase of $203,000 to complete the emergency phase. This increase raised the ceiling total to $935,000.

On April 15, ten truckloads filled with sludge waste along with a tank truck with liquid wastes, left the Drake site for ultimate disposal at CECOS Inc., New York. Rollins Environmental Services, Inc., of New Jersey received and handled the drummed wastes from the Drake site.

On April 21, 1982, the emergency phase of the clean-up at Drake Chemical was deemed finished.

Expenses for the Drake Chemical Company emergency action included approximately $162,000 for the disposal of liquids; $53,000 for the solid hazardous material; $400,000 for the tank neutralization phase; $28,500 for site security; $18,400 for disposal of cylinders and shock sensitive materials and $10,000 for the removal of laboratory materials (See Table 2).

TABLE 2

Hazardous Waste

15,000 gallons chlorinated solvents
15,000 gallons non-chlorinated solvents
5,000 gallons Trichlorophenylacetic Acid
36 drums Hexachlorocyclopentadiene still bottoms
212 tons contaminated sludge
37 Lab packs containing lab chemicals
1 drum Chlorosulfonic Acid
20 Lab cylinders
3 5-gallon pails Benzoyl Peroxide
30,000 gallons Base-neutral waters

Remedial Action

The Remedial Action program is a key portion of the Superfund mandate to clean up uncontrolled hazardous waste sites. The U.S. Environmental Protection Agency has developed a remedial planning process that ensures consistent and rational decisions at priority hazardous waste sites. There are five major constituents of the remedial planning process: (1) Remedial Action Master Plan (RAMP); (2) Remedial Investigations; (3) Feasibility Studies; (4) Selection of a Remedial Alternative, and (5) Remedial Design.[5]

A RAMP is generally prepared for sites that have been ranked as a priority on the National Priorities List and selected for remedial action. The RAMP contains information necessary for planning a strategy and for assisting in the selection of an appropriate course of action. It identifies the type, schedule, and cost of remedial planning activities required. The RAMP for the Drake site, as prepared by Roy F. Weston, Inc., has identified the following as major problems which remain at the site:

• Groundwater beneath the site is heavily contaminated with organics.
• Soils are similarly contaminated.

- Drums of chemical waste have been buried.
- An unlined lagoon contains high concentrations of organic materials.
- Buildings and on-site debris may be contaminated.
- The leachate stream emanating from the site has contaminated downstream soils.

The Remedial Investigation is conducted to assess the problems at the site and collect data necessary for solving those problems. Investigation activities are carefully planned to fill any data gaps as a result of previous studies and activities.

Some of the activities proposed as part of the Remedial Investigation at the Drake site include:

1. Prepare a site safety plan for the remedial investigation/feasibility study;
2. Obtaining subcontractors for certain jobs such as drilling wells;
3. Taking aerial photographs to prepare topographic maps;
4. Drilling monitoring wells on and near the site to study the groundwater;
5. Sampling water from proposed monitoring wells and other wells to learn the extent of contamination in the groundwater;
6. Sampling and analyzing soil and sludge, buildings and debris, and liquid from lagoons on the site;
7. Sampling and analyzing water and sediment from the leachate stream flowing through the park, from Bald Eagle Creek, and from the West Branch of the Susquehanna River;
8. Performing a study of fish and other aquatic life living in Bald Eagle Creek and the West Branch of the Susquehanna River;
9. Studying the land around the site to see if crops and other plants are contaminated;
10. Determining if drums of hazardous materials are buried on-site.

After all this work is done, the information will be studied and a site assessment will be made. This assessment will show exactly what is at the site and where contamination can be found. An assessment of the leachate stream will be made separately in order to speed up the study.

Using the results of the Remedial Investigation, several different alternatives for cleaning up the site will be developed. A Feasibility Study will be conducted for developing and evaluating these alternatives, recommending the appropriate cost-effective remedial action, preparing an environmental assessment, and preparing a conceptual design for the remedial action which is recommended. A separate Feasibility Study will look at ways of cleaning up the leachate stream. This is planned to start 2½ months before the study for the rest of the site to "fast-track" cleanup of the priority leachate stream.

Upon completion of the Feasibility Study, the recommendation plan will be transmitted from U.S. Environmental Protection Agency, Region III, and Pennsylvania Department of Environmental Resources to U.S. Environmental

Protection Agency Headquarters for approval of the alternative that represents the appropriate extent of remedy.

Upon completion of the feasibility studies, a Remedial Design will be prepared. Where U.S. Environmental Protection Agency has the lead, as in the case of the Drake project, the final design and implementation activities will be managed by the U.S. Army Corps of Engineers. Remedial Designs usually result in a set of contract documents, including plans and specifications, that describe the remedial plan in detail to allow preparation of competitive bids.

Conclusions

On February 28, 1982, the U.S. Environmental Protection Agency and the Pennsylvania Department of Environmental Resources began emergency cleanup operations after the Drake Chemical Company failed to respond to a request for voluntary cleanup. As a result of the cleanup, a fence was constructed around the site; drums, sludges, and liquids which were considered hazardous wastes were removed from the site. The cleanup was completed on April 21, 1982. On June 16, 1982, the U.S. Environmental Protection Agency's legal office recommended initiation of a civil action against Drake Chemical and against the president of that company.

Following the completion of the immediate removal operation, the concerns of the former employees of the plant and the citizens of Lock Haven centered on possible health hazards as a result of certain chemicals which Drake manufactured. The major chemical identified was Beta-napthalamine, a known bladder carcinogen.

Concerned that working in the chemical plant was health threatening, former employees formed a citizens group called Citizens and Laborers for Environmental Action Now (CLEAN) to put pressure on environmental agencies to set up health screenings for affected persons and to identify other potential hazardous waste sites which may be located in Clinton County.

In a public meeting held in May 1983, it was concluded that the Center for Disease Control, in cooperation with the Pennsylvania Department of Health, would conduct a statistical analysis to support the need for a large-scale health screening for former employees of Drake. The proposed health study will include a statistical breakdown of rates pertaining to cancer, birth defects, and other disease associated with the chemicals.

The cleanup of each uncontrolled hazardous waste site presents different environmental, operations, and health problems. In this report, an attempt was made to provide an insight into the history, the different phases of cleanup, and problems which are encountered during cleanup activities.

REFERENCES

1. Comprehensive Environmental Response, Compensation, and Liability Act of 1980. 1980. Enacted by P.L. 96-510, December 11, 1980, Stat. 2767; 42 USC 9601 *et seq.,* Amended by P.L. 97.272, September 30, 1982.
2. U.S. Environmental Protection Agency, 1982. "Superfund, What It Is, How It Works." Office of Solid Waste and Emergency Response. (WH-562A). Washington, D.C. 20460.
3. LaGrega, M.D., Gilardi, E.F., and Campbell-Loughead, J., "Remedial Action Master Plan—Drake Chemical Company Site, Lock Haven, PA," Prepared for the U.S. Environmental Protection Agency (Roy F. Weston, Inc.), December, 1982. Revised January 21, 1983.
4. U.S. Environmental Protection Agency. 1982. Environmental Response Team, Edison, NJ, "Extent of Contamination, Drake Chemical Company, Lock Haven, PA." March 1982.
5. Bixler, Brint., B. Hanson, G. Langner. Planning Superfund Remedial Actions. *In: Management of Uncontrolled Hazardous Waste Sites.* Library of Congress Catalog No. 82-0834820, Hazardous Materials Control Research Institute, 1982.
6. Pennsylvania Emergency Management Agency, *Clinton County Emergency Response Plan.* Drake Chemical Waste Disposal Site, March 5, 1982.
7. National Institute for Occupational Safety and Health. Interim Report HETA 81-248. Drake Chemical, Inc., July 1981. U.S. Department of Health and Human Services, Cincinnati, Ohio 45226.
8. Bihle, R.L., LaGrega, M.D. and Gilardi, E.F., "Clean-up of the Drake Chemical Site at Lock Haven, Pennsylvania" in *Toxic and Hazardous Wastes: Proceedings of the Fifteenth Mid-Atlantic Industrial Waste Conference* (M.D. LaGrega & L.K. Hendrian, Eds.), Ann Arbor Science Publishers, 1983.

Hazardous and Toxic Wastes: Technology, Management and Health Effects. Edited by S.K. Majumdar and E. Willard Miller. © 1984, The Pennsylvania Academy of Science.

Chapter Sixteen

Pennsylvania's Control Program for Hazardous Waste

Gary R. Galida
Chief, Division of Hazardous Waste
Pennsylvania Department of Environmental Resources
P.O. Box 2063
Harrisburg, PA 17120

The Commonwealth of Pennsylvania is today and will remain for the foreseeable future one of the largest generators of hazardous waste in the nation. The effects of past illegal or inadequate management of hazardous waste that have resulted in contamination or pollution of the Commonwealth's air, water and land resources have proven the need for a comprehensive legislative, regulatory and enforcement program to control hazardous waste.

It must be understood initially that Pennsylvania's problem with hazardous wastes is not different from that experienced by any other state in the United States and that problem is reasonably easy to define. Illegal or inadequate storage, treatment and disposal activities exist in the Commonwealth. There is a shortage of suitable facilities to properly manage Pennsylvania's wastes, and the Commonwealth is experiencing increasing public opposition to the location of new facilities that are necessary to curtail both illegal disposal and to upgrade our existing facilities.

In order to solve this problem, Pennsylvania has had to develop and bring together the legal, scientific, fiscal and personnel tools necessary to attack the consequences of yesterday's uncaring or unknowing practices, today's ongoing and complex needs for safe and proper activities and tomorrow's necessity for resolving the siting of future facilities.

NATIONAL EFFORTS

This problem has been attacked nationally by Congress through the enactment of the Federal Resource Conservation and Recovery Act, which establishes as the final phase of pollution control a nationwide comprehensive program for the control of hazardous wastes from the point of generation to the point of disposal. That program as it has been devised by Congress, is a national program with opportunities for the states to participate. States may participate if they can demonstrate that their program is substantially equivalent, and only through that mechanism can they acquire authorization to control hazardous wastes within their boundaries. But it must be remembered that this program is a Federal responsibility until states are so authorized and that, once enacted, any federal standard preempts any less stringent state requirement. States attempting to control hazardous waste must do so in a manner that is always mindful of the fact of federal oversight and possible preemption of state programs and enforcement authority.

Federal authorization is both complex and confusing as it is currently structured. This multi-phase authorization has caused states to expend considerable resources in the exercise of applying for multiple components over a period of years. If states are successful in this effort they are allowed then to administer the control program and to receive assistance and funding for those efforts.

The Commonwealth of Pennsylvania was successful in 1981 in being granted the first phase of Federal authorization, to administer a program of regulation, inspection, monitoring and enforcement for generators, transporters and storage, treatment and disposal facilities. Subsequently, the Commonwealth sought phase two authorization for the sole authority to permit hazardous waste facilities; and lastly the Commonwealth is seeking final authorization to continue to conduct these programs in a manner that is nearly identical to Federal requirements as they have evolved over a period of years.

With the achievement of authorization, the Commonwealth is demonstrating that Pennsylvania has established a control program that will allow the continued receipt of federal funding that currently provides approximately 50 percent of the resources necessary to manage hazardous waste in Pennsylvania.

LEGISLATION

Pennsylvania's Solid Waste Management Act is one of the most comprehensive and effective pieces of legislation in the United States. The Act provides for the planning and regulation of solid and hazardous waste storage, collection, transportation, processing, treatment and disposal; provides for regulation of the management of municipal, residual and hazardous waste; requires permits for operating hazardous waste storage, treatment and disposal facilities and solid waste processing and disposal facilities and licenses for the transportation of hazardous waste; and provides remedies and prescribes civil and criminal penalties for violations of the Act.

For those engaged in hazardous waste activities, the Act establishes requirements for: (a) maintenance of records; (b) proper labeling and use of containers; (c) furnishing of information on waste composition; (d) use of a manifest system of waste tracking; (e) transport only to shipper designated permitted facilities; (f) periodic reporting to the Department; (g) compliance of transport activities with State and Federal Department of Transportation requirements; (h) development of contingency plans; (i) maintenance of operation and financial responsibilities to prevent adverse effects on the public health and safety and the environment; (j) notifications and containment and clean-up of spills; and (k) design, construction and operation of facilities in compliance with all applicable environmental statutes.

Working under the broad provisions of the statute, the Department of Environmental Resources has structured a program which provides for the identification and listing of hazardous waste produced in Pennsylvania and has established standards and requirements for all those who generate, transport, store, treat or dispose of hazardous waste, and is enforcing penalty provisions equivalent to all national requirements and in excess, in many instances, of Federal requirements.

RULES AND REGULATIONS

Beginning no later than eight days after the signing of the Act into law, rules and regulations were adopted by the Environmental Quality Board in 1980 which identify the characteristics of hazardous waste and list particular hazardous waste for which all subsequent regulatory requirements were to be established.

Facing Federal preemption at the end of 1980, additional rules and regulations were adopted by the Environmental Quality Board establishing standards for those who generate, transport, store, treat or dispose of hazardous waste as follows:

A. *Section 260* established definitions and the procedures by which a generator of hazardous waste can demonstrate that a particular waste does not meet the characteristics or contain the constituents for which that category of waste was listed as hazardous in Section 261 "Criteria, Identification and Listing of Hazardous Waste".

B. *Section 262* established requirements for generators of hazardous waste including notification of activities, determinations of whether particular wastes are classified as hazardous, compliance with the manifest system, pre-transport packaging and labeling requirements, requirements for accumulations, recordkeeping and reporting requirements, exception and quarterly reporting responsibilities under the manifest system, responsibilities for discharges and spills, contingency plan requirements and responsibilities for farmers disposing of pesticides and containers.

C. *Section 263* established requirements for transporters of hazardous waste including notification of activities, licensing, bonding and insurance requirements, compliance with the manifest system, limitations on blending and mixing of waste, responsibilities for spills and discharges, safety requirements and compliance with State and Federal Department of Transportation standards.

D. *Section 264* established requirements for new hazardous waste storage, treatment and disposal facilities including notification of activities, waste analyses requirements, security and inspection standards, standards for personnel training and preparedness and prevention, handling requirements for ignitable, reactive and incompatible waste, contingency plan and emergency procedure requirements, compliance with the manifest system and recordkeeping and reporting requirements.

E. *Section 265* established requirements for existing storage, treatment and disposal facilities equivalent to those contained in Section 264, and in addition, established requirements for groundwater monitoring, closure and post closure care, financial responsibilities, and site specific requirements for the operation of land disposal, land treatment, incineration, physical, chemical and biological treatment, surface impoundments, tanks, basins and waste piles. This section also establishes the initial phase of permitting requirements including the sequence and content of applications.

F. *Section 267* established the requirements by which notification of generation, transportation, storage, treatment and disposal activities are to be accomplished.

On the basis of the requirements of these rules and regulations, the Act and the Department's program, the Commonwealth was granted the first phase of Federal interim authorization and with it the authority to solely conduct a control program for hazardous waste.

The Department's control program included (a) receipt and processing of notifications of activities; (b) receipt and processing of applications for permits for storage, treatment and disposal facilities; (c) operation of a manifest system of waste tracking; (d) conduct of compliance monitoring and inspection of generators, transporters and storage, treatment and disposal facilities; (e) investigation of spills, discharges and emergencies; (f) enforcement of rules and regulations for generators and transporters and operational standards and requirements for storage, treatment and disposal facilities.

Though that control program was effective for existing facilities and activities, delays in Federal rulemaking for permitting requirements resulted in a national moratorium on the construction of new facilities until such standards could become developed and effective.

Since the time following adoption of Pennsylvania's requirements, numerous revisions of preemptive federal standards were enacted. In addition, new federal standards for the design, construction and permitting of hazardous waste storage, treatment and disposal facilities were adopted and became effective during 1981.

Comprehensive revisions to the rules and regulations for the generation, transportation, storage, treatment and disposal of hazardous waste were adopted by the Environmental Quality Board during 1982. These revisions incorporated all changes to Federal standards enacted since 1980 and established the design, construction and permitting requirements for surface impoundments, tanks and containers, waste piles, incinerators, physical and biological treatment facilities and landfills. These revised rules and regulations made the Commonwealth eligible to achieve Phase II Authorization and able to accomplish the upgrading, permitting or closure of more than 700 existing hazardous waste facilities.

PERMITTING OF FACILITIES

Owners and operators of all facilities that manage hazardous waste are required to obtain a permit for those operations, and in obtaining a permit, they are required to meet new federal and state standards through either the upgrading of an existing facility or the construction of a new facility to meet those standards.

When Congress enacted the federal statute, it realized that not all facilities could be permitted simultaneously. Consequently, Congress established the mechanism of "interim status" which allows existing facilities to continue operation while administrative action is undertaken on applications which have been submitted.

Owners and operators apply for permits in two distinct phases: Part A applications, which were filed with the Department and the federal government in 1980 specify what activities are being undertaken, which wastes are handled and which facilities are in operation. Part B applications which were filed with the Department in 1982 and 1983 include the specific engineering, design and construction submission for the permitting of that facility.

Because of the complex system that has been established at the federal level and must be carried over at the state level to receive and process applications and to conduct public participation, a storage facility would require as long as six months to receive a permit, a treatment facility as long as nine months, and a disposal facility as long as 12 months, and beyond.

The procedures for permit issuance have basically been set by the federal government and have been supplemented by requirements of the Pennsylvania Solid Waste Management Act. Because of this, permitting of facilities is both a lengthy and an involved process. This process can range from pre-application conferences to receipt of applications in multiple phases, public notification, local government review, publication of draft permits, public comment and review, public hearings, and finally, appeals through the court system for any permit which may be granted by either the state or the federal government. This is a lengthy process that will take Pennsylvania a period of years to accomplish, but it must be accomplished if Pennsylvania is to achieve the upgrading of its existing facilities and the construction of new facilities.

ENFORCEMENT AND MONITORING

Pennsylvania's program for licensing and bonding of transporters has been successful under the provisions of the Solid Waste Management Act. Pennsylvania standards are separate from federal requirements, and no such federal requirements exist.

Pennsylvania's statute requires that all transporters must obtain a license on or before July 7 of 1982. More than 500 applications were received before that time from transporters from as many as 27 different states. All of those applications were reviewed and processed. Bonds totalling more than eight million dollars and insurance and other requirements were assessed and more than 200 licenses were issued before that date. 113 licenses were denied and more than 155 applicants withdrew from the licensing system because they could not or would not meet Pennsylvania's requirements.

In the first two years of administering the Commonwealth program, over 4500 inspections of generators, transporters and facilities were completed. More than 900 notices of violation, administrative orders and civil penalty actions were issued during the period, and compliance of all entities regulated exceeded 75 percent. Many of the remainder are under corrective action schedules.

Under the operation of Pennsylvania's manifest system, more than 300,000 manifest documents were distributed within the first two years since the system took effect in November of 1980. More than 75,000 individual manifest documents, together with quarterly reports, which detail waste characteristics, volume, where the waste was generated, transported and where it was finally received, are processed annually. Moreover 200,000 shipments of hazardous waste were tracked by the end of 1982. DER data has been computerized not only for enforcement purposes, but also to serve as the inventory of waste produced and managed to be used in the development of Pennsylvania's Hazardous Waste Facilities Plan.

For disposal facilities or any facility at which waste will remain after closure, groundwater monitoring requirements and systems have been implemented within the Commonwealth. More than 160 facilities were initially identified as potentially needing such groundwater systems. Sixty were eliminated as having to meet such requirements because of their own actions or because of subsequent changes in industrial processes. All of the remainder of those facilities currently either have adequate groundwater monitoring systems in place or are undergoing the last measures to get those systems in place.

Spill discharge and response to emergency is an additional program that the Commonwealth undertakes, and within any six month period more than 30 such spills and discharges are reported and responded to. In Pennsylvania there are more than 4000 entities that either generate, transport, store, treat or dispose of hazardous waste: 3200 are generators, over 200 are transporters and 700 are existing facilities. This seemingly large regulatory universe actually represents a considerable decrease since 1980. At the time Pennsylvania's program took effect, more than 5,500 entities speculated that they either generated, transported or managed hazardous waste. Changes in industrial processes, changes in wastes and methods handled have decreased that number by at least 1,500 during that period.

By increasing the level of staffing to approximately 200 personnel a decentralized program which allows regional administrative and technical staff to accomplish permit reviews, inspection monitoring and enforcement activities and to assist industry, local government and private citizens has been established. In addition, environmental assessment reviews of permit applications in response to the requirements of Article I, Section 27 of the Pennsylvania Constitution have been established and local government reviews of permit applications according to the Act's requirements were initiated even before the effective date of the Act.

In cooperation with the Department of Justice, a special investigative surveillance and enforcement effort against illegal activities has been undertaken since October of 1980. In the first two years under that program, 140 investigations were opened. Twelve cases involving 27 defendants were filed. Nine of those, against 24 defendants, were solved, and three remained in the court system. Total recovery during the first two year period included some $1,500,000 in fines, imprisonment of two executives and conviction with sentence pending against four additional defendants.

HAZARDOUS WASTE FACILITIES PLAN

In the future, problems that remain to be resolved include the siting of facilities in Pennsylvania. As indicated previously, there is a shortage of facilities throughout Pennsylvania and, in fact, throughout the industrial Northeast. It is not unusual for Pennsylvania waste to be transported hundreds of miles to the few sites that exist here and in other parts of the country. Pennsylvania's legislation gives it the tools necessary to safely permit and regulate and establish the new facilities that are needed. Together with these siting efforts, Department staff are working with industry daily to achieve more reuse and recovery and reduction in the kinds and amounts of hazardous wastes that are produced.

In addition to a comprehensive regulatory control program, the Solid Waste Management Act provides three mechanisms to establish the facilities necessary to manage Pennsylvania's hazardous waste. It provides that treatment and disposal facilities may be located, designed, constructed, and operated through a) the issuance of Departmental permits for construction and operation of facilities whose locations are consistent with applicable zoning and land use controls; (b) the issuance of Departmental permits for construction and operation of facilities which are to be located on leased, unused state lands; and c) the issuance of Departmental permits for construction and operation of facilities which have been issued a superseding "Certificate of Public Necessity" by the Environmental Quality Board for the location of facilities not in compliance with zoning, land use, or other controls.

In order to determine the number and type of facilities necessary to manage Pennsylvania's hazardous waste and to provide for the establishment of such facilities, the Act establishes the Department's power and authority to develop, prepare, and modify the Pennsylvania Hazardous Waste Facility Plan. Conformance to the Facility Plan is a requirement for the issuance of a "Certificate of Public Necessity".

In order to assist in development of the Plan, the Act required the formation of the Hazardous Waste Facilities Planning Advisory Committee. The Committee was to be composed of (1) private citizens, (2) public officials, (3) public interest groups, (4) organizations of substantial economic interest, including a waste treatment operator, a waste generator, local governments, environmentalists, and an academic scientist.

These committee appointments were made and the committee became functional in December, 1980. The committee was "charged" with the preparation of the Pennsylvania Hazardous Waste Facilities Plan. The committee, with the Department's concurrence, chose the assistance of a private consultant to assist in preparation of the plan.

The Act provides that the Pennsylvania Hazardous Waste Facilities Plan is to include:

1. Criteria and standards for siting hazardous waste treatment and disposal facilities.
2. An inventory and evaluation of the sources of hazardous waste concentration within the Commonwealth including types and quantities of hazardous waste.
3. An inventory and evaluation of current hazardous wastes practices within the Commonwealth including existing hazardous waste treatment and disposal facilities.
4. A determination of future hazardous waste facility needs based on an evaluation of existing treatment and disposal facilities including their location, capacities and capabilities, and the existing and projected generation of hazardous waste within the Commonwealth and including where the Department within its discretion finds such information to be available, the projected generation outside the Commonwealth of hazardous wastes expected

to be transported into the Commonwealth for storage, treatment or disposal.
5. An analysis of methods, incentives, or technologies for source reduction, detoxification, reuse and recovery of hazardous wastes, and a strategy for implementing such methods, incentives, and technologies.
6. Identification of such hazardous waste treatment and disposal facilities and their locations (in addition to existing facilities) as are necessary to provide for the proper management of hazardous waste generated within this Commonwealth.

The first product of the Facilities Plan involved the development of environmental, social and economic criteria for the siting of hazardous waste facilities, where were proposed during 1981, which were the subject of public hearing, comment and revision, which subsequently were rewritten into rule and regulation form and were proposed as rulemaking under Pennsylvania's hazardous waste requirements. These standards establish considerations of concern to local government and private citizens which have never before been considered in the permitting of hazardous waste facilities.

As proposed the siting regulations are divided into two provisions, or "phases" (the term originally used in the Department's preliminary criteria). Phase I sometimes known as the "exclusionary criteria", prohibits the siting of certain kinds of hazardous waste treatment and disposal facilities in ten areas. These areas are coastal and riverine flood areas; coastal and fresh water wetlands; oil, gas and coal bearing areas; carbonate bedrock areas; national landmarks and historic places; dedicated lands in public trust; and agricultural lands. When promulgated, the Department will have to deny permits to applicants wishing to locate in the areas described in this provision.

The second provision requires additional information and assessments from those applicants located within certain types of areas. These areas are not so risky that they are exclusionary, but they are risky enough locations to put hazardous waste treatment or disposal facilities that the applicant must make special assessments to show how and why his facility overcomes the risks normally present in these areas.

These cautionary areas are defined by geologic, hydrologic, soils, mineral, seismic, land use, transportation, population density and economic factors, and also by the list of environmental resources previously developed in satisfaction of Article I, Section 27 of the Pennsylvania Constitution. The existence of any of these factors at the proposed facility location would not necessarily cause the Department to deny a permit, but it would cause DER to require additional information to be submitted by the applicant.

In some cases, this additional information may show that the risks usually associated with such a site don't exist or have been overcome through design engineering. In other cases, the additional informaion may show that the permit must be denied because the risks cannot be reduced to an acceptable minimum.

Finally, in still other cases (economic factors, for example), the additional information will probably not affect DER's permit decision, but will inform the public and the Department what impact the proposed facility will have on the local community.

The remaining tasks of the Facilities Plan have disclosed that:

(a) Pennsylvania industries generate 5.6 million tons of hazardous waste annually and these industries manage 4.6 million tons annually on-site. The primary metal products industry generates 53% of the total. Chemical and allied products generate 35%, and the electric and electronic equipment industry accounts for 8% of the total. All the others are very small generators in comparison.

(b) Of the seventeen existing commercial hazardous waste facilities within the Commonwealth eight are expected to continue in existence; five facilities ceased operation during the time the plan was being written and four facilities may not continue operations. These facilities manage 1.3 million tons of hazardous waste annually including approximately .5 million tons from generators outside the Commonwealth. Lastly .3 million tons of waste produced in Pennsylvania is exported to as many as twenty five states each year.

(c) It's projected that by the year 2001 waste production will increase from 4.6 to 6.3 million tons managed on-site and from 1.3 to 1.5 million tons managed off-site.

(d) In addition to current recovery, treatment and detoxification and land disposal technologies within Pennsylvania alternative technologies which are applicable to wastes produced within the Commonwealth include solvent and metals recovery, incineration, aqueous waste treatment, energy recovery, land farming, deep well injection and above ground secure long-term storage.

(e) In the near term the Commonwealth faces a capacity shortfall of more than .5 million tons annually in aqueous treatment, solvent recovery and land disposal in eastern Pennsylvania and aqueous treatment and solvent recovery in western Pennsylvania and fuel blending and incineration statewide.

The Hazardous Waste Facilities Plan once adopted becomes the basis for the issuance of "Certificates of Public Necessity" by the Environmental Quality Board. The Board judges a proposed facility's conformance to the plan, its effect upon host and neighboring municipalities and many other considerations evaluated through public hearings and public comment. Once issued, if issued, the certificate will allow a facility to suspend and supersede any existing local restriction which would otherwise prevent that facility from becoming operational.

With this plan complete, Pennsylvania faces what will be one of the most serious environmental challenges of this century. That is, how does the Commonwealth convince citizens and local government that a facility should be located in their county, in their municipality, and near their own homes.

If the reasons for the current public concern, fear and opposition to the siting of facilities which include a heightened awareness based on government and media disclosures of workplace hazards, illegal or inadequate practices, pollution incidents and cleanup activities of past abuses are little understood, even less understood are actions necessary to resolve and overcome such opposition.

Current efforts towards public education, sound regulatory and enforcement programs, public participation in siting and permit decisions and appeal rights and remedies for those unsatisfied with permitting decisions can go far to resolve public concern but no state or political jurisdiction has to date conceived either a complete or successful solution to the siting dilemma.

SUPERFUND EFFORTS

The Commonwealth's control program for managing today's permitting, monitoring and enforcement problems together with tomorrow's siting difficulties is lastly balanced by Pennsylvania's efforts to provide cleanup of and remedial action for yesterday's abuses. The Federal Comprehensive Environmental Response, Compensation and Liability Act of 1980 (Superfund) was established to help address problems caused by past inadequate hazardous waste management practices and Pennsylvania's legislation enables full participation in this effort.

Superfund calls for a 1.6 billion dollar fund over five years to be used for the payment of response costs, remedial actions, and damages to natural resources caused by the release of hazardous substances. The law also authorizes temporary evacuation and housing of people threatened by such activities and incidents. Pennsylvania has identified thirty multi-million dollar hazardous waste cleanup projects that today or in the future will qualify for aid under this legislation. More than 10 million dollars has been committed to such cleanup activities through 1982 and state funding to authorize an additional 30 million dollars in Federal and matching state funds was provided during 1983. These efforts represent a reasonable start in contending with this serious problem, but EPA estimates that nationwide 22 to 44 billion dollars will be needed to address past hazardous waste disposal areas and spill problems. EPA and the Department of Environmental Resources are currently investigating as many as 500 potential hazardous waste problem sites in Pennsylvania alone.

CONCLUSIONS

Pennsylvania has established a sound, comprehensive and workable control program for the inspection, monitoring, enforcement, siting and permitting of hazardous waste facilities and activities.

That program recognizes and understands that the need to control hazardous waste is everyone's problem and that industry, government and citizens all have critical responsibilities for solving this problem, and all must work together to achieve that goal.

That program also recognizes and understands that regulatory efforts must be combined with, and enhanced by, education of all concerned, participation and responsible decision making by all concerned, and scientific, technological and sociological initiatives that in many cases must be begun by those outside the Department who are directly affected by the magnitude of this problem.

Lastly, that program provides a sound legal and technical foundation upon which those regulated or affected by the control of hazardous waste may build the science, the engineering, the commerce and the government necessary to allow Pennsylvania to meet its goals and insure that future generations do not inherit the environmental problems and consequences that must be faced today.

The Commonwealth's continuing efforts to remove pollutants and contaminants from the air and the water where they can be a source of direct contact to the public must be accomplished recognizing that once removed, these pollutants must also be ultimately managed so that they are not, in the future, discharged back into the same source from which they were originally removed. If the loop on pollution is not closed by safely managing the residues of pollution control, then the problem has not been solved. Only its consequences have been postponed.

Hazardous and Toxic Wastes: Technology, Management and Health Effects. Edited by S.K. Majumdar and E. Willard Miller. © 1984, The Pennsylvania Academy of Science.

Chapter Seventeen

Federal And State Regulation Of Hazardous Wastes In Kentucky

Ronald R. Van Stockum, Jr.
Attorney at Law
Suite 200, 539 W. Market St.
Louisville, Kentucky 40202

RCRA, the Resource Conservtion and Recovery Act, was enacted in 1976 as Public Law 94-580 and is presently codified as 42 USC 6901 et seq. A last-minute cut and paste job during an election year week-end in 1976 led the United States House of Representatives to pass the bill 396-8 and the Senate to follow suit two days later. President Gerald Ford signed it into law on October 21, 1976. It was amended in 1978 by the Quiet Communities Act (PL 96-482). New amendments to the law are currently being debated.

In Kentucky the need for this law had been acutely felt. The infamous "Valley of the Drums" containing 20,000 barrels of hazardous waste outside of Louisville, Kentucky was touted along with New York's "Love Canal" as reason for quick federal action concerning the improper disposal of hazardous waste. In the spring of 1977, Louisville's new $55 million Morris Forman Waste Water Treatment Plant was contaminated by illegally disposed pesticide procursors hexachlorocyclopentadiene (hexa) and octocholorocyclopentiene (octa). Employees at the plant became ill and the entire plant shut down for six months while it was cleaned at a cost of $2 million. As if that spill weren't enough, on February 13, 1981 a spill of hexane was dischared into the sewers of Louisville, Kentucky causing two miles of sewer lines and streets to literally explode.

Kentucky was already faced with a questionable nuclear waste disposal site and numerous truck and train wrecks causing evacuation of neighborhoods. When the federal law was enacted along the lines of the Federal Water Pollution Control Act, it gave the states the power to enforce their own laws in its stead. Kentucky quickly jumped into this regulatory arena through enactment of various statutes in Kentucky Revised Statutes (KRS) Chapter 224. Since the Kentucky law closely parallels the federal law the following discussion will highlight the elements of the federal law.

RCRA contemplated a regulatory program larger than any that EPA had undertaken before. It's "cradel-to-grave" regulation of hazardous waste was to leave no element of its handling and disposal unregulated. But it is important to note that RCRA does not ban the production of hazardous waste. The generators of hazardous waste perform a major informative step in the regulation and must see to it that all waste that they generate is properly disposed. Congress deals with prohibition through the Toxic Substances Control Act (TOSCA, 15 USC 2601 et seq) and the Federal Insecticide, Fungicide and Rodenticide Act (FIFRA, 7 USC 135 et seq).

The core of the RCRA regulatory program was to be promulgated within 18 months of the law's enactment in 1976. Because of the immense "universe" of waste to be regulated, EPA missed its statutorily set deadline. It was only after a lawsuit by the State of Illinois and environmental groups (*Illinois v Costle*, 12 ERC 1597, DDC 1979) that the first regulations were promulgated. These initial sets of regulations were filed on May 10, 1980 but were only the beginning of the flood of regulations to impact industry with the RCRA mandate.

The regulations themselves are difficult to read and understand. Fortunately, when they were first published in the Code of Federal Regulations (CFR) they contained preambles or explanatory notes. It is essential for any reader in the subject to obtain these explanatory notes and read them in conjunction with the regulation.

There are nine subtitles to RCRA. Because the law deals extensively with solid waste, not all of its sections deal with hazardous waste. Three subtitles are important:

Subtitle A-General provisions (§1001-1008)
Subtitle C-Hazardous Waste Management (§3001-3013)
Subtitle G-Miscellaneous provisions (§-7001-7009)

Chapter 40 of the Code of Federal Regulations (CFR) contains numerous parts which deal with various aspects of regulations promulgated by the Environmental Protection Agency (EPA) pursuant to RCRA.

Part 260-General
Part 261-Identification and listing of hazardous waste
Part 262-Generator standards
Part 263-Transporter standards
Part 264-Treatment, storage and disposal (TSD) facility standards
Part 265-Interim status TSC facility standards
Part 266-Interim status land disposal standards
Part 270-Permits
Part 124-Decision making

Although the Code of Federal Regulations does not contain the explanatory preambles of the Federal Register, many citations to the Federal Register are given in the CFR text.

The definition of the term "solid waste" includes both liquid and gaseous waste (40 CFR §261.2). The most important element of these definitions is that hazardous waste must first qualify as a solid waste.

There are many twists in the definition of solid waste. For example, garbage, refuse and sludge are always assumed to be discarded and are always solid waste. Other waste material becomes solid waste only if *discarded*. The term "discarded" contains two very important exclusions which remove the following materials from the definition of hazardous waste.

1) materials which are "used, re-used, reclaimed, or recycled." (40 CFR §261.2(c)).

2) Materials "being burned as a fuel for the purpose of recovering usable evidence."

Thus hazardous materials which otherwise meet the definition of hazardous waste are excluded on the basis of not being solid waste.

Part 261.4 lists further exclusions of waste material from the definition of solid waste:

1. Domestic sewage
2. Clean water act point sources (NPDES)
3. Irrigation return flows
4. Nuclear by-products
5. In-situ mining waste

Part 261.4 also lists certain solid wastes which could normally fall into the category of hazardous waste but are excluded by regulation.

1. Household waste
2. Farm agricultural and animal waste used as fertilizers.
3. Mining overburden
4. Fossil fuel waste from power plants.
5. Drilling fluids
6. Certain chromium wastes.

Hazardous waste is defined as a solid waste which is covered by one of the four characteristics or is found on any of four lists. The four characteristics and a complex toxicity test determine which compounds are found on the lists. Additionally, there is an Appendix VIII to Part 261 which contains numerous compounds which have been shown to be toxic, carcinogenic, mutagenic or tetrogenic and which, if present in a waste, will cause the entire waste to be listed. The four characteristics and four lists will be discussed in a more thorough manner below.

The characteristic of *ignitability* is applicable to a solid waste which in liquid form has a flash point less than 140 °F or if a solid, "burns so vigorously and persistently that it creates a hazard when ignited." Compression gases can also be termed ignitable (40 CFR §261.21).

The characteristic of *corrosivity* is applicable to a solid waste which has "a Ph less than or equal to 2 or greater than or equal to 12.5 or corrodes steel at the rate greater than 6.35 mm per year at 130°F" (40 CFR §261.22).

The characteristic of *reactivity* is the least specific test of all. It is linked to such qualitative standards as "undergoes violent change without detnoating" or "reacts violently with water" (40 CFR §261.23). In my experience as a environmental prosecutor in the State of Kentucky, this was the most difficult standard to enforce.

The characteristic of *EP toxicity* is determined by using a specified extraction procedure to see if any of 14 metals or pesticides will leach out of the waste (40 CFR §261.24).

The four lists of hazardous wastes contain those wastes which contain the toxic constituents found in Appendix VIII of part 261; exhibit any of the four characteristics previously discussed; or meet the definition of acute toxicity which is found in the definitional section (40 CFR §261.11).

The first list is labelled *"Hazardous Wastes from Non-specific Sources."* On this list, thirteen generic groups of compounds used in a variety of industrial processes are listed and given "F" numbers (40 CFR §261.31). For example, F004 lists, "the following spent non-halogenated solvents; cresols and cresylic acid and nitrobenzene; and the stillbottoms from the recovery of these solvents."

The second list is labelled *"Hazardous Wastes from Specific Sources"* (40 CFR §261.32). On this list there are 87 specific types of waste generated by eleven general categories. These compounds are given "K" numbers. For example, K062 is found under the "iron and steel" category and is described as "spent pickle liquor from steel finishing operations."

Elements on the non-specific and specific sources lists are included because they contain hazardous constituents found in Appendix VIII. Appendix VII tells you specifically which of those constituents are found in each "F" or "K" category. In the examples used above, Appendix VII lists "cresols and cresylic acid, nitrobenzene" for "F004." For K065 Appendix VII lists "hexavalent chromium, lead."

The final two lists are pure chemical products which have been discarded, are off specification or spill residues (40 CFR ᴷ261.33(e) and (f)). These compounds are not specifically listed if in a mixture and such circumstances are dealt with in the so-called "mixture rule" (CFR§261.3).

Those wastes which show *acute* toxicity as defined in the definitional section (40 CFR §261.11) are given "P" numbers and listed in the (e) paragraph of 40 CFR §261.33. As of the date of this publication, there are 123 of them.

The (f) paragraph of 40 CFR §261.33 contains the fourth list. It contains those compounds identified as *toxic* wastes (Appendix VIII wastes) or exhibits any of the four hazardous waste characteristics. These compounds are given "U" numbers and as of the date of this publication total to 247.

Wastes are continuing to be added to these lists as test data is generated on more and more substances. Some of these wastes have been delisted as more evidence is gathered (40 CFR §§260.22). For example,one original category included in the specific source list was "leather tanning finishing." Seven separate "K" numbers were listed for various chrome wastes generated by the industry. After reviewing industry petitions to delist the waste, a distinction was made whereas only hexavalent chromium wastes were to be listed. The tanning industry produces tri-valent chromium wastes and, therefore, the entire "Leather tanning finishing" category was dropped from the list. The timing of this delisting was of particular interest to the author who had filed the first hazardous waste action in Kentucky on the basis of the original listing. Needless to say, that aspect of the suit was quickly settled.

The mixture rule was first discussed in the May 19, 1980 regulations (45 FR 33095). It states that a solid waste which contains a mixture of non-hazardous and listed hazardous wastes is considered hazardous in its entirety unless the listed hazardous waste was listed on the basis of meeting one of the hazardous waste characteristics. In the latter situation, the solid waste is only deemed to be hazardous waste if the entire mixture exhibits one of the characteristics. If the mixture contains waste listed on the basis of toxicity or acute hazardness then the entire mixture is deemed hazardous waste. Through 40 CFR §260.22 industry may petition the EPA to delist specific mixtures which do not exhibit toxic or acute hazardous characteristics as a whole, even though the mixture contains hazardous waste listed because of those criteria.

It is interesting to note that a mixture of most hazardous wastes into wastewater subject to §§402 or 307 (b) of the Clean Water Act does not result in the entire mixture becoming hazardous waste if the hazardous constituent is generated from "de minimis" losses or lower in concentration than that specified in 40 CFR §261.3(a) (2) (iv).

Two small sections of the hazardous waste regulations have profound influence on the vast number of waste generators which fall within the "universe" of regulated waste.

The "small quantity generator" is addressed in 40 CFR §261.5. Without an exclusion addressed specifically to them, the burden and expense of complying with the hazardous waste regulations would put many small establishments out of business and swamp the EPA with paperwork. A difficult question involved the amount of hazardous waste a generator could produce in one month without breaking a threshold limit and becoming subject to full regulation. For most hazardous waste, this threshold limit was set at 2,200 pounds or less of hazardous waste a month. Generators producing less than this amount of hazardous waste must identify it as such, but need not manifest it and may dispose of it at any facility which is permitted by the state to handle industrial hazardous waste. Of course, if in any one month the small quantity generator produces more than 2,200 pounds, the entire amount becomes hazardous waste subject to full regulation under RCRA.

The mixture rule is applied to small quantity generators in a fashion which will allow them to mix less than 2,200 pounds of hazardous waste with a non-hazardous waste and still not be subject to regulation, even if the resulting mixture is greater than 2,200 pounds (40 CFR §261.5(h)). Of course, if the entire mixture meets any of the four hazardous waste characteristics, then the generator becomes subject to full regulation under the statute.

The 2,200 pound threshold limit is applicable to all hazardous wastes, except ones on the 40 CFR §261.33(e) list, which are listed for acute toxicity. For these substances, the small quantity generator threshold limit is 2.2 pounds per month.

A second short, but important, regulation deals with "special requirements for hazardous waste which is used, reused, recycled or reclaimed" (40 CFR §261.6). This provision, barely taking up one-half of one page, effectively excludes a large segment of industry which would otherwise be regulated. Yet this wide-open recycling exclusion is necessary if the intent of RCRA is ever to be carried out. Section 1003 of RCRA states:

The objectives of this act are to promote the protection of health and the environment and *to conserve valuable material and energy resources...* (emphasis added).

It is because of the need to provide an impetus and economic motive towards conservation and provide an impetus and economic motive towards conservation and recycling that the provisions of 40 CFR §261.6 were developed. This section exempts from RCRA regulation those wastes which are used, re-used, recycled or reclaimed. Considering the cost associated with RCRA compliance, it is easily seen that recycling waste makes the handling of hazardous waste much less expensive. New concepts, such as "waste exchanges," have sprung up across the country where hazardous waste is traded to those companies who can use the waste as a component of their production.

However, with what one hand gives...the other takes away. After creating this large recycling exemption, EPA closed it again. 40 CFR §261.6 states that any hazardous waste otherwise exempted by this section is still covered by RCRA regulation while it is transported or stored.

Through the exemption of 40 CFR §261.6 many high energy waste streams are being recycled, blended into fuel and otherwise being utilized as fuel for boilers and incinerators. In the 1983 Federal legislature much debate was centered on closing this exemption and regulating more fully the blending and burning of hazardous waste. One can expect the EPA in the future to examine closely whether or not the use, reuse, recycling or reclamation of hazardous waste is *legitimate and beneficial.* Without these two attributes, the utilization of these wastes becomes truly treatment and disposal in the guise of reclamation.

The Federal EPA recognized the magnitude of the task dictated by RCRA. Instant and complete regulation of the industry was not feasible. Therefore, EPA created an interim category for facilities that were engaged in treatment storage and disposal (TSD) of hazardous wastes on August 18, 1980. This category is called "interim status" and is only applicable to TSD facilities in existence on August 18, 1980 and who filed a notification of hazardous waste activity by that date. An important further criterion was the filing of the first part (part "A") of a comprehensive permit application by November 19, 1980. In August, 1980 many industries and TSD facilities were unaware that such a massive set of regulations was about to descend upon them. If they missed the August, 1980 notification date, they were disqualified from "interim status" and were operating without a permit. The harshness of this position caused EPA to allow these unpermitted industries to operate under an "interim status compliance letter" (ISCL). Through this mechanism, EPA states that these TSD facilities may continue to operate if they comply with all the regulations of a facility with "interim status." Although EPA thus binds itself, these ISCL facilities are not immune from Citizen suits for failure to obtain a permit.

It is contemplated by EPA that all of the interim status facilities will eventually be called upon to file a second part (Part B) of their permit application and be fully regulated under the final facility regulations found in 40 CFR §264. EPA estimates that it will take as many as ten (10) years to call for and review all of the interim status permits. Any new facilities must file both part A and part B of the permit application and be immediately required to comply with the more stringent final facility standard found in 40 CFR §264.

Section 3008 of RCRA details the comprehensive federal power of enforcement. Simply stated, penalties range as high as $25,000 per day for civil violations of the requirements under the act. Furthermore, a facility's permit can be suspended or revoked for such violations.

Criminal sanctions are applicable against anyone who knowingly, either,

a) Transports hazardous waste to a non-permitted facility;

b) Treats, stores or disposes of hazardous waste without having a permit or in violation of any material condition of a RCRA permit.

c) Makes false statements or representations on documents.

d) Knowingly destroys or conceals records concerning hazardous waste.

Those sanctions include up to two years imprisonment and $50,000 fine for each day of violation.

Anyone committing *knowing endangerment* shall be subject to a fine of not more than $250,000 and imprisonment for two years. *Knowing endangerment* involves violation of RCRA standards where the violator "knows at that time that he thereby places another person in imminent danger of death or serious bodily injury." The 1980 amendments include detailed language concerning state of mind, definitions and defenses in determining whether a person is guilty of knowing endangerment.

In November of 1982, the United States Justice Department created an environmental crimes unit. In the largest environmental criminal action to date, the A. C. Lawrence Leather Co. and various corporate officers and employees were charged with various violations of RCRA, Section 3008(d) (3),

42 U.S. C. 6928(d) (3). The individuals involved entered pleas and received fines totaling nearly one half million dollars. The corporation was fined $150,000, sentenced to five years probation and ordered to repay the Federal Government $238,420 it received in federal construction grants. (U.S.V. A.C. Lawrence Leather Co., et al., No. 82-0037-01-06-L and No. 83-0007-01-05-L (D.N.H. 1983).

The miscellaneous provisions section of RCRA contains several noteworthy provisions. Section 7001 is entitled "Employee Protection" and details the protection afforded informants. Section 7002 entitled "Citizen suits" allows for citizens to sue the government or industry for enforcement of the act and receive costs regardless of the outcome. Section 7003 is entitled "Imminent Hazard" and gives the EPA immediate injunctive authority to restrain conditions presenting, "an imminent and substantial endangerment to health or the environment."

Several Federal Courts have held that the "Imminent Hazard" section of RCRA (§7003) is retroactive to past disposal practices at abandoned waste sites.[1] However, this interpretation is not uniform across the nation.[2] Furthermore, a federal court in Pennsylvania held that section 7003 of RCRA is not applicable to "non-negligent past off-site generators."[3] This controversy is likely to continue until the Supreme Court of the United States speaks to the issue. It is likely that further prosecution for past handling and disposal of hazardous waste will be brought under the Comprehensive Environmental Response, Compensation and Liability Act (CERCLA) which deals with abandoned disposal sites more directly.

RCRA contains a provision similar to the Federal Water Pollution Control Act (FWPCA) which allows individual states of the nation to enforce their own hazardous waste program in lieu of the Federal program (§3006 RCRA). This transfer of regulatory enforcement will only be made to states which have enacted "substantially equivalent" programs. The transfer is accomplished in two phases described as "interim authorization." Phase I gives the states authority in dealing with identification, listing, generator and transporter standards and facility interim status standards. As of the date of this article, 36 states, including Kentucky, have been granted Phase I of interim authorization.

Phase II of interim authorization is broken down into 3 components. Component A deals with tanks, containers, surface impoundment, storage and waste piles. Component B deals with incinerators and Component C deals with land disposal facilities. The State of Kentucky has received Phase II interim authorization for Components A and B. As of the date of this article, only Mississippi has received component C of the Phase II interim authorization. The State of Kentucky has developed land disposal regulations and is awaiting Component C authorization.

Section 3006 of RCRA contemplates that states will eventually receive final authorization to enforce their program for hazardous waste regulation in lieu of the federal program. Because of the statutory language, Kentucky must qualify for and receive final authorization by January 2, 1985.

Because of language in the Federal Act, Kentucky statutes must be, and are, "equivalent" to their federal counterparts for final authorization. KRS 224.864 incorporates the Federal standard for classification of wastes in total:

The criteria and lists promulgated by the Department (Kentucky) shall be identical to any such criteria and lists proposed or promulgated by the U.S. Environmental Protection Agency pursuant to P. L. 94-580 (RCRA).

There are, however, some significant departures from the Federal Statute. KRS 224.855 (5) gives a veto power to the governing bodies of counties and cities wherein a permit is sought for land disposal of hazardous wastes. The governing body must take into account social and economic impacts, but the absolute nature of the veto is expressed by the statute which allows the governing body to take into account "psychic" costs.

KRS 224.866 is a cornerstone of Kentucky's statutory control of hazardous waste. It contains several additions to the federal scheme.

The first paragraph states that in no case shall a permit to construct or operate a hazardous waste disposal facility be issued unless it can be "integrated into the surroundings in an environmentally compatible manner."

The first paragraph of KRS 224.866 also contains this controversial language, "The Department may consider *past performance* in this or related fields by the applicant." This concept of "past performance" is unusual. It opens the door, legally, for a broad factfinding proceeding which can consider past behavior. The rules governing such an inquiry were not described. Therefore, it can be expected to generate much litigation.

The second paragraph of KRS 224.866(2) contains a ban on the land disposal of liquid, incompatible, or reactive hazardous waste in the state. Unlike the Federal law, which allows such disposal, Kentucky prohibits it until the waste is solidified. The issues concerning the landfilling of liquid hazardous waste continues to be debated at the federal level. It is interesting to note that states such as California are moving to ban the land disposal of any hazardous waste for which an economically feasible treatment process is available to render it non-hazardous.

Paragraph 3 of KRS 224.866 deals with financial responsibility. Federal regulation dictates that TSD facilities demonstrate financial responsibility for sudden accidental occurrences at the level of $1 million per occurrence and an annual aggregate of $2 million. Kentucky adds to this federal standard by demanding that disposal facilities demonstrate additional excess coverage of $10 million annual aggregate.

An interesting issue is raised by paragraph 4 of KRS 224.866. This statute provides for termination of responsibility of the owner of a hazardous waste disposal facility after 20 years. The statute states that such persons,

Shall be responsible for the post-closure monitoring and maintenance of the permitted facility for a minimum of thirty (30) years after closure of the facility and may apply *to the Department for termination of the responsibility* for post-closure monitoring and maintenance at any time after the site has been closed for at least thirty (30) years...The Department shall determine either that post-closure monitoring and maintenance of the site is no longer required in which case *the applicant shall be relieved of such responsibility.* (Emphasis added)

A question arises as to who would be responsible for a disposal facility if problems arose after thirty years and after the state has relieved the owner of responsibility. Under KRS 224.866 it appears that the state assumes responsibility after the release.

An intriguing difference between state and federal regulation of hazardous waste is found in KRS 224.868 entitled, "Special Waste." This section was originally codified in Kentucky in order to mimic a similar section in proposed federal regulations. The Federal counterpart was later amended and eliminated while the Kentucky statute remained. At first glance it appears to create an intermediate class of waste somewhere between the hazardous and non-hazardous classification yet unregulated by the bulk of hazardous waste regulations. Closer inspection, however, reveals that what appears exempt is again brought within the sphere of regulation. The interesting result of this categorization is that disposal sites of special waste are exempt from the local veto embodied in KRS 224.855. Although only certain mining and wastewater treatment wastes are specifically enumerated in the statute, other "high volume and low hazard" wastes may be designated special wastes.

There are no fully permitted land disposal sites for hazardous wastes in Kentucky.* With the memory of such disposal disasters as the Louisville sewer explosions and "Valley of the Drums" so clear, the local veto becomes an almost insurmountable barrier to the siting of a land disposal facility. Because of this situation, a statute was passed in the 1982 General Assembly setting up an "integrated waste treatment and disposal facility siting board." (KRS 224.2201, et seq.)

The siting board was given the mandate to site *one* "regional integrated waste treatment and disposal demonstration facility." This one site can be placed regardles of the local veto. The type of site to be chosen must 1) receive wastes from more than one county, 2) more than one person and must 3) utilize multiple treatment and disposal technologies which must include a secure landfill and a high technology incinerator. These requirements, plus a 5% gross receipts license fee make the "regional integrated waste treatment and disposal demonstration facility" a difficult project for which to qualify or desire.

The State of Kentucky has compiled statistical data for the production of hazardous waste in Kentucky for the year 1981. The following is a list of the four counties which produce the most hazardous waste:

---

* There are five interim status landfills, two interim status land treatment operations and two interim status surface impoundments.

County	No. of Generators	Haz. Waste Tons	% of Total Waste Produced in State
Marshall	5	8,111,067	90.41%
Carroll	3	481,190	5.36%
Hancock	4	158,677	1.77%
Jefferson	64	65,073	0.73%
State total	226	8,971,806	100.00%

Of the 8,971,808 tons of hazardous waste produced in Kentucky, 99.53% is treated or disposed of within the state. Of that quantity, almost all is rendered non-hazardous.

The following is a list of the six types of hazardous waste produced in the greatest quantities within the state.

Waste No.	Type of Waste	Haz. Waste Tons	% of Total Waste Produced
D002	Corrosivity characteristic	5,66,170	63.04
K071	Brine purification muds from the mercury cell process in chlorine production	1,845,698	20.57
D009	E. P. Toxicity-mercury	658,109	7.33
D007	E. P. Toxicity-chromium	628,664	7.00
F005	Spent non-halogenated solvents	86,059	0.59
K062	Spent pickle liquor	32,300	0.36
	from steel finishing operations	8,907,000	98.89%

REFERENCES

1 . *U.S. v Price*, 523 F. Supp. 1055 (D. NJ 1981); *U.S. v Solvents Recovery Service*, 496 F. Supp. 1127 (D. Conn., 1980); *U.S. v Reilly Tar and Chemical Corp.*, (D. Minn., 1982, 12 ELR 20954).

2 . *U.S. v Waste Industries*, 13 ELR 20286 (E.D.N.C. DEC. 1982).

3 . *U.S. v Wade*, 546 F. Supp. 785 (E.D. Pa. 1982).

Hazardous and Toxic Wastes: Technology, Management and Health Effects. Edited by S.K. Majumdar and E. Willard Miller. © 1984, The Pennsylvania Academy of Science.

Chapter Eighteen

Western European Hazardous Waste Management Systems and Facilities, A Survey

John E. McClure, Jr. B.S., M.B.A., C.P.G.
Director, Waste Management Services
GRW Engineers, Inc.
801 Corporate Drive, Lexington, Kentucky 40503

When the U.S. Congress passed the Resource Conservation and Recovery Act (RCRA) of 1976, it specified requirements for an integrated management system for the control and safe disposal of hazardous industrial and commercial wastes in Subtitle C of the Act. This law, which signaled the beginning of a major new national policy, will eventually lead toward the adoption of waste management practices in the United States similar to those used in Western Europe since the early 1960's.

These European systems incorporate several basic components: advanced waste management technology; waste reduction economics; creative financing; and regulation of wastes. This chapter will briefly examine several such systems, as represented by facilities in Denmark, West Germany and France. Other hazardous waste management facilities in these countries and Austria were also visited by the author to investigate the degree to which technological resources have been applied to the control of these wastes.

GENERAL CHARACTERISTICS OF TYPICAL EUROPEAN SYSTEMS

In general, a typical Western European hazardous waste treatment facility will include the following components:
- A plant covering approximately 20 acres (more if a landfill is on-site)
- Facilities for secure pretreatment storage of a variety of wastes, including sufficient storage for up to one year's waste input

- Facilities and equipment to:
- Thermally treat and/or destruct organic wastes
- Chemically treat and neutralize inorganic wastes
- Recapture energy be means of waste-heat boilers to produce steam or electric power
- Access to secure chemical landfill, either on- or off-site for disposal of post-treatment and post-reclamation residues
- Complete analytical laboratory facilities and modern safety and maintenance facilities
- Sophisticated plant management, security, contingency, and data management systems.

A facility may also incorporate materials reclamation equipment, such as organic solvent distillation units, oil/water emulsion separators ("crackers"), or chemical precipitation units for dissolved metals. Additionally, commercial facilities may select particular combinations of waste streams for resale to others (waste exchange) or to effect mutual neutralization through mixing (synergistic waste chemistry).

Such a facility will be the designated receiving point for wastes generated in its service area (called a "waste shed"). The facility may either be the required disposer for a region, the facility "of last resort" for a nation, a jointly used facility for a group of local industries, or a private installation serving a commercial market or a group of waste generators within a manufacturing company. In the case of a national or regional unit, environmental regulations may require all generators in the service area to use the facility; in the case of joint venture or private facilities, the environmental agencies may require the use of the facility through their permitting apparatus. In all cases, each load of waste received is identified, its origin is known, and it is generally planned for as part of normal facility operations.

With these systems, there is extensive knowledge of what type and volume of waste is being generated by whom, which techniques may be used to make hazardous waste as secure as possible, and where the generator must ship it for treatment of disposal.

The treatment and disposal plants, which form the most visible parts of these systems, are designed to serve a particular geographic area of waste generator group, and they are limited to accepting those wastes for which they are equipped. This acceptance limitation, which has been achieved through a mature regulatory system, industry/agency co-operation, and public and corporate conscientiousness, has prevented dangerous accumulations of untreated wastes (which has occurred in Elizabeth, New Jersey, Seymour, Indiana, Louisville, Kentucky and other parts of the United States) from being a problem. This cooperation and trust among government, industry, and the European public has been a key element to the generally good performance of these systems.

HOW RCRA RELATES TO EUROPEAN SYSTEMS

RCRA's required waste management practices are patterned from Western European techniques and, thus, will ultimately lead management of hazardous industrial wastes in the United States toward the European model. Resistance to new waste management practices in the United States is partially due to the speed with which this change is being imposed by the statutory and regulatory timetables of RCRA and the U.S. EPA.

Western Europe has a long history of reliance on incineration as an acceptable means of solid waste (that is, garbage) disposal. One incineration equipment supplier, VonRoll, Ltd., of Switzerland, has built a total of 157 solid waste incineration plants around the world since 1957, and 17 plants worldwide for the treatment of industrial wastes since 1963, illustrating the time which municipal and hazardous waste incineration has had to become accepted elsewhere in the world. This time span, when compared to the relatively short time since the passage of RCRA in 1976 and the promulgation of the first RCRA regulations in 1980, gives an idea of the speed at which change has been required of U.S. industry.

The technology which will ultimately be used by U.S. industry may differ from that presently in use at the Western European plants discussed in this chapter because of technological advances. However, similar systems of management will be used, since they are demonstratively effective in Europe and allow industrial production and expansion while protecting the public and the environment at the Same time.

The Western Europe systems discussed in this article describe four basic types of facilities: Kommunekemi A/S, the central treatment facility for Denmark; Sondermullplatz Mittelfranken, a regional West German facility that serves Middle Franconia (a region of Bavaria); Sidibex, a commercial French facility used by a group of private industries near LaHavre, and the in-house waste treatment facilities of the BASF chemical plant at Ludwigshafen, West Germany. The specific waste treatment and disposal techniques vary between facilities, but the general management practices used at each one are similar in many ways. Some other facilities which are listed to further illustrate regional differences and similarities technologically and operationally are: SARP Industries's commercial facility at Limay, France; the Hessische Indutriemull (H.I.M.) state owned facility at Biebesheim, West Germany; and the Entsorgungsbetriebe Simmering (E.B.S.) joint city and commercial facility at Vienna, Austria.

KOMMUNEKEMI

Danish national statutes and regulations governing industrial waste, passed in 1973 and 1974, assign a dual responsibility for waste management to the community in which the waste generator is located as well as to the generator. The mechanism for this is a requirement that the host community certify that wastes generated within its jurisdiction can be properly treated or disposed in the community's waste management system. Should this not be possible, the community and the generator are required to send any wastes beyond the capability of the community facility to a more advanced treatment facility.

Kommunekemi is located in the town of Nyborg on the island of Funen. The plant, which came on stream in 1973, was established to provide treatment for industrial and other wastes which were too difficult for community and industrial facilities.

The Kommunekemi facility was established through the creation of a non-profit public corporation, Kommunekemi A/S, which was funded by low-interest government loans; it is owned by the Danish communities. Since general Danish policy is for government enterprises to be priced competitively with private industry, the repayment period for the loans was set short enough to ensure that Kommunekemi disposal charges were competitive with those of private facilities in Europe. This also ensures the public that hazardous waste generation is not "subsidized" by low disposal costs.

Kommunekemi serves the entire country of Denmark through a system of 21 major collection substations and some 250 small collection points. These substations, located near major points of waste generation throughout the country, receive and temporarily store industrial and commercial wastes. The wastes are then shipped to the facility by rail cars owned by Kommunekemi. All wastes received at the collection stations and at the plant are accompanied by an identifying manifest similar to that required by RCRA. The wastes are sampled and tested as necessary in the Nyborg plant laboratory when they are received. Wastes are shipped in a variety of containers, from tank cars to small cans,

TABLE 1
Waste Received by Kommunekemi
in 1979

TYPE	VOLUME (tonnes)
Waste Oil	12,000
Halogenated Wastes	4,000
Solvents	8,000
Organic Chemicals	13,000
Inorganic Wastes	8,000
Pesticides	1,000
Miscellaneous	4,000
Total	50,000

although most nonbulk waste is shipped in standard steel drums. Non-organic wastes, such as metal-plating sludges, are shipped in reusable, skid-mounted containers furnished by Kommunekemi. The approximate types and quantities of wastes received by the plant in 1979 are shown in Table 1.

Waste Treatment Processes

Processes and techniques used to treat the wastes are: oil and solvent recovery, with subsequent sale of the recovered oil; blending of compatible wastes for storage and treatment; rotary kiln incineration of solid, liquid and pasteous (a European term for pastry wastes) organic wastes; and chemical treatment, neutralization and precipitation/filtration of inorganic metal-cleaning, -coating and -plating wastes. The contaminated water from the waste-oil recovery process is injected into the kiln to remove its organics and to cool the incinerator gas stream before it enteres the waste heat boiler. The primary storage capacity for organic solvents prior to incineration was approximately 6.2 million gallons in 1980.

Organic Wastes - The basic equipment used for incineration of organic wastes is a two-chamber incinerator: the primary chamber is a rotary kiln operating at up to approximately 1300 °C and the baffled secondary chamber operates at approximately 950 °C. A waste-heat boiler is in line immediately after the secondary chamber, reclaiming thermal energy (at an estimated rate of 22 metric tons [tonnes] of steam per hour) from the hot gas stream before it enters an electrostatic precipitator to clean the exhaust gases. The total residence time for the gas stream is seven seconds, with two seconds in the primary chamber. A rotary kiln was chosen for the primary chamber because of its capability for handling a wide variety of waste feedstock types, including whole steel drums of solid wastes. The average feed rate in 1978 for the kiln was approximately 5.41 tonnes per hour or about 35,000 tonnes per year.

A second, small incineration system, equipped with a caustic scrubber for exhaust gas cleanup, was being used in 1980 to burn strongly halogenated solvents at a rate of approximately 2500 tonnes per year. This unit was only capable of handling liquids through its vortex burner; however, and was scheduled to be phased out when a second, full-sized incinerator, equipped with a caustic scrubber, was brought on line in 1982.

Inorganic Wastes - The inorganic treatment unit treats spent pickle liquor (acid solutions used to clean metals) and chromate and cyanide coating and plating wastes by chemical precipitation to produce a filter cake, consisting of 42 percent calcium sulfate with a pH of 9.5. Heavy metals contained in the cake are in a stable, insoluble form suitable for deposit in clay-lined cells or pits at a nearby chemical landfill, which is also used for the incinerator slag and fly ash. The ultimate plan for the inorganic precipitate is reclamation of the stabilized heavy metals (chrome, mercury, cadmium, etc.). Therefore, landfilling of these stabilized wastes is considered a temporary storage method (until reclamation is economically feasible) rather than ultimate disposal.

Environmental Control

The environmental compatibility of the plant is monitored in a number of ways, technically and institutionally. The Director General of the Danish National Agency of Environmental Protection is on the Board of Directors of Kommunekemi. The entire plant operation is audited twice each year for the agency by an independent, third party laboratory, performing complete tests of incinerator exhaust gas content, slag and ash analysis, and other parameters. The inorganic treatment plant effluents are tested weekly for a number of parameters, including a bioassay, and the precipitation runoff from the entire plant is collected, treated and tested prior to release. Finally, and probably of most importance, the entire plant operation is conducted in an open, full-disclosure manner that has prevented undue public concern from arising over its existence. This approach, coupled with the positive benefit of the energy from the incinerator provided to the community of Nyborg in the form of district home heating through a recirculating, hot-water pipeline system, has created an atomsphere of acceptance on the part of the community which has been a key factor in the success of the Danish facility.

Facility Costs and Disposal Charges

The total capital investment in the plant, including equipment, buildings, land, and improvements was estimated at $17 million (in 1973 dollars). Using a 1.8 inflationary factor, the estimated 1980 replacement cost for the total plant was about $30.6 million. The disposal prices charged by the facility during 1979 averaged $116.55 per tonne, with drummed organic wastes in solid or semisolid form being disposed at $286.75 per tonne. These prices reflect the inclusion of transportation costs as well as direct operating and loan repayment costs. Since repayment of the original loan was near completion in 1980, the new incineration system that came on stream in 1982 replaced and raised the original loan repayment cost, thereby, ensuring that Kommunekemi's waste disposal prices still remain competitive with the private disposal industry elsewhere in Europe.

SONDERMULLPLATZE MITTELFRANKEN

The Sondermullplatze Mittelfranken (SMM) hazardous waste management facility in Schwabach, West Germany near Nurnberg was established in 1968 to serve Middle Franconia, a region of Bavaria. The facility is owned and operated by a non-profit administrative association made up of nine cities, three communities, and seven counties or rural areas in Middle Franconia and, in addition, serves parts of the districts of Upper Franconia, Upper Pfalz, and part of the federal state of Baden-Wurttemberg. The total area served is approximately

TABLE 2
Sondermullplatz Mittelfranken
Waste Treatment Processes

Method of Treatment	Volume (tonnes per year)
• Directly landfilled without further treatment	44,000
• Decantation or emulsion separation	27,500
• Neutralization/precipitation	22,000
• Incineration	16,500
Total	110,000

25,000 square kilometers, which is about 10 percent of the Federal Republic of West Germany.

Waste Treatment and Disposal Processes

This facility uses a secure chemical landfill, waste-oil separation, inorganic chemical treatment, and incineration to process an estimated 111,000 tonnes of wastes per year. A general breakdown of wastes by treatment method is shown in Table 2.

All wastes are tested upon arrival at the facility (which does not operate its own transport system) in its laboratory. Parameters for which analyses are performed include: cyanide; nitrite; nitrate; chromium; pH; copper; iron; nickel; lead, cadmium and other heavy metals; free oil; hydrocarbon emulsification; heat value and falsh point; and halogen content. The normal time required for testing each load of waste is approximately 10 minutes, thus, allowing a prompt decision on the subsequent routing of the waste through the plant.

The types of wastes which are directly landfilled without further treatment are limited. Typical examples are: metal hydroxide sludges with greater than 30 percent solids content; light metal (such as magnesium) machine shavings; dry cleaning-solvent sludge; oil-contaminated soil; and lead- or mercury-bearing wastes in which the metals are in an insoluable form. All other wastes are treated or incinerated prior to disposal in the landfill.

The treatments employed range from simple mechanical dewatering of wastes which otherwise are landfillable to multistage chemical processes for handling highly toxic contaminants such as cyanide, arsenic, and pesticides.

The incinerator is a two-chamber unit, the primary chamber of which is a rotary kiln, followed by a baffled secondary chamber. Since the volume of oil-contaminated water is rather large at this facility, the operating temperature of approximately 800 °C in the primary chamber increases to approximately 950 °C in the secondary chamber due to oil/water injection at that point. The exhaust gases are then cooled to 300 °C by water sprays in an enclosed conditioning tower before being introduced into an electrostatic precipitator for cleaning prior to release. There is no waste-heat recovery on this unit.

The landfill, which was the earliest part of the facility to be established (the treatment plants and incineration unit were built four years later), was developed in a clay pit located in the outskirts of Schwabach. The original capacity of the pit was 1 million cubic meters, with a 15-meter depth. The landfill has a surface area of approximately 67,000 square meters or 14 acres. One of the primary reasons for selecting the site was its water table, which is 30 meters below the original land surface. Additional protection for ground water was provided before the landfill was put in use by the addition of a compacted clay liner and an underground drainage system to collect any leakage or leachate. Additionally, six monitoring wells were installed around the perimeter of the pit to monitor ground water quality.

The present operation plan of the landfill limits the area of its working face, or active disposal area, to only 100 square meters at any one time. All other portions of the landfill are covered by one meter of soil in order to prevent rainwater from infiltrating the buried wastes. The wastes are deposited in lifts, or layers, approximately seven meters thick, sandwiched between the one-meter soil layers. Currently, disposal is taking place at approximately the original ground level. Operators of the facility plan to extend the life of the landfill by another 10 years, however, by continuing to dispose of wastes in soil-covered lifts (or layers) to a final height of 18 meters above ground level.

Environmental Control

The facility is responsible for its own environmental control. SMM laboratory and technical personnel periodically test all emissions and plant treatment effluents, as well as monitor ground water samples from the wells. Additionally, sheep which are pastured on the portions of the landfill that are covered and vegetated, are bioassayed on a regular basis for contaminants and other possible problems.

This self-monitoring is augmented by further testing conducted by the Bavarian State Authority for Environmental Protection and other governmental offices, since these agencies have been made responsible for the technical supervision of the facility by the regulations which established SMM.

Facility Costs and Disposal Charges

The original cost of the total facility was $16.5 million, with the incineration unit costing slightly more than $3 million (in 1972 dollars). Current revenues are almost $5 million annually; however, these may be expected to increase in the future due to inflation, as well as for new loan repayment costs for additional equipment. A new incineration system is planned to go on stream in 1984 with a capacity of 22,000 tonnes per year, thus raising the total annual incineration capacity of the facility to 37,000 tonnes.

Since the facility is operated as a non-profit public utility owned by an association of local governments within the region, disposal charges are based on the treatment cost for each type of waste and the general operating overhead of the facility. Also, as the facility is non-profit, any excess revenues are distributed among the member governments to benefit the public, and any revenue shortfall is made up by assessments to the members. The member governments also have a statutory "obligation to use" the facility, further emphasizing the governmental connections of the facility.

Basic disposal charges in force in 1980 are described in Table 3. Additional fees are assessed for extra treatment requirements, labor, chemicals, or equipment use.

Additional charges are assessed for wastes with high concentrations of toxics, extreme pH levels, and so forth. In 1980 these costs ranged from $16.50 to $99.00 per tonne. Extra labor and treatment plant usage were charged at $13.75 per hour, and at $22.00 per hour, respectively.

SIDIBIX

The regional hazardous waste management facility at Sandouville, near LeHarve on the west coast of France, is an example of a joint venture waste management project among an area's private industries, with local and regional government agency encouragement.

The need for adequate hazardous waste treatment and disposal capabilities in the industrial region of the Lower Seine Basin near LeHarve was recognized in 1973 by the Mine Service Agency of Rouen and the Agency of the Seine Basin in Normandy, agencies responsible for pollution control in that region of France. Based on the results of waste generation surveys conducted by the local industries, it was found that a majority of the industries in the area were petroleum or petrochemical based and that an incineration unit with energy recovery would best serve the waste management needs of the group.

A consortium, named the Society of Regional Study for the Protection of the Environment (SERPE), was formed by 20 companies with plants in the area, including those of such American-connected companies as Ashland Oil, Esso Standard, Esso Chemical, Firestone, Goodyear, Mobil and Shell.

TABLE 3
Sondermullplatze Mittelfranken
Disposal and Treatment Charges

Method	Cost (base price per tonne)
Landfilling without treatment	$24.75
Landfilling with mechanical treatment	37.75
Basic chemical or mechanical treatment	35.75
Special treatment (arsenic, cyanide, etc.)	148.50
Thermal treatment (incineration)	115.50

This consortium coordinated the design and construction of the facility at San-douville and formed the group that, presently oversees the activities of the facility in conjunction with the Chamber of Commerce and Industry of LeHavre and other civic groups. Day-to-day operations of the facility are carried out by a contract operator, Sidibix, which is a subsidiary of the Compagnie Generale Des Eaux of Paris.

Construction of the plant, which incorporated equipment to separate oil emulsions and neutralize inorganic wastes as well as incinerate organics, was completed in the spring of 1977. Since the wastes generated in the area were found by the SERPE survey to contain minimal amounts of chloride and sulfur (.04 percent and .6 percent, respectively) and those with recoverable heat value represented approximately 75 percent of the total anticipated, incineration with waste-heat recovery is the primary waste disposal method used at the facility. The resultant heat is recaptured to produce steam which is sold to neighboring industrial buyers as well as to generate power for in-plant use.

The facility accepts approximately 65,000 tonnes of waste annually, 50,000 tonnes of which are incinerated and the remainder physically or chemically treated. The geographic origin of the wastes is: 60 percent from the city of LeHavre; 38 percent from the community of Lillebonne; and 2 percent from the city of Rouen. Nineteen of the original founding companies send a waste stream to the facility to be incinerated, which totals about 35,000 tonnes annually with the balance of burnable wastes coming from nonmember generators.

Waste Treatment and Disposal Processes

The hazardous wastes which are accepted are divided into 11 categories for management and pricing purposes: seven for incineration of organics and four for physical and chemical treatment of inorganics. These wastes are categorized as follows:

Incineration Wastes
- Barrels of solidified wastes requiring manual handling, compressed blocks of wastes requiring cutting
- Barrels of heavy pasteous wastes or other difficult wastes,
- Loose solids, unusual sludges or pasteous wastes
- Bulk solids or regular sludges
- Liquids in barrels
- Bulk irregular liquids
- Bulk regular liquids

Physical and Chemical Treatment Wastes
- Acids
- Neutral wastes or weak bases
- Hydrocarbon-aqueous mixtures or emulsions
- Strongly alkaline wastes

The average chemical composition of the burnable waste stream received at the facility is: carbon, 65 percent; hydrogen, 9.9 percent; water, 14.2 percent; chloride, .04 percent; sulfuric acid, .01 percent; sulfur, .6 percent; and inerts, 11.2 percent. The average heat content of this waste stream is about 5,000 therms (19,800,000 Btu) per tonne, with a density of about 1 tonne per cubic meter.

The facility incinerator is a two-chamber unit with a rotary kiln primary chamber and a secondary chamber with liquid injection. It operates normally at 900 °C in both chambers. This operating temperature has been found to give good refractory life as well as adequate destruction of the waste stream. In order to ensure waste destruction at this temperature, the management exercises tight control over incoming wastes in two complete laboratories on-site, as well as tracking wastes by means of a manifest with each load. Wastes chosen for incineration must be compatible with the operating temperature of the kiln and with the capabilities of the electrostatic precipitator, the only control on exhause emissions. This input analysis control over the waste stream ensures a feed to the kiln that is low in chloride and sulfur and that produces an environmentally acceptable emission.

Wastes are received and stored at the facility according to type: containers in a covered stockade area; bulk liquids in storage tanks; and bulk semisolids or solids in concrete pits. Wastes are also fed into the kiln according to type: barrels by an elevator and chute; and bulk wastes from the concrete sludge pits by an overhead crane bucket.

The kiln is 4 meters in outside diameter and 12 meters long and accepts a feed rate of about 3 tonnes per hour. Solid residue (slag and clinker) is discharged from its lower end and is quenched in a water tank. Because of the tight control exercised over the wastes which are accepted, the incinerator clinker is considered nonhazardous and is used for road construction fill in the region.

The exhause gas from the kiln enters the secondary chamber at 900 °C and that temperature is maintained by four additional burners injecting liquid wastes at an approximate rate of 1.4 tonnes per hour. The interior dimensions of the secondary chamber are 4 meters wide by meters long by 13 meters high; it is steel framed with .5 meters of refractory lining.

The gas stream leaves the primary chamber at about 900 °C and, then, enters an eight-pass, waste-heat boiler. It cools to about 280 °C by the time it leaves the boiler and enters the precipitator. The boiler produces approximately 40 tonnes of steam per hour at a pressure of 40 bars and a temperature of about 320 °C.

The steam is then delivered at this temperature and pressure to a 700-kilowatt (kW), counter-pressure turboalternator and, then, is routed into the steam-distribution pipeline at a pressure of 20 bars and a temperature averaging 275 °C. The current produced by the turbo-alternator is 280 volts at 50 hertz (Hz). This enables the plant to be supplied with 31 kW of power per tonne of steam produced.

The steam-delivery pipeline is insulated and incorporated a condensate return line; the steam is distributed to customers located to 5 kilometers from the facility. The steam is sold on the basis of a guarantee of temperature and pressure, rather than flow. These characteristics have been optimized at 16 bars and 200 °C based on the results of a technical feasibility and user needs study. It was determined in the initial planning of the facility that the sale of energy as steam would be more cost effective than purchasing the additional capital equipment for power generation. In addition, the area electric utility also is able to offer surplus power at a lower price than Sidibex.

Physical and chemical treatment of wastes at the facility is directed toward producing an environmentally acceptable effluent by removing excessive contaminats from non-burnable wastes. Wastes are determined to be non-burnable by the facility's laboratory if either their heat content is too low (such as that of oil/water emulsions), thus, affecting the combustion process in the incinerator, or if they contain contaminants (such as sulfur, lead, and so forth) in concentrations that are too high for safe injection in the incinerator under its normal operating conditions.

Oil/water emulsions are separated (or cracked) by means of flotation and skimming after heating and the addition of de-emulsification chemicals. The separated oil is burned by injection into the secondary chamber of the incinerator. The separated water is then used in the preparation of lime solutions for use in the acid waste neutralization treatment system, with any hydrocarbons remaining in the water being removed by the flocculation process accompanying the neutralization.

Acid wastes and highly alkaline wastes are also blended to achieve neutralization to the extent that is possible, given the mix of wastes received at the facility. Further pH adjustment to cause flocculation or precipitation of metallic ions in solution is accomplished by the addition of the lime described above. After the chemical separation is complete, the decanted effluent is given a final treatment using an activated carbon filter before it is released into an inspection basin and ultimately into a local canal. A set of three inspection basins is used in order to allow the contents of any one basin to receive further treatment if any of the following effluent discharge permit requirements are not met:

Temperature	$\leq 30\,°C$
pH	5.5 to 8.5
Suspended solids	≤ 30 mg/1
Biological oxygen demand (5 day)	≤ 30 mg/1
Chemical oxygen demand	≤ 120 mg/1
Hydrocarbon content (Hexane extraction method)	≤ 5 mg/1
(Total)	≤ 20 mg/1
Chrome (hexavalent)	≤ 0.5 mg/1
Lead	≤ 0.1 mg/1
Fluoride (monovalent)	≤ 15 mg/1

Sludges from the chemical separation process are filter pressed and pelletized, and then landfilled off-site, since they are now considered to be chemically neutral.

Environmental Control
The environmental regulation of the waste disposal activities at the facility is provided by the Mines Service Agency of Rouen, which is the primary permitting agency. Waste storage and handling design criteria for the facility are generally similar to those used in the area's petroleum refineries, as are the emergency equipment and training given to the personnel.

The Mine Service Agency also monitors the performance of the incineration emissions in accordance with the following standards:

Particulates	.15 grams per normalized cubic meter (normalized at 7 percent carbon dioxide)
Chloride	1.1 percent
Sulfur	2.6 percent

Additionally, other impacts, such as noise, are also monitored. The Sandouville Facility works closely enough with the regulatory agencies that a permanent office is set aside in the facility's administration building for the agencies' use.

Disposal Charges
Disposal prices charged by the center are shown in Table 4.

The pricing structure at Sandouville is based on the type of wastes received; however, an additional factor is the heat content of waste. Disposal prices for high-heat content wastes are discounted in order to encourage the flow of energy-rich feedstock to the incinerator and boiler. Additionally, the proceeds from the sale of steam are distributed to the 19 companies of the consortium in proportion to the quantity and heat value of the wastes they send to the facility. The type of wastes received, thus, also serves as an allocation method for steam distribution among the member companies who are also users.

BASF AKTIENGESELLSCHAFT, LUDWIGSHAFEN

The principal manufacturing plant of the BASF Group, a large multinational, integrated chemical manufacturing company, is in Ludwigshafen, West Germany. The city, and the plant, lie on the Rhine approximately 50 miles south of Frankfurt. The plant covers approximately 2½ square miles, employs about 50,000 people and produced sales of 12.49 billion deutche marks in 1980, 45% of the total group sales. The plant manufactures over 6000 different products from approximately 300 different production lines and in 1979 generated 210,000 tons of industrial waste while manufacturing 6.5 million tons of product.

110,000 tons of this waste stream, consisting of simple ignitable liquids and solids with a low degree of hazard, were co-fired in the plant's three conventional steam power stations. 98,000 tons of the balance was incinerated in kilns at the plant's hazardous waste incineration facility and 2,000 tons of highly hazardous wastes were either incinerated at sea or disposed of at a facility located in 700 foot deep salt mine at Herfa-Neurode, West Germany.

Hazardous Waste Incineration Facility

The present waste incineration facility at Ludwigshafen has been developed over the period since 1960 and now consists of six rotary kiln units and a converted moving grate calcining furnace. These units have a combined waste incineration capacity of 109,000 tons per year and a total steam generating capacity of 113 tons per hour from waste heat boilers. The facility, which is located on the main plant property, and feeds its generated steam into the plant steam supply lines, primarily incinerates industrial wastes from the Ludwigshafen and other BASF plants. The only exception is a small amount of outside hospital and laboratory wastes which is incinerated as a public service.

Controlled incineration of BASF wastes has been under development at this site since 1956, initially utilizing a vertical calcining kiln (phased out in 1970) and a moving grate calcining furnace constructed in 1960. Municipal solid waste from the city of Ludwigshafen was originally co-fired in the moving grate furnace with BASF wastes; however, this was discontinued in 1969 when the furnace was modified to increase its capacity and improve its performance in incinerating the growing waste stream from rapidly increasing chemical production.

Additional capacity was actually needed prior to 1969, to incinerate thick, pasteous wastes and contanerized solids with a higher heating value (9,000 to 18,000 BTU/lb) than those wastes typically burned in the moving grate unit. Also, feeding problems from the effect of the pasteous wastes on the moving grates had made a new design necessary. Therefore, two additional units which were built in 1963 consisted of two chambered kilns with rotating primary chambers. These units were modified from existing industrial furnace designs by

TABLE 4
Sidibex Disposal Rates

Wastes	Rates (Per tonne)
Sulfuric acid wastes	$91.00
Hydrochloric acid wastes	65.00
Chromic acid wastes	150.00
Burnable wastes	65.00 to 100.00

theoretical calculations to accommodate the waste streams causing difficulty in the other units. The general design requirements were that the kilns be able to successfully burn liquids, pastes, and contanerized and bulk solids in varying mixes and with high heating values.

VonRoll Ltd. of Zurich supplied the two units which were equipped with waste heat boilers, and were designed to accept 7,700 tons of wastes per year per unit. The heat output of each unit was 33,000,000 BTUs/Hr and the boiler would produce 9.3 tons of 500 °F steam at a pressure of 260 lb/sqin.

These units have been in service since 1963 and considerable knowledge on practical methods and techniques for incinerating difficult wastes has been obtained during their operation. For instance, the purposeful addition of a fluxing and slagging material to the waste feed was found to create a thick molten slag "blanket" or annular skin which completely covered and protected the refractory lining of the rotating primary chamber. Prior to this discovery, there had been rapid chemical and mechanical erosion of the refractory due to waste reactions with the fire brick and mechanical stress caused by the solids. Discovery and use of the slagging technique, and a change in refractory formulation, greatly increased the service life of a kiln before re-bricking.

Another problem was encountered during the early period of operations which led to an important design change. Salts which were contained in the wastes were found to volatilize during combustion and be carried in the hot gas stream to the boiler tubes, where they would condense into salt cakes which would build up and ultimately clog the gas passages through the boiler tube bundles. The original boiler designs had been standard ones and had not anticipated utilizing such a complex combustion gas stream.

Two boiler design changes were implemented during this period which have since become widely recognized: the water tube spacing has been increased to help prevent clogging and to facilitate cleaning, and some tube bundles were rebuilt to incorporate self-cleaning devices and modular construction allowing their removal for maintenance without shutting down the kiln.

Additional waste incineration capacity was again needed by 1969, so two more rotary kiln units were constructed, by VKW, at that time. The basic configuration of a rotating primary chamber, a stationery secondary chamber, and a waste heat boiler was retained, but a number of changes were made which reflected the operating knowledge which had been grained through the operation of rotary units #1 and 2.

Further expansions in waste management capacity were necessary to meet production increases, so an additional rotary kiln unit was constructed the same size as numbers 3 and 4, in 1974. This unit was constructed by Krauss-Maffei and connected to a central electrostatic precipitator which also serves units 1, 2, 3, and 4. The emissions from the modified moving grate unit (unit "Zero") are controlled by a separate electrostatic precipitator, but both precipitators discharge into a single 120 meter high cluster stack.

Air pollution control by dry electrostatic precipitators is not effective when the waste stream consists of high chlnoine or high sulfur content materials, therefore, a sixth rotary kiln unit was added in 1978 which was equipped with a two stage wet caustic scrubber to remove HCL, HF, and SO_2 from the exhaust gases. This unit was built by Babcock Krauss-Maffei based on a design developed by BASF engineers which utilized the experience gained from operating the earlier units.

The scrubber is an adjustable throat rubber lined unit, furnished by Lurgi, but also incorporating BASF design concepts. It utilizes water with pH adjusted to 6.8-7.0 with NaOH which is recirculated except for a 10 to 15% bleed-off. The bleed-off stream is treated with additional caustic outside of the main recirculating loop in order to prevent excess generation of CO_2 in the system. The adjustable venturi is ring shaped and incorporates a pressure drop of between 500 and 700 mm of water to handle a gas stream of approximately 70,000 m³/hour.

Kiln number 6 has twice the capacity of the other rotaries, and can accommodate 24,000 tonnes of waste annually. It operates at approximately 900 °C in the primary chamber and normally 1200 °C in the secondary chamber except when very refractory halogenated hydrocarbons are burned. The waste heat boiler on the unit delivers 28 tonnes of steam per hour at 18 bars and 270 °C. The rotation of the primary chamber is held below 1 rpm in order to increase retention time of the wastes and to maintain the slag blanket which protects the refractory lining.

A unique feature of the entire BASF facility is the containerization of the wastes, which are transported from their points of generation in the plant almost entirely in special containers in a liquid or semi-liquid form. The containers are specially designed for each waste stream and utilize special materials, pressurization and heating coils as required. The sizes range from a rail tank car to 1,000 gallon skid mounted tanks. The tanks are connected directly to the kilns and the contents can be forced out by nitrogen or steam pressure directly into the kiln. Some solids are fed directly in their shipping drums, but this container system allows lthe bulk of the wastes to be injected through various nozzles at a more controlled rate. Feed control is the key to maintaining the optimum combustion and destruction conditions in the incinerators.

Another feature of the container system is that is allows all wastes to be stored at the generation points, therefore, the only wastes at the incinerator facility are usually those scheduled for particular days operation.

BASF has been successful in developing a waste management system which operates in an urban environment with an extremely high degree of environmental safety while furnishing a substantial amount of the energy required for the entire chemical plant.

CONCLUSIONS

As can be seen from the brief descriptions of these four facilities, considerable variation exists in the methods of establishment and operation from plant to plant. While these differences may appear to be substantial, the general systems of waste management used at the facilities have many similarities.

The author has visited several other European hazardous waste management facilities, including: Entsorgungsbetriebe Simmering (E.B.S.) in Vienna; Hessische Industriemull (H.I.M.) in Biebesheim, near Frankfurt; and SARP Industries in Limay, near Paris. An examination of these other facilities only reinforces the main premise of this chapter that Western European waste management systems have many similarities, even though the details of the physical plants and their organization may differ. The diversity between the plants stems from the differing disposal requirements of the various areas.

The general type of hazardous waste management system used in Western Europe is one approach for managing hazardous wastes in a safe, technically sound and economically acceptable manner which also has government and public approval. The European systems are representative of the types of systems to be expected in the United States as a result of RCRA requirements and, therefore, a detailed knowledge of them should be productive.

ACKNOWLEDGEMENTS AND BIBLIOGRAPHY

Erbach, G. 1980. Hessische Industriemull GMBH, Wiesbaden, West Germany, private communication.

Fade, F. S. and Leroy, J. B. 1978. Industrial Wastes and Energy, T. S., L'Eau Vol. 73, No. 1, pp. 21-29.

Fauquan, J. C. 1980. Sibidex, LeHarve, France, private communication.

Gontard, B. 1980. SARP Industries, Limay, France, Private communication.

Greisbaur. 1980. Sondermullplatz Mittlefranken, Schwaback, West Germany, private communication.

Jourdan, B. 1980. Compagnie Generale Des Eaux, Paris, France, private communication.

Kristensen, A. 1980. Kommunekemi A/S, Nyborg, Denmark, private communication.

Lauridsen, J. 1980. Danish National Agency of Environmental Protection, Copenhagen, Denmark, private communication.

Pirzer, R. W. 1980. BASF, Ludwigshafen, West Germany, private communication.

Scharsach, A. 1980. VonRoll Ltd., Zurich, Switzerland, private communication.

Womann, Heinz, 1971. Experiences With Industrial Waste Incineration at BASF, "Energie", Vol. 23, No. 11.

Womann, Heinz, 1979. New Waste Incineration Facility With Wet Chemical Scrubber, *Chemie-Technick,* Sonderdruck 8 p., 487-492, Heidelberg.

Portions of this chapter which were published earlier in the Dames & Moore *Engineering Bulletin* of March, 1981, are used here with the kind permission of that firm.

Invaluable advice and assistance in the preparation of this manuscript was provided by Diana B. Ratliff, who has colaborated with the author may times previously on this subject.

Hazardous and Toxic Wastes: Technology, Management and Health Effects. Edited by S.K. Majumdar and E. Willard Miller. © 1984, The Pennsylvania Academy of Science.

Chapter Nineteen

Management of Hazardous Materials within an Industrial Facility

Charles L. Fraust, Raymond G. Dinsmore, and Barbara M. Russell
Western Electric Company
Allentown, PA 18103

There are currently over 58 thousand different chemical compounds in use within industry, an estimated ten times the number used in 1970.(1) With this large growth rate in chemical usage, it is evident that even the best managed industries may be confronted with associated chemical emergency situations at one time or another. No matter what precautions are taken, the possibility remains that things can still go wrong.

Probably the single most important factor that distinguishes management of hazardous materials in an industrial facility from management in other settings is the innate ability to know what materials are on hand, their properties, how to properly employ them, and the proper means of responding when problems occur. Unquestionably, the most important requirement in responding to a release or spill anywhere is the determination of the identity and properties of the material. Until this information is obtained, effective response will be severely hampered. Response efforts undertaken before completely understanding a problem are ill advised and may lead to improper actions which could exacerbate a situation to the extent of injuring people, damaging property, or even endangering life.

The proper use of hazardous materials may be divided into four areas:
1. Certification of production chemicals
2. Management of in-plant usage
3. Management of wastes generated
4. Contingency Plan

Each of these areas will be discussed independently and broken down into its component elements. The end result will be a total program that will enable "cradle to grave" management and insure proper use of hazardous materials.

CERTIFICATION OF PRODUCTION CHEMICALS

The initial step to effective management of hazardous materials is the development of an inventory of materials in use. Such an inventory should include, as a minimum, the material name, user area and location, quantities used, total quantities on hand at each user location, and container sizes. This information will serve as a starting point for developing an in-house data base. This data base should include items such as physical properties, hazard classifications, problems peculiar to the materials, chemical imcompatibilities, required environmental controls, required personal protective equipment, spill and clean up procedures, disposal procedures, fire fighting techniques, and any other information deemed appropriate to satisfactorily manage the use of this material. Key to establishment of a data base is obtaining a Material Safety Data Sheet, MSDS, from the material supplier. Much of the information required to manage the use of a hazardous material will be given on the MSDS form. However, the amount of information provided varies from supplier to supplier from complete, detailed data to very sketchy information, particularly for proprietary formulations. When dealing with sketchy information, several alternatives are available. First, the supplier can be contacted for further information. If the information is proprietary, the supplier may be willing to divulge the identity of specific hazardous components. Another approach is to provide the supplier with a list of known hazardous materials and require the supplier to certify if these are present or not. If these approaches fail, a last resort is to have the material analyzed either in-house, if such capability exists, or at a vendor analytical laboratory. Under no circumstance should unidentified formulations be used unless, as a minimum, proper assurance is obtained that the materials are not hazardous.

The MSDS form should be viewed as a starting point from which a localized information system is developed. The MSDS form can at best provide a general idea of appropriate control measures. Specific requirements depend on such things as the way a material is used, quantities involved, in-house response procedures, and applicable regulations.

It is a worthwhile practice to maintain a reference file of texts which deal with properties of hazardous materials. Many such texts are now available and the choice should depend on suitability for individual needs. (2)(3)(4)(5)(6)(7)(8) Access to services such as Toxline and Medline is also quite valuable.

A critical stage in the control of hazardous materials is when new chemicals are introduced as part of new processes or modifications to existing processes. Research and development are examples of areas which are prime candidates for introduction of new processes, new chemicals, and, possibly, new management problems. As such, it is important for environmental control groups to develop working relationships with company groups engaged in these types of activities. Likewise it is important to interface with process engineering groups so that when new processes are evaluated for possible use, an assessment of the hazard potential may be included. An effective means of prescreening new processes is the establishment of internal or intercompany evaluation committees. Such committees should meet regularly to evaluate promising new technologies, new processes, and new chemical formulations for their potential impact on a facility and its environs. This review procedure affords environmental control personnel, who may not be fully acquainted with new developments, the opportunity to evaluate potential problems, make recommendations, develop necessary control strategies, and include the new process in the facility management program. Failure to provide this review and evaluation can lead to delays in implementation of a new process, purchase of equipment unsuitable for use at a facility, and an inability to control byproducts.

Critical issues addressed in the review may include an overview of the process, chemicals and chemical reactions, possible byproducts, quantities, physical and chemical properties of chemicals used or produced, compatibility with materials of construction, known toxic effects, waste disposal procedures, environmental controls, personal protective equipment and how to handle material losses, such as spills. If possible, analyses should be performed on prototype or pilot equipment to substantiate effectiveness of controls and to identify and quantify any byproducts.

MANAGEMENT OF IN-PLANT USAGE

As indicated earlier, a large number of chemicals are used in industry. The handling of these chemicals involves the potential for material loss. Provisions to reduce this potential should be designed into storage, transport, use and disposal of all chemicals used in manufacturing processes.

Materials that have been approved for manufacturing use should be stored in a chemical warehouse upon receipt. This should, preferably, be separate from the manufacturing facility and designed specifically for chemical storage. Chemicals should be segregated categorically to limit the opportunity of mixing incompatible materials in the event of an accident. Procedures to follow in the event of an emergency should be established. Items such as drum overcoats, spill clean-up materials and personal protective equipment are necessities in a chemical storage area.

Compressed gases require separate segregation and storage. Cylinders under pressure represent the storage of a considerable amount of energy. Failure to properly store and secure cylinders may lead to rapid loss of the contents. This may be the result of a cylinder toppling, overheating the cylinder, or a cylinder valve failure. If not secured, rapid loss of the contents can launch a torpedo-like projectile. The contents of a cylinder may be corrosive, toxic, flammable, or even explosive. Like liquids, compressed gases must be segregated in storage by their hazardous properties. While no industry standard presently exists, as a minimum it is a good practice to separate corrosives, toxics, and flammables.

Chemical transport is a critical operation and great care should be exercised to avoid accidental spills. Small glass continers should be protected by using non-breakable carriers. Stainless steel safety cans may be used with flammable solvents. When dealing with larger quantities, such as 55-gallon drums, transporting them in leak-tight cabinets can prevent spills onto the shop floor.

Compressed gas cylinders should be transported through manufacturing areas using vehicles designed for safe handling. Cylinders should never be rolled except to move them from the conveyance vehicle to the storage cabinet. All cylinders should be leak tested prior to delivery to production areas and again leak tested after use. Cylinder caps should be maintained in place until the cylinder is safely secured in its storage cabinet and ready for installation of the CGA connector.

Labeling of all chemicals is an important aspect in hazardous material management. While most chemicals have some type of identifying label when they are received, it is good practice to attach another label indicating hazard information such as health, fire and reactivity.

This is especially true for proprietary materials, where the name may not give an indication as to the associated hazards. One labeling system which is frequently used in industry is the hazard alert symbol. Originally developed by the National Fire Protection Association,(9) this label identifies numerically the relative hazard grading for health (toxicity), fire (flammability) and reactivity (instability) of the chemical. The ratings range from zero to four with four being the most hazardous and zero being the least hazardous. The symbol provides a quick, easy indication of the types of hazards associated with a chemical.

As earlier indicated, before any chemical is used in production it must be reviewed and approved by the environmental engineering organization and the potential hazards of its use identified. The Material Safety Data Sheet is helpful for this purpose. To provide manufacturing personnel with the information necessary to work safely with the chemical, a summary fact sheet can be provided. The fact sheets may be compiled into a manual which is given to each production group. The fact sheet should describe the chemical and identify the effects of over-exposure. It should also identify the particular spill cleanup materials required and the appropriate personal protective equipment. The necessary cleanup material, along with personal protective equipment, should be located strategically within the building. One means of ready access is a portable spill cart which can be rapidly deployed to affected areas.

Adequate chemical storage space must be available in the production area. Again materials must be separated by hazard category. Normally storage cabinets are designated for acids and bases, flammable and non-flammable solvents and special materials. Chemicals stored within a production area should be in ventilated, fire-resistive cabinets and/or safety containers. Solid materials should always be placed on the top shelves to avoid possible reactions if a bottle of liquid is broken. Combustible material such as boxes, towels or clothing must be kept separate from these storage areas.

Suitable space should be made available in production areas to store compressed gases. Where appropriate, ventilated gas cabinets should be provided, in particular for toxic and corrosive gases. Materials deemed particularly hazardous for use inside a manufacturing facility should be deployed outside the building and piped to the work position. As a general rule, it is advisable to deploy only the minimum quantity of material suitable to perform an operation without causing excessive cylinder changes.

Operations that release toxic airborne contaminants or have the potential for such release must be controlled with local exhaust systems. Local exhaust ventilation is generally provided at wet chemical stations, equipment using solvents, and hazardous gas systems. These localized exhaust systems provide the necessary air movement to capture and exhaust toxic materials. If the contaminant is highly toxic or corrosive, air cleaning equipment such as a wet scrubber may be required prior to discharge. Periodic inspections of exhaust systems is necessary to ensure adequate velocities are maintained.

Hazardous compressed gases must be protected at their point of use. In addition to ventilated storage cabinets, heat detectors and sprinkler heads are recommended to insure that in the event of a fire or a condition of high heat, response personnel are notified and water deployed to keep the cylinder cool. Failure to maintain cool temperatures can result in pressure buildup ultimately leading to gas release via the overpressure - over-temperature relief disk. Hazardous gas cylinders should also be provided with adequate distribution systems. Key elements include appropriate regulators, a high pressure shutoff valve, welded fittings, and flow limiting devices.

Although local exhaust ventilation is utilized for hazardous chemicals, air monitoring to insure compliance with OSHA Permissible Exposure Limits should be performed on an on-going basis. Monitoring should also be employed during spill response to insure that response personnel are not exposed to hazardous materials. There are several techniques that are available for measuring airborne contaminants.[10] The type of sampling method used will depend on the nature of the contaminant and the reason for taking a sample.

Personal protective equipment is required for the majority of chemical operations. Written guidelines on the selection, use, and maintenance should be provided. Safety glasses, chemical splash goggles, gloves, aprons, protective sleeves, and respirators are just some of the items that are used routinely. Specific hands-on training in the wearing of personal protective equipment is extremely important.

Education and training in health and safety should be given to all employees prior to starting a job. Subjects such as chemical safety, fire protection, compressed gases, and personal protective equipment should be included in this basic training. This material can be presented to employees by various means: written manuals or booklets, video tapes, slides or movies to name a few. The shop supervisor or other designated individual should also review the appropriate health and safety procedures for a specific process with each new employee.

MANAGEMENT OF WASTES GENERATED

Most industries today are aware that proper waste disposal must be factored into management's philosophy. The cost for disposal of hazardous materials, like the cost of labor and raw materials, is simply one of the costs of doing business, and must have management attention at the outset.

When first looking at a plant's hazardous waste production, it is important to separate these by-products into several categories based on characteristics, the most important elements in a plant's hazardous waste management program are the development, characterization, and segregation of these waste streams. Once this is done, it remains fairly straightforward to select the proper disposal or reclamation alternative.

Disposal Alternatives

There are a number of alternatives available when deciding what to do with hazardous material by-products. These alternatives can be arranged in a descending order of desirability as follows:

1) Recovery, Recycle, Reuse - By far the best alternative is to reclaim the materials for reuse either within the plant or by another company. It is indeed true that one man's waste can be another's raw material, although some measure of clean-up is generally required. Many chemical companies and other independents now provide waste recycle services available to industry. These services have become widely available for chlorinated and some other solvents. More recently recycle companies have become interested in other types of waste materials. Waste exchanges are also becoming very popular. These statewide or regional organizations provide listings of waste streams wanted by industry so they can be matched with companies having waste streams available. In this manner both industries involved as well as the general public benefit.

2) Recovery of Energy from Waste - Many waste chemicals and oils have the potential of providing valuable heating capacity for other industrial processes. For example, cement kilns have been commonly used to recover heat and, at the same time, destroy hazardous waste. Numerous examples exist in the literature where chemical wastes are successfully rendered non-hazardous while providing a low cost fuel in the operation of a cement kiln.(11)

3) Treatment (Detoxification) - Where it is impractical to recycle a hazardous waste for either its chemical value or heating potential, it becomes a require-

ment to treat it so that it is no longer hazardous. Treatment may be as simple as chemical neutralization to bring the pH value out of the "hazardous" range. More often, additional steps such as oxidation, reduction, chemical coagulation, or incineration will be necessary. In any case the decision must be made whether to treat on-site or to use a reputable contract disposal firm. Incineration, although currently accounting for only about 6%(12) of the total wastes disposed, will most likely be a primary disposal technique in the next decade. Its main advantage is in its ability to break down the hazardous components of many organic wastes while leaving relatively little residual material to deal with.

4) Disposal - The treatment techniques described above also involve some amount of disposal, whether it means discharging neutralized water to a receiving stream or emitting combustion products into the air. In these cases, however, the materials disposed of are either considered harmless or of such a small quantity as to be insignificant. The term disposal as used here means long-term storage of the untreated waste within the environment. Historically, this has generally been accomplished in three ways: ocean dumping, deep well injection, and landfilling. Landfilling remains in widespread usage today and certain areas of the country having a suitable geologic structure still specialize in deep well injection. Ocean dumping of hazardous wastes is no longer a common practice. The problem with any of these techniques is that the material remains in the environment in its hazardous state and the chance always exists that these materials might find their way to surrounding ground water. To say the least, these techniques remain controversial and their use should be based on careful evaluation of all options.

Shipments of Hazardous Wastes Off Site

Several references have been made to the decision whether to treat on-site or to use contract disposal. When off-site shipments of hazardous wastes are made, new responsibilities fall onto the industry which under the Resource Conservation and Recovery Act, RCRA, is considered the waste "generator." In particular, generators must comply with hazardous waste manifest programs. All states are required to enforce the Federally mandated hazardous waste manifest system. This system is intended to assure that all waste shipments are correctly handled, transported and disposed of by a permitted disposal facility.

From the waste generator's standpoint, it is important to keep several things in mind with regard to the manifest system. Trained, responsible people should supervise shipments of waste materials. Detailed records must be kept of all shipments indicating the date the wastes were picked up and when the manifest confirmation copy is expected to be received in the mail. The supervisor should ask to see the driver's license card or permit before loading the truck. All transporters of hazardous waste must be licensed to do so.

Periodic inspections should be conducted at all TSD (treatment, storage, disposal) facilities that an industry deals with. This keeps the lines of communication open and could possibly alert an industry to potential problems which a TSD may be facing. A generator could well have to share the costs of cleaning up a poorly run TSD facility.

CONTINGENCY PLAN - PLANNING FOR THE UNEXPECTED

It should be clear that given the quantities and varieties of chemicals employed by industry, the nature of by-products produced, and the potential for accidental release of these materials to the environs, some program was needed to organize and direct efforts to minimize effects of unexpected losses. It is with this in mind that the Rules and Regulations promulgated under RCRA provided that all generators of hazardous wastes must develop and implement a contingency plan.(13) Other Federal regulations and many State laws also require industries, which deal with hazardous chemicals and oil, to establish a preparedness and contingency plan of one type or another.

Although different formats and guidelines are available, the primary objective of such a plan is to develop an organizational structure and response programs to deal with emergency situations (fire, explosions, spills, gas leaks and/or discharges) which could result in environmental degradation or endangerment of public health and safety. An effective contingency plan will serve to mitigate the consequences of these types of unforeseen situations.

Any contingency plan can be broken down into two main parts: 1) prevention, or steps taken to avoid or mitigate an emergency situation and 2) preparedness, the ability to quickly react in emergency situations that occur despite efforts at prevention.

Prevention

Hazardous material losses in industry generally occur because of one or a combination of the following: inadequate design of equipment, poor housekeeping, inadequate maintenance of equipment, operator error, or sabotage. Conveyance and processing systems associated with hazardous materials must be designed with the expectation that equipment components will eventually fail. Consequently, back-up systems and containment structures should be incorporated into the original design. High level alarms and shut-off devices should be a part of all tanks containing potentially polluting liquids. These tanks should be surrounded by dikes or an impervious area pitched to a containment vessel. Thought must also be given to the compatibility of containment vessels with the liquids they are likely to see. High temperature conditions caused by an exothermic chemical reaction or simply heat from the sun may render an otherwise satisfactory tank liner material totally useless.

In designing a facility which processes or stores hazardous materials or wastes it is helpful to analyze each segment of the process, determine the consequences of an equipment malfunction, and formulate an alternative plan to deal with the situation. This is crucial in the design of any gravity flow treatment process where the operator has little control over what may be coming down the pipe and the rate at which it is coming. By formulating alternative plans in the design phase, the proper back-up equipment can be provided from the start.

Great care must be used in siting facilities which contain hazardous materials. Tanks and containment structures should be placed above the 100-year flood plain. There is very little that can be done to contain such materials in a flood situation. Containment structures such as dikes or troughs should be provided for all exposed piping and areas where trucks make chemical deliveries. A diked chemical delivery pad is beneficial for two reasons. First, it would naturally contain any spills which occur during the transfer of a material from a bulk delivery truck. Such an area would also serve as a "safe spot" to which a leaking truck can go in an emergency situation. The material would thus be contained until proper clean-up methods could be effected.

Some prevention measures are administrative in nature, the best example being good housekeeping. Although often overlooked, proper housekeeping plays an important role in prevention of accidents which could result in injuries, damage, or pollution incidents. Good housekeeping is a state of mind which management must convey to its employees through company policies and by example. Most good housekeeping practices are obvious but an additional administrative policy might be to limit the amount of hazardous chemicals stored on site, thereby reducing the chance, or at least the severity, of a pollution incident.

Lack of equipment maintenance is often the reason for a pollution incident. It is easy to overlook back-up systems such as containment dikes or standby pumps which are rarely called on to perform. Therefore, it is vital to have in place an inspection and preventive maintenance program which will check out these systems regularly. Back-up equipment should be used as the primary system on a regular basis, to assure its effectiveness. Routine inspections by third parties (neither designer or operator) will also help pick up potential problems.

Underground tanks should be periodically examined for leakage to insure that materials are not leaching into the ground water. Similar checks should be made on underground pipelines where possible. Daily inspections should be conducted around areas where hazardous wastes are stored in tanks or drums.

Where an industry's storm runoff drains directly into a river, lake or estuary it is wise to set up a daily inspection procedure on the receiving stream. A visual inspection may pick up traces of oil or foam, indicating a leak situation in its initial stages. Samples should be taken for quick tests such as pH and dissolved oxygen. An effective inspection program should include an inspection log with monthly reports on results and infractions.

The fourth major factor in prevention of a hazardous material incident is employee training. Most industries do a good job of initial orientation into a newly designed manufacturing process or emergency response procedure. Problems generally develop when new employees are brought into an existing operation and receive no training or at best an on-the-job education from the previous operator. In this manner key information may not be passed on, or worse, erroneous information given. Supervisors must coordinate training of new employees with the persons responsible for training. The training organization should develop the proper training aids for dealing with hazardous materials and emergency response and schedule sufficient training hours. Examples of key target groups to receive training are: 1) fire brigade, 2) chemical operators and delivery personnel, 3) security force, 4) spill clean-up teams, 5) waste treatment operators, and 6) shipping personnel.

The final area of prevention to be addressed is security. Sabotage of an industrial facility although infrequent can result in significant pollution incidents. Sabotage usually takes the form of mindless acts of vandalism and are preventable by the implementation of strict security measures. Heavy-duty fencing should surround all significant portions of the industrial complex with appropriate warning signs posted. Any water valves serving the sprinkler system must be located within the confines of the secure area and locked open. Critical areas of the plant should be surrounded by fencing provided with tamper alarms. In a large facility, it may be appropriate to provide closed circuit video suveillance of plant entrances and other key areas.

Preparedness

It was mentioned initially that regardless of the effort directed at prevention, accidents will occur. When they do, the industry involved is generally in the best position to provide the necessary information and coordinate the response. A group within the facility must be designated to formulate a response plan and coordinate its implementation. A large company may find it advantageous to organize a committee composed of persons familiar with various aspects of the plant's operation.

The state of a plant's "preparedness" to deal with an emergency situation can be broken down into two main categories: communications and resources.

The category of communications can be further broken down into internal communications and external communications. Most industrial plants are already equipped with fairly sophisticated internal communication systems for day-to-day operations. So it is only a matter of planning to adapt the system to serve the needs of emergency response. Even plants with only single shift operations generally have at least a night watchman on duty. One area of the complex should be designated the primary communications center where someone is always stationed. Ideally this would be located near the main entrance to the plant. This center would be the point where all emergency calls come in from locations around the plant. All telephones within the plant should be marked with this emergency number. First aid, fire brigade, spill teams, etc., should be equipped with two-way radios or pagers which can be immediately accessed via the communications center. All automatic alarms should also be fed to this point.

This center should be the point from which all emergency calls are made for outside assistance. Persons manning the desk would have all emergency telephone numbers available and possibly direct phone or alarm links to the municipal fire departments and ambulance service.

An on-site plant emergency coordinator must be designated ahead of time along with appropriate back-ups. After the guard on duty makes the original decision to summon aid, all major decisions should come from the emergency coordinator or his designee. The coordinator will be in the best position to make consistent decisions because of his familiarity with all aspects of the contingency plan.

A plant spill or fire which has the potential of affecting persons or property outside of the plant involves communications of another sort. Advanced warning of possible water contamination or similar problems should be given to any neighboring industries or residents, municipal water treatment plants and public works offices. Additionally, the appropriate State, local or Federal regulatory agencies should be contacted as well as the U.S. Coast Guard Response Center where navigable waterways are involved.

The other aspect of a plant's "preparedness" is the resources it has available to it to deal with an emergency. These resources may be in the form of people, equipment or simply information. Large industries may be fortunate enough to have their own medical department and trained response personnel on site. However, even smaller plants can generally find employees who have experience and training acquired outside the plant with volunteer fire departments or ambulance corps.

Where needed talents are not available, outside sources must be tapped. Besides the local fire company, police and ambulance, a plant should have outside clean-up specialists and spill response contractors available for its use.

Emergency response equipment must be readily available for personnel to use. Examples of types of equipment which should be readily accessible within the plant are: 1) various types of portable fire extinguishes, 2) fire hoses or standpipe connections, 3) personal protective clothing, 4) face shields, 5) self-contained breathing apparatus, 6) spill clean-up absorbent materials, 7) neutralizing chemicals, 8) portable pumps, 9) containers such as 55-gallon drums, 10) air purifying respirators, and 11) emergency first aid supplies.

The final resource which an industry should have available is information; information on what types of chemicals are used within the facility and how they may react with other chemicals, water, or even air; information on fire fighting and spill clean-up techniques recommended; and information on the plant sprinkler system, valve locations, and main electrical disconnects. The responding fire department will be knowledgeable on how to fight fires but may not be familiar with unique situations which occur within an industrial facility. The plant emergency coordinator must be able to quickly familiarize outside assistance with the nature of the problem and any particular hazards associated with it.

Other information which an industry should have available includes assistance "hot lines" such as Chemtrec, regulations regarding spill reporting and pollution incidents, plus various other chemical handbooks and response guide books. Smith provides a comprehensive listing of available information sources in his book, "Managing Hazardous Substances Accidents."(14)

SUMMARY AND CONCLUSIONS

The key to effective management of hazardous materials at an industrial facility is to plan ahead. While no one can foresee and prepare for all circumstances, much can be done to insure that all reasonable precautions have been taken. This includes familiarity with materials used, proper controls, minimizing available quantities of hazardous materials, development of loss response capability, contingency planning, and, training of plant personnel who use, handle, dispose of, or respond to losses of hazardous materials. Such a program requires considerable effort to develop and even more effort to effectively implement. Furthermore, programs must be reviewed and updated on a regular basis. Each person having a specific area of responsibility in hazardous materials management must be totally familiar with his part. A means must be provided to insure continuity as personnel change or leave jobs. Finally, and of key importance, all response personnel must be well trained and well drilled so that, if the need arises, they are prepared to act swiftly and effectively to address any hazardous material problem that may occur during the normal course of production.

REFERENCES

1. Office of Toxic Substances, U.S. EPA, Washington, D.C.
2. Lewis, Sr., R. J. and Tatken, R. L., Editors, 1980, Registry of Toxic Effects of Chemical Substances, 1979, Edition, Volumes I and II, National Institute for Occupational Safety and Health.
3. Proctor, N. H. and Hughes, J. P., 1978, Chemical Hazards of the Workplace, J. B. Lippincott Company, Philadelphia, Pa.
4. Hawley, G. G., 1981, The Condensed Chemical Dictionary, Van Nostrand Reinhold Company, New York, N.Y.

5. Windholz, M., Editor, 1976, The Merck Index - An Encylopedia of Chemicals and Drugs, Merck & Co., Inc., Rahway, N.J.
6. Clayton, G. D. and Clayton, F. E., Editors, 1981, Patty's Industrial Hygiene and Toxicology, 3rd Revised Edition, John Wiley & Sons, New York, N.Y.
7. Sax, I. N., 1979, Dangerous Properties of Hazardous Materials, 5th Edition, Van Nostrand Reinhold Co., New York, N.Y.
8. Braker, W. and Mossman, A. L., 1980, Matheson Gas Data Book, Sixth Edition, Lyndhurst, N.J.
9. National Fire Protection Association, 1969, Recommended System for the Identification of the Fire Hazards of Materials, NFPA No. 704M, Boston, Ma.
10. Air Sampling Instruments for Evaluation of Atmospheric Contaminants, 5th Edition, 1978, American Conference of Governmental Industrial Hygienists, Cincinnati, Ohio.
11. Lauber, J. D. 1982, Burning Chemical Wastes as Fuels in Cement Kilns, Journal of the Air Pollution Control Association, Volume 32, No. 7, p. 771.
12. Gregory, Robert C., 1983, Incineration-A Viable Waste Treatment Alternative, Hazardous Materials and Waste Management, Volume 1, Number 2, pp 33-37.
13. Title 40, Code of Federal Regulation, Part 264, Subpart D and Part 265, Subpart D.
14. Smith, A. J., Jr., 1981, Managing Hazardous Material Accidents, McGraw-Hill, pp. 139-160.

Hazardous and Toxic Wastes: Technology, Management and Health Effects. Edited by S.K. Majumdar and E. Willard Miller. © 1984, The Pennsylvania Academy of Science.

Chapter Twenty

Management and Handling of Chemical Waste: An Industrial Approach

P. M. King, J. D.[1] and J. W. Osheka, M.S.[2]

[1]Director, Environmental Affairs
Environment, Health and Safety Department

[2]Environmental Engineer
Industrial Chemical Division
PPG Industries, Inc.
Pittsburgh, PA 15272

The purpose of this paper is to review the philosophy and specific components of an industrial approach to the management and handling of chemical wastes. The starting point for such a program must be the statutory and regulatory requirements and the program must include such additional items as the company's management believes are necessary to protect the public health and environment, and to minimize the firm's long-term exposure for potential liability for such waste. The approach described does not represent the specific program of PPG Industries or any one firm in all respects, but rather contains an array of components that may be adapted to a given organization. While the dimensions of the chemical waste generated in the United States each year are not precisely defined, chemical waste is, without a doubt, a major environmental concern from scientific and technical aspects, as well as from legal and public relations standpoints.

The management and handling of chemical wastes is regulated extensively by federal and state laws and regulations. The Resource Conservation and Recovery Act of 1976 ("RCRA"), 42 U.S.C. 6901, established the federal statutory framework for the regulation of hazardous waste. The program involves five basic components:

1. an identification and listing program for hazardous wastes;
2. standards of care for all who deal with hazardous wastes;
3. a manifest system and recordkeeping program;
4. a permitting and enforcement program; and

5. federal delegation of responsibility to states with programs at least equivalent to the RCRA program.

The regulations implementing RCRA begin at 40 CFR 260.

Pennsylvania also has adopted an extensive hazardous waste management program in the Solid Waste Management Act, 35 P.S. 6001. In contains elements similar to the first four components of the program under RCRA. Pennsylvania regulations implementing the Solid Waste Management Act can be found at 25 Pennsylvania Code Chapter 75, Sub-Chapter D: Hazardous Waste.

In addition, the Comprehensive Environmental Response, Compensation and Liability Act of 1980 ("CERCLA" or "Superfund"), 42 U.S.C. 9601, establishes certain requirements dealing with releases of hazardous substances, including chemical waste, into the environment.

Much has been written about these statutory and regulatory programs. The basic components of the statutes and regulations are incorporated in the approach to be described here.

ORGANIZATION/PHILOSOPHY

Having an adequate number of competent professionals is critical to designing, implementing and adjusting, as necessary, an industrial program to effectively manage chemical wastes.

PPG Industries, a diversified manufacturer of chemicals, coatings and resins, glass and fiber glass, has established a matrix approach to this management challenge.

Personnel are available in several corporate staff departments to assist in the design and implementation of the program. These areas include, as necessary: Environment, Health and Safety, Law, Transportation, Risk Management, Finance, Government Affairs and Public Relations Departments. These functrions are responsible for identifying the legislative and regulatory requirements and long-range trends in waste handling and management, assessing associated risks and identifying ways to manage such risks; complying with financial responsibility requirements and assisting with public relations aspects of waste management, as well as such other regulatory concerns as packaging, labeling and shipment.

At the operating division level, environmental affairs and manufacturing professionals are charged with the designing and implementing of the program as it applies to the plants within their responsibilities. At the plant level, the manufacturing and environmental affairs staffs implement the program for each facility. Divisional personnel must assure that each process generating waste is reviewed and an appropriate program implemented to ensure proper handling. Logic dictates that different programs well may be required for different operations. Flexibility must be built into the overall program so that each component has a suitable waste management method.

PPG's philosophy in the management and handling of chemical wastes can be briefly described as follows: First, reduce the volume of generated waste to a practical minimum so that the problem is no larger than necessary; second, recycle or reclaim waste generated to further reduce the management problem and obtain an economic return for the material; third, reduce the volume and/or hazard of the waste prior to disposal; fourth, employ the proper safeguards in selecting an appropriate disposal option; and finally, at a minimum, operate in compliance with all applicable laws and regulations.

IDENTIFICATION AND RESOLUTION OF A CHEMICAL WASTE PROBLEM

The obvious first step in the process of management of chemical waste is the identification of a given material as a hazardous waste requiring additional care beyond that accorded solid waste which is not hazardous. RCRA defines a "hazardous waste" in Section 1004(5) as:

"A solid waste or combination of solids wastes, which because of its quantity, concentration, or physical, chemical or infectious characteristics may:

(A) Cause, or significantly contribute to an increase in mortality or an increase in serious irreversible or incapacitating reversible, illness; or

(B) Pose a substantial present or potential hazard to human health or the environment when improperly treated, stored, transported or disposed of, or otherwise managed."

Basically, a hazardous waste management program must address those wastes listed in an applicable state program or in the Federal Regulations at 40 CFR Sections 261.31, 261.32 or 261.33; or any waste which exhibits characteristics of ignitability (a liquid with a flashpoint under 140°F), 40 CFR 261.21; corrosivity (aqueous material with a pH of 2 or less or 12.5 or more), 40 CFR 261.22; reactivity (normally unstable or violently reacts with water), 40 CFR 261.23; or toxic (according to an extraction procedure method), 40 CFR 261.24.

The burden is on the generator of a material to determine whether a solid waste is hazardous. This material may be identified as hazardous by regulatory definition or considered as such by the generator. Once a water is classified as hazardous, all parties who deal with the material thereafter must comply with the applicable regulatory requirements, beginning with accumulation, storage and recordkeeping requirements, as well with as any other aspects of the firm's chemical waste management program.

With this background, let us examine some of the waste management techniques available to industrial firms recognizing that not all of the techniques will be available to all firms.

MANAGEMENT OF WASTE GENERATION

Waste Generation Reduction

Once a waste stream from a particular process is determined to be hazardous, the best management solution is to examine the process to determine whether it can be redesigned to operate more efficiently, whether work practices can be improved to optimize efficiency or reduce waste or whether raw materials can be substituted to solve the problem which may include the presence of contaminants. This review could include a chemical process engineering study of the operation to improve its stochiometric performance. Such a study could suggest the need for control or instrumentation improvements such as adding computerized control systems to reduce the possibility of process upsets. It also could show the need for worker training programs to instill in employees the importance of their activities on the environmental aspects of process performance.

The scope of the problem of reduction of generated waste varies dramatically when examining an existing facility as opposed to a new operation. An existing facility may have inherent space limitations or existing process configurations that may limit or preclude otherwise available alternatives. There may also be a significant economic impact in attempting to retrofit a plant.

With regard to a new facility, many waste reduction considerations can be factored into the facility design, engineering and construction, so that a more cost-effective result can be achieved. In addition, a new facility may allow the selection of a more advanced basic process which already incorporates many of these waste-reduction concepts.

One possible aspect of waste reduction which also facilitates reuse or recycle of waste is that of waste segregation. If hazardous wastes can be appropriately segregated from solid non-hazardous wastes or from other categories of hazardous waste, the total volume of hazardous material will be reduced or made more manageable by eliminating the contamination of non-hazardous material. This can be as simple as separating plant trash from chemical waste or as complex as sewer separation or repiping of process areas to achieve the desired result. The key is to start the examination at the initial point of waste generation, and follow the process from there, examining each alternative as it becomes apparent.

Recycle/Reuse

Once a given waste stream has been isolated in theory, other possible uses can be considered for the material which would otherwise become a waste. A waste may be something that does not fully meet the customer's specifications, perhaps it can be reprocesed or blended to comply with specifications, or sold to that customer or another at a lower price without the requirement of a capital investment.

At this point, it may be appropriate to consider all available reclamation alternatives without regard to whether or not they can be performed on-site. The analysis may be considered as another "make or buy" decision similar to those often performed with raw materials. All aspects for each alternative such as operating economics and capital costs, permitting constraints, public opposition and the ability to control the process and any products thereof would be taken into account. The peculiar environmental aspects of the decision become apparent, however, in consideration of other factors. These include potential liability for an off-site disposal facility once a waste is comingled with material from other sources.

Reduction of Volume/Hazard

After all possible alternatives have been evaluated for recycle or reuse of a given waste stream, the material should be examined with respect to the reduction of its volume or the nature of the hazard presented by the material.

Volume reductions can be achieved by techniques such as dewatering sludge or the separating of multi-phase streams. The resulting liquids can be treated in the plant's waste-water treatment operation or sent to a publicly owned treatment works, with the remainder of the material handled as a chemical waste. If the resulting sludge is organic in nature, incineration may be a viable alternative or used as a secondary fuel in industrial boilers or furnaces. Solidification with cement kiln dust or fly ash may be an alternative if the sludge material is inorganic.

Other physical and chemical treatment methods to solidify or chemically fix a material to reduce its hazardous nature also can be used. These processes are described in greater detail elsewhere in this paper.

SELECTION OF A DISPOSAL ALTERNATIVE

Once a waste has been reduced to its lowest volume and least hazardous state, a method of final disposition must be selected. Many factors influence this decision, including the technical limitations of available processes, economic considerations, permitting requirements for on-site alternatives, public relations and transportation concerns.

Treatment Technologies

Regardless of reduction and/or recycling programs, waste will continue to be generated by the chemical industry. Management of the material, of course, depends upon its volumes and characteristics. However, the best management approach will be one that is the most environmentally acceptable.

A variety of technologies and waste-management alternatives exist today. Selection of an appropriate treatment or disposal method is a multi-faceted process.

Physical, chemical and biological treatment are designed to reduce the level of hazard or minimize volumes required for ultimate disposal. Physical treatment, such as, filtering or solidification, can significantly reduce the volumes required for further treatment or disposal. Chemical treatment, which may include neutralization, oxidation/reduction, ion exchange and chemical fixation, are designed primarily to reduce the hazard associated with the waste. Biological treatment involves the utilization of microorganisms to decompose organic wastes. These processes are the most widely used and cost effective techniques for the treatment of industrial waste waters. Available physical and chemical treatment processes are listed on Table 1.

TABLE 1[1]

Physical and Chemical Treatment Processes

Physical Processes or Unit Operations	Liquid-liquid extraction (including liquid membranes and supercritical technology)
Stripping—air or steam	Reverse osmosis
Suspension freezing or freeze crystallization	Ultrafiltration
Adsorption (on carbon, resin, or other absorbers)	Crushing and grinding
	Cryogenics
Centrifugation	
Distillation (including steam distillation)	*Chemical Techniques or Unit Processes*
Electrodialysis	Calcination and sintering
Evaporation	Catalysis
Filtration	Chlorinolysis
Flocculation, precipitation, and sedimentation	Electrolysis
	Hydrolysis
Solid-liquid extraction or dissolution	Microwave discharge
	Neutralization
Flotation and foam fractionation	Oxidation
High-gradient magnetic separation	Ozonolysis
Ion exchange (solid-liquid or liquid-liquid)	Photolysis
	Reduction

[1] National Materials Advisory Board, National Research Council, Management of Hazardous Industrial Waste: Research and Development Needs; Report of the Committee on Disposal of Hazardous Industrial Wastes. National Academy Press, 1983, page 30.

Waste that cannot ultimately be treated requires the selection of a proper disposal alternative. One such alternative is thermal destruction. The technology of thermal decomposition has progressed during the past several years. Wastes that once were landfilled now are being utilized for their energy values in incinerators, or industrial furnaces and boilers. In addition to recovering their energy value, thermal destruction of most organic streams has proven to be a successful disposal option.

Properly designed incinerators can destroy harmful constituents of these streams while also significantly reducing the volumes of material requiring further management.

Waste streams for which there are no other management alternatives must be secured in environmentally sound systems. Secure, well designed landfills, which are properly operated and monitored, presently are the best way to handle these materials. The primary advantage of a secure landfill is its ability to isolate potentially hazardous materials from the surrounding environment. The effectiveness of a successful landfill operation depends primarily on its design. Proper management and placement of materials within the landfill reduces the potential for liquid migration out of the facility. Liquids within the landfill can be collected for treatment by a properly designed leachate collection system. However, the liner system serves as the ultimate barrier between the landfill and its surrounding environment. The design of the liner usually requires materials of natural and/or synthetic components. Depending on geologic and hydrologic conditions, natural liners consist primarily of recompacted clays. Artificial liners usually are constructed of impermeable plastics or rubber materials.

Once it becomes full, the landfill is generally capped with the same impervious liners to prevent leaking and to seal the landfill's contents from the environment. The landfill's internal design usually is complemented by a monitoring system. This is accomplished either by a ground-water and/or leachate monitoring network.

Selection of a particular hazardous waste management control alternative, whether on or offsite depends on a number of factors. Since there are a variety of technologies from which to choose, the least expensive or most convenient alternative may not be the best in the long-term. Although initial cost is important in selecting a particular process, several other factors must be taken into account to determine the overall desirability of a particular alternative. These factors include the availability of resources, degree of hazard, volume of waste generated, physical characteristics, long-term disability and necessary personnel training as well as the potential for harm to the environment or the public.

Contractor Utilization

Selecting an independent contractor to manage a portion of the waste is as critical as proper handling of the material. Retaining a contractor who is unqualified and irresponsible could result in significant environmental harm and expose the generator to significant liability. This liability includes potential responsibility for clean-up and remedial actions for the specific site involved, third party suits for liability, and possible criminal prosecution by federal or state agencies.

To minimize this exposure, PPG utilizes a basic program to apply a degree of control over its waste contractors. The key to this program is the development of an approved contractor and waste disposal site list. PPG will not select and continue to use a site unless it has been inspected on a periodic basis by environmental affairs personnel and has received approval for specific chemical waste streams and disposal options. The key to this inspection process is using objective criteria to evaluate each potential contractor and specific site. Although initial inspections can establish the suitable physical aspects of the operation (permits, design, etc.), the evaluation of the sites managerial attitudes usually requires several visits to be fully appreciated. Some firms believe that it is advisable to limit the number of contractors and sites used in order to better control the risks associated with the use of an independent contractor. Even these firms often maintain more than one useable site to protect against immediate problems should a primary site be shut down on short notice by a breakdown, weather problems or the like. Tables 2-5 contain elements of such a contractor evaluation program. As such, these materials are intended to provide a guide to the development of an appropriate program rather than set forth a rigid approach.

TABLE 2

General Criteria Standards

1. *General Management Considerations*
   a) *Experience of Management and Personnel*—A particular site is only as good as the people who operate the facility. Personnel should have the experience, training and attitude to ensure proper handling of waste materials. This involves a thorough understanding of the regulations and how to operate the facility accordingly.
   b) *Federal, State and Local Permits*—Does the contractor have current federal, state and local permits? Copies of the documents should be obtained and reviewed. In addition, the applicable agencies should be contacted to verify that the permits are current.
   c) *Relationship with Regulatory Agencies*—Has the contractor had any problems interacting with the applicable agencies? This information can be verified through any observed governmental action such as notices of violation, inspection reports, information requests and subpeonas, as well as interviews with regulatory personnel.
   d) *Public Attitude*—Does the contractor have a good rapport with the public? Sponsorship of community and public awareness programs would demonstrate civic responsibility.
   e) *Financial Background*—With the potential increased liability being imposed upon waste handling, it is essential to establish that the contractor financially stable. This can be determined by a review of financial statements, annual reports, and/or credit ratings. It is also important to verify that the contractor carries proper insurance. This can be ascertained directly through the insurance firm.

2. *General Operating Consideration*
   a) *Geographic Location*—What is the approximate mileage from the source to the contractor's facility? The population density surrounding the waste site and the transportation corridors are additional concerns.
   b) *Transportation*—If the contractor is to provide transportation, is its transportation record good? Using a contractor's transportation services has the potential to reduce the legal exposure of the generator by transferring responsibility sooner. Is the contractor using a subcontractor to provide transportation? If so, what is the subcontractor's record?
   c) *Safety Considerations*—Are proper clothing and respiratory protections being used to reduce personnel exposure?
   d) *Waste Limitations*—Can all waste materials be handled? Is care being taken to segregate waste types to avoid dangerous chemical interactions?
   e) *Pre-Disposal and Pre-Treatment Practices*—If necessary, are proper pre-treatment and/or pre-disposal practices being used? If so, are the handling methodologies acceptable?

TABLE 2 (Continued)

General Criteria Standards

f) *Laboratory and Research Facilities*—To ensure proper handling and management of the waste, are adequate sampling and analytical protocols being followed? Facilities should be properly equipped with practical instrumentation and trained personnel to carry out the testing.

g) *Recordkeeping*—Does the contractor maintain an adequate recordkeeping system? Are files accessible and retained accordingly to regulatory requirements? Is appropriate information made readily available to the generator?

h) *Storage Facilities*—Are the storage facilities designed in accordance with safety and regulatory specifications? Is sufficient storage space provided for incoming shipments in the event of a shutdown of the operating unit? Large waste inventories could potentially reflect substandard operations.

i) *Management Practices*—Does the facility have appropriate and adequate analysis, training and emergency preparedness plans?

j) *Design Life*—How long does the company expect to continue present operations? What provisions have been made for projected control after closure?

In addition to these general criteria, the facilities also should be inspected for the specific technologies being used. Design and construction specifications should be evaluated for landfills, incinerators and treatment/storage/reclamation facilities. Specific design and operating criteria would encompass the following:

TABLE 3

Landfill Evaluation Criteria

1. *Geological and Soil Characteristics*—Critical design criteria of any landfill are its geological location, soil composition and risk of seismic activity. The type of soil determines the potential for leachate migration. Location of a landfill on a geological fault or fracture could increase the potential for liner failure.

2. *Hydrology*—The depth of the local groundwater table is important. Also, the nature and use of the surrounding groundwater system—perched/flowing, brackish/fresh—must be considered.

3. *Proximity of Surface Water*—What is the potential for surface run-on/run-off or flooding? Does the local topography promote proper run-on/run-off control? What type of treatment is used for contaminated run-on/run-off of water?

4. *Climate*—Ground may freeze in colder climates, thus causing operational difficulties including frost heaving which may affect the integrity of the liner or the cap. Odor and fugitive dust problems may occur in hotter climates. Significant quantities of rainfall may present concerns in wetter climates.

5. *Liner Design*—Is the liner design based on sound engineering? Is the composition synthetic, natural or both? If it is natural, have compaction studies been performed to ensure integrity? When the liners are artificial, are the liner materials compatible with the wastes?

6. *Leachate Collection System*—Is the leachate collection system designed to promote proper dewatering of the landfill? How is this system managed and the leachate treated?

7. *Segregation of Materials*—Is the chemical composition of the waste evaluated for proper segregation of materials? Does the landfill design assure segregation of interacting materials?

8. *Cover/Cap*—After daily operations, are the waste materials covered with inert media? Once it becomes full, is the landfill capped with an impervious liner to provide proper run-on and run-off?

9. *Groundwater/Leachate Monitoring Detection Systems*—Is a groundwater/leachate monitoring system installed to monitor liner failure and leachate migration? How is the system managed and analysis performed?

TABLE 4

Incinerator Evaluation Criteria

---

1. *Mixing and Blending Operation*—Are proper precautions taken to minimize waste reactions during mixing and blending operations?
2. *Incinerator Design*—What is the design? Do design parameters meet regulatory requirements?
   a. *Capacity*—Are waste inventories comparable to design capacities?
   b. *Down Time*—Has the unit been out of operation for unusually long periods?
   c. *Feed System*—Are feed systems designed to enhance incineration?
   d. *Combustion Temperature*—Does operating temperature ensure proper combustion.
   e. *Residence Time*—Is proper residence time allotted for complete combustion? How is residence time calculated?
   f. *Emergency Shutdown Controls*—Are proper warning and emergency shutdown systems in place? How are they verified?
3. *Pollution Control Equipment*—Is the design of the pollution control equipment based on sound engineering?
4. *Performance Testing*—When and what types of tests have been performed on the control equipment? Who conducted the sampling and analysis?
5. *Management of Control Equipment Residue and Ash*—How are control equipment residuals managed? Are the residuals landfilled, if so, where? If the residue is claimed to be nonhazardous, what is the waste analysis plan? Has the residue been officially "delisted" under federal or state law?

TABLE 5

Storage/Treatment and Reclamation Facility Criteria

---

1. *Operation*
   a. *Storage Facilities*—Is the storage facility designed in accordance with safety and regulatory specifications? Is sufficient space provided for present inventory?
   b. *Treatment Facilities*—Are proper treatment safeguards folowed to reduce any adverse effects from chemical interaction of the waste? Have treatment methodologies being employed been demonstrated effective?
   c. *Recycling/Reclamation Facilities*—Is there an incentive to recycle the material? What would be the legitimate use for the materials being reprocessed? Is the proposed reuse beneficial?
2. *Disposition of Waste*—How are waste materials generated from these operations ultimately disposed? Are similar disposal controls being practiced as presented in these protocols?

Upon the completion of this evaluation, the site then can be approved or rejected for the corporate site list. This approved site list then can be circulated to in-house personnel who are responsible for the day-to-day selection of disposal contractors and sites for disposal of specific waste streams. An essential element of the selection procedure is to preclude the sending of a chemical waste stream to a facility that is not on the approved list for such stream without the prior written approval of appropriate environmental officials.

Waste Contract Program

If a material is to be handled offsite by an independent contractor, it is critical that an effective industrial waste disposal agreement be entered into by the parties. The purpose of this agreement is to clearly identify the relationship and expections of each party with respect to the waste to be handled. Like any other commercial transaction, many of the elements of the agreement are obvious but in light of the potential residual liability after the contractor has performed its services, many additional considerations must be examined. The traditional practice of using general or "blanket" purchase orders for all contractor services is simply inappropriate for hazardous waste disposal.

PPG Industries has attempted to approach this problem by entering into national contracts with major waste disposal firms so that specific agreements need not be negotiated for each plant or each waste stream or shipment. This process contemplates that PPG and the contractor will negotiate and arrive at an agreement which sets forth the basic relationship between the parties in all respects except for specific volumes to be shipped and the prices to be paid for the contractor's services. These latter items are negotiated by specific plants that have responsibility for waste management. In this way, PPG and the contractor can be assured that each of its individual concerns are addressed in a comprehensive way in one document. This document then can be incorporated by reference into specific purchase orders for disposition of given waste streams.

This program does not eliminate the need to negotiate specific waste-disposal agreements for certain waste streams or for plants where disposal services are not available from those firms with whom PPG has national relationships. However, it reduces the number of those special contracting situations.

Key elements to be addressed in the Industrial Waste Disposal Agreement are set forth in Table 6.

TABLE 6

Key Elements of an Industrial Waste Disposal Agreement

1. Parties - Who is ultimately responsible for performance
2. Scope of Work/Nature of Work
3. Disposal Charge and Billing Procedures
4. Title, Risk of Loss and Demurrage
5. Means of Disposal - Selection by Whom and Alternatives
6. Disposal Indemnity - To continue after performance of Agreement
7. Governmental Permits
8. Force Majeure
9. Insurance - Type amount and allocation of cost
10. Disposition of Rejected or Non-conforming Material - Controlled by whom
11. Contractor's status as independent contractor
12. Confidentiality of proprietary information
13. Prohibition or Limitation of Assignment and Subcontractor

The contract for a waste management or handling service essentially should define the parties who are engaging in the relationship, should identify what is to be done, at what cost and when and should specify the responsibilities of the parties if something goes wrong. It is critical for the contract to determine the point at which title to the waste is transferred to the contractor and the extent to which the contractor assumes present and future liability for the waste.

Special attention should be paid to subsidiaries or affiliates and possible assignment or subcontracting for a portion of the work. As with any supply relationship, it is important to know who is performing what aspect of the work and examine that operation as closely as any supplier would be scrutinized.

What is the waste to be handled? It is especially important to adequately define the waste recognizing that there may be variations in the specifications. It should be recognized that the waste by its nature is a heterogeneous material. Accordingly, the contractor must impose some reasonable limits on the range of material which it can accept. The handling of material that the contractor cannot accept must be considered. In addition, consideration must be given to defining the desired result if the contractor has mixed a portion of the material with waste from another source.

With regard to disposal and billing charges, until proper notification that disposal or treatment has been completed, payment of the invoice should be withheld. Who has title to the material at each point and who has responsibility for risk of loss or delay in shipment must be carefully examined.

The contract obviously should address the method of disposal as well as possible alternatives if there the material cannot be disposed of by the first alternative.

A key feature in the agreement should be the disposal indemnity. Through the disposal indemnity, the contractor recognizes and assures responsibility for his performance of the agreement. This responsibility includes the time after expiration of the contract during which the waste remains under the control of contractor, such as in a landfill situation. A related feature is the idea of insurance to support the indemnity obligation.

Finally, it may be appropriate to address the confidentiality of the precise nature of the waste, if such information could be used by a competitor to identify a proprietary technological aspect of the firm providing the waste. This concept is complicated by requirements to manifest the waste and otherwise disclose information to governmental bodies in permitting applications.

Obviously, other typical contractual terms must be addressed and included in the waste-disposal agreement. However, the key fact is that such aggreements are not merely normal supply relationships. Environmental expertise (technical as well as legal) must be included in the management team contracting for the service.

CONCLUSION

The management and handling of hazardous waste must be performed so as to avoid harm to the environment or people who may be expected to come in contact with the material. This objective must be acccomplished in a way that compliments the goals of the organization to optimize its production of saleable products, while effectively controlling the cost of chemical waste management and disposal. These goals can be achieved best by having a defined chemical waste management program that takes into account the current state of the laws and regulations regulating such activity as well as the nature of the firms business and the evolutionary nature of technology in the area of chemical waste management.

Hazardous and Toxic Wastes: Technology, Management and Health Effects. Edited by S.K. Majumdar and E. Willard Miller. © 1984, The Pennsylvania Academy of Science.

Chapter Twenty-One

Benefits of Managing Hazardous Wastes

Ann Fisher, Ph.D.
Economist
U.S. EPA, 401 M St., S.W.
Office of Policy Analysis (PM-220)
Washington, D.C. 20460

During the past decade, there has been increased concern about unwanted toxic substances that are either by-products of various manufacturing processes or the ultimate wastes of discarded products. This concern led Congress to pass the Resource Conservation and Recovery Act (RCRA) in 1976, and the Comprehensive Environmental Response, Compensation, and Liability Act (CERCLA, or Superfund) in 1980. Charged with the responsibility for these laws, the U.S. Environmental Protection Agency (EPA) has issued several regulations for implementing them.

One component of the complex rule-making process is determining how stringent a regulation must be to protect human health and the environment. Since there are various degrees of protection, we need information on the benefits and costs of different levels of controlling hazardous wastes so we can decide whether the last dollar we spend on such control is really worth it.

Using well-established methods and data from experience with existing facilities, we can estimate fairly accurately the costs of various types of facilities for treating and disposing of hazardous wastes. But predicting the benefits of managing hazardous wastes is more difficult.

This chapter explores why measuring the benefits from restricting toxic wastes is important. It defines benefits and describes alternative approaches that have been used or suggested for measuring them, explaining the advantages and limitations of each, and summarizing what we can expect in the near term with respect to their accuracy and applicability.

RATIONALE FOR MEASURING THE BENEFITS OF ENVIRONMENTAL REGULATIONS

To ensure the economic efficiency of government regulations, President Reagan signed Executive Order 12291 on February 17, 1981 (1). According to this order, before taking final action on major rules, government agencies must prepare regulatory impact analyses (RIAs) and submit them to the Office of Management and Budget for review. Each RIA must state the need for the proposed regulatory action, quantify its benefits and costs, value them in dollar terms to the extent possible, and compare them with the benefits and costs of alternative ways of solving the problem. It must also discuss the unquantifiable or nonmonetizable benefits and costs of the proposed regulation and compare their importance with those valued in dollar terms.

EPA has responded to the spirit as well as the letter of the President's order by drafting *Regulatory Impact Analysis Guidelines* for its offices to follow when developing new regulations (2). The document stresses the importance of both performing sensitivity analysis for parameters and estimates that have substantial uncertainty, and considering distributional issues and nonmonetizable and unquantifiable effects along with efficiency in the decision process.

DEFINING AND MEASURING THE BENEFITS OF ENVIRONMENTAL REGULATIONS

Since the benefits of environmental regulation typically are not exchanged in markets, they are more difficult to measure and value than ordinary commodities. A simplified view for an ordinary market commodity is shown in Figure 1. The demand curve, D, shows the quantity that will be purchased at each alternative price. It assumes that other factors influencing demand—such as incomes, other prices, and tastes—are held constant. For ordinary commodities, such demand curves can be estimated from data on actual sales. As would be expected, consumers purchase more of the commodity at lower prices. If the market price is P_o, consumers will purchase Q_o units, and their costs will be price times quantity, or OP_oAQ_o.

Points on the demand curve show consumers' willingness to pay for each additional unit of the commodity. When they purchase Q_o units, their total willingness to pay is OP_jAQ_o. The excess of willingness to pay over total expenditures (P_oP_jA) is called consumer surplus.

For an ordinary market commodity, a person must purchase the amount he wants to consume. Some commodities, however, are *public goods* in the sense that if they are provided at all, they automatically become available to a large number of people. National defense, radio and television signals, and environmental quality are examples of public goods. Rather than purchasing

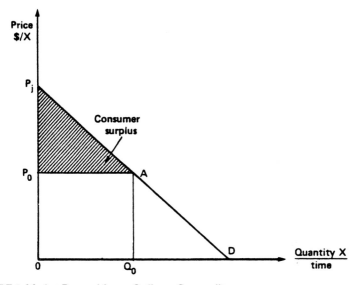

FIGURE 1. Market Demand for an Ordinary Commodity.

public goods directly, the costs of these commodities tend to be covered by taxes or by higher prices for ordinary market goods. Since the price of an environmental good to an individual consumer—e.g., reduced risk from hazardous waste pollution—typically is zero, consumer surplus equals total willingness to pay for such goods. Thus, the change in consumer surplus resulting from a change in a particular environmental commodity provides an appropriate measure of its benefits.

Unfortunately, estimating consumer surplus for an environmental commodity is less straightforward than for most goods. This is because it usually is easier to define a market good (e.g., a quart of strawberries) than an environmental good (e.g., the benefits of toxic waste management), and because there are no observed sales of the environmental good to serve as a basis for its demand curve.

A first step to get around this problem is to define the benefits of hazardous waste management in terms of physical units. Preventing toxic waste pollution avoids damage to human health, plants, animals, and materials.

In some cases, we can trace the release of a particular pollutant to its impact on ambient environmental quality, to the resulting exposures and adverse effects, and ultimately to what people would pay to avoid those effects. Possible mitigating behavior, such as using bottled water and receiving medical care, makes this chain of effects difficult to model. In other cases, insight on willingness to pay for hazardous waste control can be inferred from consumers' purchases of goods that are somewhat related to the environmental commodity of interest. However, preliminary results for measuring the benefits of hazardous waste management using this approach have been discouraging. Thus, particularly when there is substantial uncertainty as to the magnitude or types of pollution impacts, it may be more feasible to examine the value people place on reducing the discharge itself.

Two basic approaches are used to evaluate the benefits of proposed regulatory actions. The first measures the reduction in physical damages to people and ecosystems. The second examines society's valuation of the regulatory action. This valuation may be in terms of asking how important those physical effects are. It also may be in terms of a more general question of how important is a reduction in the risks from hazardous waste pollution.

MEASURING REDUCTIONS IN ADVERSE PHYSICAL EFFECTS

The most common method for measuring the benefits of environmental regulations examines the resulting reductions in adverse physical effects to people, plants, animals, and materials. Often called the damage function approach, this technique requires substantial modeling and availability of data, and a number of judgments about how to handle various uncertainties. The damage function approach relies on four basic modeling components: facility failure, transport and fate, exposure, and effects.

Facility Failures

The first requirement for the damage function approach is the ability to model expected failure in hazardous waste facilities. This modeling is needed for alternative technologies within a particular facility (such as clay v. plastic liners for landfills) and for alternative types of facilities (such as landfills v. incinerators). Given the relatively short time since hazardous wastes have been recognized as a significant problem, this often means that predictions must be made for new technologies where experience is quite limited. Predicting facility failures is complicated by the need to consider a long time frame, since leaks may not occur for many years, and because there may be a substantial lag between exposure and any adverse effect. Nevertheless, good information is available for predicting some facility failures.

Transport and Fate

If a facility has a failure, it is necessary to find out what happens to the contaminants released. But predicting how far and how fast the contaminants will move—through surface water, groundwater, air, or some combination of these—depends on information about hydrogeological and meteorological conditions at the site. Any reactions that occur among contaminants, or between contaminants and the soil, also can affect movement. Such reactions may yield compounds that are either more or less dangerous than the toxic materials originally disposed of in the facility. Although the transport and fate of substances released from hazardous waste facilities can be modeled, such efforts are expensive and time consuming.

Exposure

The next step is to model the exposure of population groups and ecosystems to the contaminants. The sources of exposure vary—water ingestion, inhalation, skin contact—and actual doses will depend on the concentration of the pollutant and its pattern over time. Information about whether the exposure is to one contaminant or to many can help in predicting effects (which is the final step). Policy choices about how to reduce exposures may be influenced by how exposure from the facility's failure is related to other sources of exposure to the same contaminants. Reasonably good information is available for modeling exposures.

Effects

The last step in predicting physical damages requires the ability to model the effects of potential exposures on people and ecosystems. That is, toxicity analyses, or dose-response functions, are applied to expected exposure levels. Estimating a dose-response function is complicated by the fact that a given level of airborne exposure may not yield the same effects as the same level of exposure through food or drinking water. Because EPA's primary mission is to protect human health, it has worked more on predicting dose-response functions for humans than for plants, animals, and materials. Although some information is available for nonhuman damage from exposure, the emphasis here is on the four major approaches for predicting human health effects (3).

Epidemiological Studies

Epidemiology usually is defined as the empirical, statistical study of the relationships between human diseases and their causes. Scientists and regulators tend to regard evidence on actual human effects as more transferable to the general population than information from laboratory animal experiments, short-term in vitro tests, or structure-activity analysis.

Unfortunately, epidemiological studies typically have numerous shortcomings. Although clinical studies with controlled exposure levels are available for a few substances, ethical considerations and recent legal restrictions severely limit this avenue of research. Instead, most epidemiological studies gather and analyze data from "the real world," and so have all the problems of poor design whenever using nature's experiments.

For example, in a cohort study, groups with differing exposure to a suspect substance are followed over time to determine differences in response. The expense of keeping track of each member in the cohort tends to make sample sizes small. This means it is hard to detect relatively small changes in health effects as related to specific causes.

In case-control studies, a group with a disease is compared with a control group without it to see if exposures differ for the two groups. It may be difficult to find many people—in either group—for whom exposures can be estimated with any confidence. Even for people with exposure information, the data may be too limited to account for all confounding influences—such as smoking, diet, exercise, age, sex, race, and income. However, since one of the groups has been identified to have the disease in question, the case-control method is more useful for studying rare effects than the cohort approach where samples sizes would have to be much larger for rare effects to show up.

No matter which design is used, epidemiological studies rarely yield unequivocal answers about the effects of toxic substances on human health. To get high enough exposures (either through a few large doses or through low dosing over a long time) to lead to a statistically significant increase in the health effect, occupational exposure studies often are used. Since job holders are healthier than the general population, dose-response functions based on occupational exposures may understate the risk for susceptible groups exposed to much lower expected doses from failures at hazardous waste facilities.

In epidemiological studies, synergism and exposure to multiple hazards often make sorting out the effects of single substances difficult. At the same time, people exposed to hazardous wastes usually are exposed to a mixture of toxic substances. Simply adding the risks calculated on individual components of the waste stream will give a misleading risk estimate if some of the components are more (or less) potent in the presence of others in that waste stream.

On balance, positive results in an epidemiological study give strong evidence that a substance is harmful, even if the data are insufficient to give a detailed dose-response function. On the other hand, negative results often are weak evidence of a chemical's safety, because of small sample sizes, inaccuracies in exposure data, and the inability to account for all confounding factors. A final limitation of epidemiology is that no data exist for new substances that are or will be entering hazardous waste disposal sites.

Laboratory Animal Experiments

Tests on animals avoid the ethical issues of human experimentation. They also reduce the influence of confounding factors and the time needed to observe responses since test animals have shorter natural lifespans. However, they introduce new uncertainties.

Because of the difference in size and metabolism, animal doses must be converted to equivalent human doses. This still leaves a question about whether test animals react the same way humans do. For substances that we know adversely affect both humans and animals, actual human response may differ from that predicted on the basis of animal response by one or two orders of magnitude. This may reflect synergism with other substances in the more realistic setting for people.

Negative results in animal tests do not prove that humans will not react to the substance. For example, arsenic is a known human carcinogen, but has failed to cause cancer in laboratory animals. However, scientists think substances that adversely affect animals are likely to harm humans, even when no epidemiological evidence is available for comparison.

Because of the expense of conducting animal experiments with very large samples, high test doses are used to increase the chance that a harmful substance will show an adverse effect. It is then necessary to extrapolate from the high test doses and observed responses to predict responses at the low doses expected from exposure to hazardous wastes. Typically, several plausible dose-response functions fit the observations equally well, but imply widely differing risk levels at the expected low exposures. These dose-response functions predict responses that differ by several orders of magnitude, compared with the response varying by one or two orders of magnitude depending on the conversion of animal to human doses.

Short-term (in vitro) Tests

The high cost and time required for animal experiments spurred the development of several short-term tests. Rather than studying the animal over its lifetime, these tests observe bacteria, mammalian cell cultures, or small organisms for a few days to a few weeks after exposure to a suspect substance.

One argument is that the selected cells in the laboratory Petri dish may not react as the whole animal (or person) will. They may not reflect the target tissues of a substance under study. On the other hand, the Petri dish can be viewed as the worst possible case because the cells cannot excrete the substance or use other parts of the whole organism to combat its effects.

Most of the short-term tests identify specific types of mutagens and carcinogens, so that a battery of tests is more likely to identify a potential human carcinogen. Relatively few human carcinogens have been positively identified, so there is limited information available for judging the concordance between short-term tests and human responses. The concordance is imperfect for those short-term tests and lifetime animal studies that already have been compared.

For noncancer endpoints, the few existing short-term tests are expensive and time consuming, negating some of the advantages claimed for others. The scientific community believes that short-term tests can support existing animal or epidemiological results and indicate the need for additional study. However, they are not definitive by themselves.

Structure-Activity Analysis

In a structure-activity analysis, the molecular structure of a suspect substance is compared with those for which toxicity and metabolic pathways are well known. This is a new field, so the data base is relatively weak for some classes of chemicals. Most scientists believe this approach may be useful to support further testing, but are unwilling to accept it as definitive in identifying toxic substances. Although there is some evidence that short-term tests can be used to develop dose-response functions, more research is needed to determine whether structure-activity analysis can predict responses for alternative doses.

Benefits and the Damage Function Approach

Estimating the reduction in physical health effects has substantial appeal as a way to measure the benefits of managing hazardous wastes. Despite the many uncertainties already discussed, this approach can yield estimates of the number of lives saved or illnesses avoided. These estimates could be for a worst-case scenario, or for the most likely scenario. In some instances, recognizing the uncertainties may yield such wide ranges that the estimates have little use. Substantial research is under way to reduce these uncertainties. However, some of them are more likely to be resolved than others.

Since the damage function approach by itself measures only changes in risk, another difficulty arises when the results are to be used to choose among alternative policies. In a simple example, suppose two hazardous waste management programs each would cost $10 million. Program A would save 10 lives and avert 50 cases of kidney disease. Program B would save 5 lives and avert 100 cases of kidney disease. Which program should we choose? If multiple types of illness are included, the decision becomes even more complex.

The way out of this dilemma requires independent valuation of damage function estimates. Alternatively, an entirely different approach may determine society's willingness to pay for reducing hazardous waste risks in general, without requiring details on the specific numbers and types of physical effects involved.

MEASURING SOCIETY'S VALUATION OF CONTROLLING HAZARDOUS WASTES

Valuing Physical Effects

By estimating how important reducing physical damages is, we can ask whether the predicted reduction in risk is worth the cost. There are precedents for assigning values to health effects (2), (4). For example, morbidity can be valued in terms of its associated medical costs. This approach may understate an individual's willingness to pay to avoid the illness because it ignores pain and suffering. On the other hand, it may overstate the person's willingness to pay because insurance reimburses many of the medical costs. The insurance coverage may be offset by loss of earnings if the person is not paid for sick leave.

Most of the estimates for valuing the risk of mortality come from studies of wage premiums in risky occupations (4). Although all of these studies have shortcomings, they show that workers require $400 - $7,500 in additional annual income to voluntarily accept increased annual job-related risks of death of about one in a thousand. A regulation that reduces risk of death to each of 1,000 people by .001 would save one life in expected value terms. The increased safety would then be worth $400,000 - $7,500,000. Even though this is an a priori analysis of the value of safety to 1,000 people, among whom one person—unidentifiable in advance—is expected to die, rather than an ex post measure of compensation for a particular death, many decision makers are uncomfortable with the idea that death (or illness) can be assigned a dollar value. When this reluctance to value health effects is combined with the difficulty of estimating the changes in physical damages, other approaches to measuring society's valuation of the benefits of managing hazardous wastes seem worth considering.

Indirect Approach: Housing Values
   One way to infer society's demand for a public good is to examine how market transactions for ordinary goods might reflect it. Housing is the commodity that has been studied in relation to hazardous wastes. The rationale behind this approach is that if people prefer to avoid risks of exposure to hazardous wastes, then they will pay more for a house located farther away from a hazardous waste facility than they will for the same house located closer to the facility.
   Public Interest Economics (PIE) analyzed Pleasant Plains, New Jersey, a small community where an illegal dumping operation led to groundwater contamination (5), (6). When it learned of the contamination, the city rapidly provided municipal water to the households with wells in the contaminated zone. Although these residents complained that well water tasted better, and they had to pay part of the cost to hook up to the municipal water supply, there have been no further contamination problems. PIE also studied Andover, Minnesota, a small town threatened by contamination from an old asphalt-lined dump. Apparently, residents had little perception of a potential problem, so it is not surprising that proximity to the dump was not a significant determinant of property values in Andover.
   PIE examined the characteristics of houses, including their distance from the dumping site, along with selling prices before and after the contamination. Preliminary results for Pleasant Plains showed the expected "rent gradient" (i.e., higher housing prices further from the site) after the contamination. However, when PIE collected additional data, the rent gradient disappeared.
   This result has many possible explanations. For example, distance may not be an adequate proxy for risk. Or PIE may have omitted important housing characteristics, so that some confounding influence overwhelms the risk effect. Perhaps buyers and sellers perceived a very low level of risk. Or few sales may have occurred between the problem's discovery and correction. Finally, concern about hazardous waste risk may be such a relatively small component in the decision to purchase a house that its effect simply gets lost in the statistical "noise."
   As part of a larger project to compare alternative methods for measuring the benefits of controlling hazardous wastes, EPA is sponsoring another property

values study. Faculty from Harvard's Kennedy School of Public Policy are examining housing transactions in the suburban Boston area to see if proximity to a hazardous waste site affects property values, and to compare any such effects with those of an ordinary landfill. Results will be available in 1984. So far, however, the one completed study shows little promise for using this approach to measure the benefits of controlling hazardous wastes.

Direct Approach: Contingent Valuation Surveys

Another way to measure the benefits of managing hazardous wastes is to define a hypothetical market and use a survey to ask people about their willingness to pay for alternative levels of control. Variants of this method are all classified as contingent valuation surveys.

Benefit estimates from such surveys can be affected by several types of bias (7). These include: *strategic bias,* where the respondent tries to influence the outcome; *information bias,* where results are affected by the types and amounts of information available to each respondent; *hypothetical bias,* where respondents to not react as they would in an actual market situation; *instrument bias,* particularly for the payment vehicle and for the starting points chosen to initiate bidding; and *sampling, interviewer,* and *nonrespondent* bias.

Evidence on the importance of these biases is mixed, but most writers have found strategic bias to be small. Information bias can be minimized by making sure that all respondents have access to the same relevant information, and hypothetical bias is reduced if the contingent situation is realistic. Payment vehicle bias is reduced if the chosen vehicle seems appropriate to the respondents. Sampling bias can be minimized by designing the sample to reflect the target population, and interviewer bias can be minimized by training and using professional interviewers. Non-respondent bias is reduced as the response rate rises. The usefulness of each contingent valuation study can be evaluated in terms of the above considerations.

As with the property value studies, the contingent valuation approach has the advantage that we need not estimate the number of each type of health effect associated with the different control levels. Instead, we can ask respondents about their perceptions of risk associated with hazardous waste facilities and their willingness to pay to reduce that risk.

During 1982, Cummings, et. al. used a preliminary contingent valuation survey for valuing toxic waste regulations (8). Rather than asking about willingness to pay for reducing risks of exposure to hazrdous wastes, they asked respondents to consider a hazardous waste containment policy itself. Because our knowledge of expected exposures and expected effects for given exposures is limited, a containment policy acts as a hedge against uncertain health risks. Willingness to pay for the regulatory action will depend on respondents' perceived probabilities of damages from exposure, and their perceived probabilities of containment (or, conversely, exposure) with and without the policy. More research may show when such a regulation would be an adequate surrogate for the change in risk. (For example, respondents' bids may have been influenced by their perceptions of how effective government actions are.)

This study examined several important questions, such as the effect on willingness to pay of the amount of information people have about toxic wastes, whether people focus on the commodity "hazardous waste containment policy" or on a more general good "environmental safety," and whether bids are smaller when people are asked about the spending category that would be reduced so that they still could meet their budget constraint. An attempt was made to determine whether alternative sample designs affected survey costs, and results were compared across three cities to see if regional differences emerged.

Because of the exploratory nature of the Cummings, et al. project, few conclusions can be drawn with any confidence. Sample sizes were small, and response rates were very low, so that the responses may not be representative of the target groups. Wide variability in the willingness-to-pay amounts for these small samples limits the usefulness of the estimates. In addition, as problems surfaced in the questionnaire, it was modified so that cross-city comparisons are difficult to interpret.

The most important conclusion from this study is that the contingent valuation approach indeed may be useful in evaluating the benefits of hazardous waste regulations. Their report implies that the commodity to be valued should be carefully defined, and that more information is needed on how to measure risk perceptions and how to communicate changes in risk.

In a more recent attempt to grapple with the difficulties of valuing the benefits of controlling hazardous wastes, Desvousges and Smith stress the importance of accounting for people's perceptions of the joint probability of an adverse effect given some exposure (9). They recognize the difficulties suggested in some of the psychology literature with respect to choices involving small probabilities. Even so, they conclude that the contingent valuation approach offers the most promise for measuring the benefits of controlling toxic wastes, because the commodity to be valued can be defined in terms that are meaningful to respondents.

In response to this conclusion and as another component of EPA comparison study mentioned earlier, Desvousges and Smith are developing a refined contingent valuation instrument for use in the Boston area. This effort is viewed as a second-generation attempt. If the results are as promising as the early indications, a third-generation project will be designed to develop estimates that can be applied to regions and the nation, rather than just to individual sites. Results from the second-generation study should be available in 1984.

CONCLUSION

Given the extreme visibility of toxic waste issues, and the limited resources available for resolving these problems, it is crucial to have a reliable way of determining which regulatory actions are most important. Traditional economic analysis tells us to examine both the benefits and the costs of proposed actions, to choose the most efficient ones.

In physical terms, the main benefits of controlling hazardous wastes are reduced health risks. With sufficient time and money we can predict a facility's failure and how the hazardous substances travel through the environment. However, even when our information on exposure is reasonably good, predicting health effects may be difficult, especially because of uncertainties in extrapolating from high test doses to the much lower doses likely after a facility's failure and in converting animal doses to human equivalents.

Although progress is slow because of high costs and the long time frame needed for animal and human studies, more information about dose-response functions is being generated all the time. Future advances in short-term tests and structure-activity analysis may lower the costs and time involved in predicting health effects for expected exposure levels.

To balance the reduced health damages against the regulatory costs, we need to value health effects. We can use the broad range of value-per-statistical-life estimates from occupational studies or the medical-costs-avoided approach. In some cases, the estimates of physical damages and the broad ranges for valuing health effects may yield benefits estimates that are highly uncertain. As an alternative, we can find a more direct estimate of the public's willingness to pay for reducing health risks. Although the only completed study to use this direct approach in a toxic waste context was exploratory, more refined work is in progress. To the extent that new results can be extrapolated to similar situations, they will be useful as a complement to whatever information is available on physical health effects for a proposed regulatory action.

* The views presented are the author's alone and do not represent the views of the U.S. Environmental Protection Agency.

REFERENCES

1. *Federal Register,* February 19, 1981, 13193-13198.
2. U.S. Environmental Protection Agency, *Guidelines for Performing Regulatory Impact Analyses,* draft, various dates. A June 22, 1982 version appeared in the *Environment Report,* July 16, 1982, pp. 385-392.
3. Fisher, A. 1982. The Scientific Bases for Relating Health Effects to Exposure Levels. *Environmental Impact Assessment Review.* 3/1: 27-42.
4. Violette, D. M., and L. G. Chestnut. 1983. *Valuing Reductions in Risks: A Review of the Empirical Estimates,* U.S. Environmental Protection Agency, Washington, D.C., EPA-230-05-83-002.

5. Adler, K. J., R. C. Anderson, Z. Cook, R. C. Dower, and A. R. Ferguson. 1982. *The Benefits of Regulating Hazardous Waste Disposal: Land Values as an Estimator,* research report, Public Interest Economics Center, Washington, D.C.

6. Cook, Z., K. Adler, A. R. Ferguson, and M. J. Vickers. 1983. *The Benefits of Regulating Hazardous Waste Disposal: Land Values as an Estimator.* March draft research report, Public Interest Economics Foundation, Washington, D.C.

7. Schulze, W. D., R. C. D'Arge, and D. S. Brookshire. 1981. Valuing Environmental Commodities: Some Recent Experiments. *Land Economics.* 151-172.

8. Cummings, R. G., H. S. Burness, A. F. Mehr, W. S. Walbert. 1983. A Policy Bid Experiment for Valuing EPA Regulations on Hazardous Waste Disposal. *In: Experimental Approaches for Valuing Environmental Commodities: Vol. II,* draft, W. D. Schulze, R. G. Cummings, D. S. Brookshire, eds., U. S. EPA Grant #CR808-893-01, Washington, D.C.

9. Desvousges, W. H., and V. K. Smith. 1983. *The Benefits of Hazardous Waste Management Regulation: Technical Report.* Research Triangle Institute, Research Triangle Park, N.C.

Hazardous and Toxic Wastes: Technology, Management and Health Effects. Edited by S.K. Majumdar and E. Willard Miller. © 1984, The Pennsylvania Academy of Science.

Chapter Twenty-Two

Empirical Estimation of Hazardous Waste Disposal Damage Costs

Teh-wei Hu, James A. Weaver, and D. Lynne Kaltreider**

** The authors are Professor of Economics, The Pennsylvania State University, University Park, Pa. 16802, Manager, Economics Studies, Engineering-Science, Inc., and Research Assistant, Institute for Policy Research and Evaluation, The Pennsylvania State University, respectively.

For a government or society to enact environmental policies that will protect human health and the environment, it is important to know the costs and benefits of implementing these policies. The disposal of hazardous wastes has become an issue of concern both with regard to the environment and human health. However, current available literature does not contain sufficient empirical information on either the environmental damage costs resulting from various waste disposal practices or the measurable benefits of improving the environment, much less the effects of waste disposal practices on human health. A recent study entitled, *Costs of Environmental-Related Health Effects,* conducted by the Institute of Medicine of the National Academy of Sciences (January 1981), points out that the methodology for estimating improvement in human health as a result of improvement of the environment is still in its infancy. The chairman of the study committee, Professor Kenneth Arrow, a Nobel Prize Economist, has assessed the state of the art of estimating costs or benefits of environmental-related health effects as follows:

Pollution or other man-made alterations in the environment affect health through a two-stage process: (1) the pollution enters an ambient medium to which human beings are exposed; (2) the exposure causes a response in terms of health effects. For comparison with production costs needed for decision making, a third step is needed, an equivalence between health effects and costs in terms of resources. The study quickly found that data on all three stages were sadly lacking, and the bulk of the report concerns itself with analysis and recommendations for better information. However, our discussion of the first stage, the impact of sources of environmental hazards on human exposure, must be regarded as inadequate, due to lack of time and of relevant expertise on the committee. Further study of this area is of high priority.

The purpose of this paper is to illustrate the procedures for estimating the economic costs incurred by society vis-a-vis the disposal of hazardous wastes. At the same time, the paper provides an opportunity to examine the status of data inadequacy and make recommendations for better data collection in the future in order to provide a more adequate date base for examining the relationship between waste disposal and its environmental health effects.

The original intent of this study was to estimate the damage costs resulting from coal waste disposal sites. However, because many of these disposal sites are located on utility plant sites, there is no public disclosure of the disposal effects, and consequently no data for estimating damage costs are available. As a result, this paper will focus on disposal sites that contain some of the same elements as coal wastes as a basis for estimating damage costs.

This paper first identifies the existing data on estimating hazardous waste disposal damage costs (Section II). Section III discusses the conceptual methodology for estimating damage costs. Empirical examples of damage cost estimation are presented in Section IV. The data limitations and recommendations for estimating environmental data costs conclude this paper (Section V).

II. *Review of Previous Studies on Hazardous Waste Disposal Damage Costs*
A. *Identification and Description of Data Sources*
The information provided in this review comes from a variety of materials. The majority are U.S. Environmental Protection Agency (EPA) publications and reports from EPA files. In preparing this paper, we reviewed EPA case studies of damage incidents as well as background documents and reports and a draft economic impact analysis of the Resource Conservation and Recovery Act (RCRA) prepared by EPA.

A review of these materials revealed basically three categories of data presentations. Each will be described here briefly.

1. *Detailed reports.* EPA has issued a series of "current reports on solid waste management." This series includes three case studies (June 1976) of damages resulting from leachate from municipal land disposal sites. Each of these reports provides (a) a history of the development of the disposal site (including type of operation, type and amount of waste disposed of, design and description of the site), (b) damage assessment (including damages, litigation, remedial action, characteristics of the leachate), and (c) damage costs (in the most complete of the studies, this is broken down into tangible direct damage costs, intangible direct damage costs, avoidance costs, corrective costs, and administrative costs). In each case where well water was tested, a complete breakdown of the chemical constituents found in the water is provided with clear indication of which elements exceeded U.S. Public Health Service (PHS) or EPA drinking water

standards. These reports provide perhaps the most complete data available on solid waste disposal damage costs for particular incidents. Also useful are three other reports in this series entitled *Hazardous Waste Disposal Damage Reports.* Each of the three contains three case studies (for a total of nine incidents) which attempt to look at the extent of the damages, broken down into personal, environmental, and economic damages, along with the cause of the problem, type and quantity of waste, source of waste, and status.

2. *Brief descriptions of damage incidents.* The majority of these are found in two EPA publications—*Damages and Threats Caused by Hazardous Material Sites* (January 1980) and *The Prevalence of Subsurface Migration of Hazardous Chemical Substances at Selected Industrial Waste Land Disposal Sites* (October 1977). The former publication includes brief descriptions (abstracts) of about 350 hazardous waste sites where damages have either already occurred or are expected to occur. According to the EPA preface to this volume, these abstracts "briefly describe the site, the toxic pollutants involved, the media or resource affected and the damages" (p. vii). What becomes clear upon examining these abstracts is that there is little information upon which an economic cost/benefit analysis can be based. The information, as would be expected from abstracts, is simply too sketchy.

The latter publication includes descriptions of 50 industrial waste disposal sites (both landfills and lagoons) in 11 states. Each of these sites was sampled for subsurface contamination. The sites are not identified by name in this report; only their regional location within the state is provided. The important information provided by this report is the amounts of various hazardous chemical constituents found in the monitoring wells, soil cores, or springs. A good deal of geological data is also provided for each site. However, there is no indication of the effects of the hazardous elements in terms of damage to human health or the environment nor is there any estimate of clean-up costs for these sites.

This latter report, however, does provide some summary data concerning the migration of hazardous materials. For example, the investigation found heavy metals (excluding iron and manganese) at 49 sites and confirmed that they had migrated at 40 of these sites. Selenium, arsenic, and/or cyanide were detected at 37 of the 50 sites; at 30 of these sites, these substances had migrated. Using project criteria,[1] the investigators were able to confirm migration of at least one hazardous constituent at 43 of the 50 sites. Finally, at 26 of the 50 sites, it was found that hazardous inorganic constituents in the water in at least one of the monitoring wells exceeded the EPA standards for drinking water quality.

---

[1]Project criteria included the following: (1) "one or more hazardous constituents must be detected beyond the boundary of the waste deposition area"; (2) "the concentration of the hazardous substance must exceed the concentration of the same substance in water from a background well or other background ground-water source"; (3) "all wells used to evaluate a site must tap the same aquifer"; and (4) "based on an interpretation of geohydrology and overall ground-water chemistry, the data must identify the landfill or lagoon under study as the source of the inorganic or organic substance" (p. vi).

In addition to the two volumes just described, references were found to approximately 60 other damage incidents associated with the disposal of solid wastes. These are found in several EPA background documents (see references) prepared by EPA in connection with the drafting of regulations to implement RCRA.

3. *EPA files.* Approximately 18 individual case studies (briefs) of damage incidents were reviewed. The amount of detail provided in these files varies. Some are potentially as useful as the reports in Category 1 above. Many are little more useful than those in Category 2. The majority fall somewhere in between. These files frequently have the advantage of offering estimates of the clean-up costs (Level I and Level II mitigative measures).

B. *Limitations of Sources*

The materials reviewed have a number of limitations. Their major problems are identified and discussed in this section.

1. *Failure to provide complete list of toxic or hazardous substances.* It is not unusual to find in these materials statements like the following:

To date, U.S. EPA and the New York State Health Department have identified at least 82 different chemical compounds in the landfill (*Economic Impact Analysis*, p. 4).

This report on the Love Canal Incident goes on to note that 11 of these compounds are "actual or suspected carcinogens, and 1 is a well-established human carcinogen" (p. 5). However, only 8 substances are then listed, and there is no indication of the other 3 nor of the 71 other chemical compounds found in this landfill. The purpose of this review is to identify damage incidents involving substances which are also found in coal waste disposal. Without a complete list of substances for the various incidents, this cannot be accurately done.

2. *Failure to provide quantification of hazardous or toxic substances.* With the exception of one EPA publication, *Prevalence of Subsurface Migration of Hazardous Chemical Substances at Selected Industrial Waste Land Disposal Sites* (1977), the materials reviewed provide little quantitative data on the amount (ppb. or mg/1 for instance) of toxic or hazardous chemical constituents found in a particular waste disposal site. In order to empirically estimate the damage costs resulting from hazardous waste disposal, it is necessary to be able to link quantities of toxic or hazardous chemicals with clean-up costs. This cannot be done without quantification of the substances involved.

3. *Failure to provide data on time period involved.* In the majority of these damage reports, there is incomplete information about the timing of the damage. In other words, it is often not clear when the toxicity began. Therefore, it is difficult to determine over what period the damage existed. In some cases, the date that a particular landfill was established is noted, but it is not clear whether the toxic or hazardous substance was present from the beginning and/or whether the degree of toxicity remained at a constant level throughout the period of the

landfill's existence.

4. *Failure to provide sound data on substance impact.* It is noted in at least one of the EPA documents that further tests are necessary to determine the link, if any, between the presence of a toxic substance in an area and the disease rate of a local population. The majority of the reports reviewed are descriptive rather than analytical. Such "further testing" has not been done. There have been few scientific experiments, using control groups, for example, to establish causal links. Without such precautions, it is difficult to establish causality.

5. *Failure to provide cost data.* The lack of cost data in these materials is a major limitation. When costs are provided, they are frequently the *estimated* costs of clean-up, i.e., the costs of "mitigative measures," rather than actual costs. *Actual* cost data are lacking in terms of clean-up costs, costs associated with incidence of disease, costs associated with the impact of a toxic substance on the environment (including not only air and water pollution but damage to crops, vegetation, etc.). The most commonly provided data in these materials concern the effect of a particular hazardous or toxic substance on water quality but there are few data on human health effects. Cost data are sometimes provided with regard to the costs incurred in hooking up to a public water supply or buying bottled water, etc. to replace contaminated drinking water. As noted, clean-up costs are also frequently provided, although they tend to be projected costs rather than actual costs.

C. *Review of Damage Costs Reports*

The purpose of this literature review was to gather data which would enable us to estimate the damage costs in relation to the disposal of coal wastes. Table 1 highlights the 14 cases which contain potentially useful data for our analysis. Our criteria for including a study on this table were the following: (1) at least one of the elements involved in the incident had to be a major chemical constituent of coal wastes. The following constituents comprised this list: antimony, arsenic, beryllium, cadmium, chromium, copper, chloride, lead, mercury, nickel, selenium, silver, and zinc. (2) Some data had to be provided on the amount of the chemical substances involved in the particular incident. (3) Some damage estimates had to be provided in at least one of the following areas: (a) water quality, (b) human health, (c) environmental impact, and (d) clean-up costs.

III. *Quantification of Previous Damage Cost Studies*

A. *Valuation of Life and Health*

Life and health are sometimes considered goods of infinite value. This is especially true when people consider their own life and health and those of their loved ones. People often are willing to expend as many resources as they can to save their own life. On the other hand, from a societal point of view, resources are scarce, and it is often an unavoidable decision whether to allocate resources

TABLE 1

Summary of Hazardous Waste Disposal Damage Reports
Limited to Coal Wastes Related Chemical Constituents

Site	Toxic Substances	Time Period	Water Quality	Damage Results			Clean-up Cost
				Human Health	Environment		
1. Perham, Minnesota (EPA/530/SW-151) disposal of grass-hopper bait	arsenic trioxide (in grasshopper bait) subsurface migration	bait burial 1934-1936 illness report-ed 5/72 and next 10 weeks	arsenic up to 21,000 ppb. (vs. 50 ppb., U.S. Public Health Service Standard) in drinking water	arsenic poisoning 11 persons, 2 required hospitalization gastrointestinal symptoms severe neuropathy for one person	contamination of soil and ground water		discontinued use of contaminated well $3,000 to install public water supply estimated cost of removing and dis-posing of contamin-ated soil-$25,000
2. Byron, Illinois (EPA/530/SW-151) industrial waste disposal	cyanides (365,050*; arsenic (60), cadmi-um (340), chromium (17,200), phenols (8), lead, mercury, copper, iron, zinc *maximum concen-tration detected in run off water, parts per billion (ppb.)	1974, dump-ing ceased 1972	unacceptable levels of lead and mercury in 46 wells unsafe level of cad-mium in 1 well		3 dead cattle con-tamination of surface and ground waters destruction of wildlife destruction of local vegetation destruction of stream-dwelling organisms		$250,000 by 1 prop-erty owner for clean-up and monitoring Level I mitigative measures estimated at $150,000 Level II mitigative measures $625,000

TABLE 1 (Continued)

Site	Toxic Substances	Time Period	Damage Results			
			Water Quality	Human Health	Environment	Clean-up Cost
3. Rocky Mountain Arsenal between Denver and Brighton, Colorado (EPA/530/SW-151.2) (+ update June 1980) groundwater contamination in unlined holding pond for storage of chemical wastes	aldrin and dieldrin (insecticides) endrin, arsenic, diisopropyl-methylphosphate (DIMP), dicyclopentadiene (DCPD)	1943-1957	contamination and temporary abandonment of 64 wells arsenic found in well water along with chloride, the herbicide 2,4-D in 12 sq. mi. area (1965) chloride concentrations 200 ppm used as indicator; concentrations ranked as high as 3,000-4,000 ppm		damage to 6½ square miles of farmland, sugar beets, pasture grass, alfalfa, corn, and barley (1975): 30 sq. mi. contamination of groundwater: aldrin (30ppb) endrin (40ppb) dieldrin (40ppb) DCPD (4ppb)	$1,000,000 for the construction of waste injection well which subsequently had to be abandoned because of earthquakes federal government has paid out $165,000 in damages cost estimates to clean up area about $25 million
4. Crosby, Texas (near Houston) (EPA/530/SW-151.2) dumping of approx. 70 million gallons of industrial wastes into a sand pit	cadmium, zinc, manganese, iron	mid 1960's-1971	discontinued use of 26 wells for 1½ years abnormally high concentration of cadmium, zinc, iron, and manganese	nausea, sore throats, headaches, from offensive irritating odors	destruction of local vegetation	$2,000 fine in 1970 $500 fine in 1971 $5,000 fine + ceding of 22A to state fines resulted from company continuing to dump, not from clean-up as none was undertaken
5. Southington, Connecticut (EPA/OSW/ May 2, 1980) incineration of solvent recovery sludges	lead, zinc	1974	contaminated company's well and groundwater		contaminated soil, high levels of lead air contamination	3 of the city's 6 wells closed

TABLE 1 (Continued)

Site	Toxic Substances	Time Period	Water Quality	Human Health	Environment	Clean-up Cost
				Damage Results		
6. Woburn, Massachusetts (EPA *Economic Impact Analysis* pp. 16-17)	chromium, arsenic, lead, zinc disposed of in area + volatile organics + chlorinated organics	no date	groundwater contamination suspected	age adjusted cancer death rates 13% higher than statistical-ly expected from 1972 -present studies being under-taken to correlate health data with en-vironmental sampling		minimum to clean up site estimated at $1.25 million ($650,000 for dis-posal of material and $600,000 for disposal of "hot spots")
7. Chester, Pennsylvania (EPA/430/9-80/004) p.208, p.35 + Background Doc. 5/2/80 pp. 23-24 industrial plant disposal, fire	chromium, copper, nickel, lead, volatile organics and aromatic hydrocarbons	no date	chromium, copper, nickel, and lead in excess of drinking water standards found via water sampling pro-gram			
8. Neville Island, Pennsylvania (EPA/430/9-80/004) pp. 33-34, p.218 + Back-ground Doc. 5/2/80 p.22 industrial waste dumping	benzene, phenois, parathion, cyanide, mercury, coal tar residues	dumping in 1950s		high rate of health problems among 26% of workers building a public park at the site including eye irritation and blood in urine		park closure with monitoring would cost $150,000 to $250,000; removal of contaminated waste to rebuild the park would cost $7-$24 million
9. South Charles City, Iowa Background Doc., 5/2/80 pp. 17-18 chemical manu-facturing com-pany dumping	arsenic, phenois, orthonitro-alinine, nitrobenzene, estimated 27 million cubic feet of chemical sludge	1953-1957	toxic chemicals found in Waterloo drinking water, 50 miles away			estimated cost of removal of toxic wastes $20 million

Site	Toxic Substances	Time Period	Water Quality	Damage Results Human Health	Damage Results Environment	Clean-up Cost
10. Montague, Michigan (EPA/430/9-80/ 004), pp. 16 & 20; Background Doc. 5/2/80 p. 129 dumping of chemical wastes	asbestos, flyash, brine, pesticides (dioxin)	1950-1970 (1957-1979)	private wells contam- inated		groundwater contamination and contamina- tion of White Lake (flows into Lake Michigan)	purge well system to be installed to inter- cept contaminated water clean-up cost estimates: $15 million to $300 million $15 million plan accepted
11. New Hanover County, N.C. (EPA/430/9-80/ 004) pp. 29, 198 industrial waste landfill	lead, tetrachloroethy- lene, benzene, trich- lorethylene, 1,2, -dichloroethane, methylene chloride	1972-1979	lead found in levels hazardous to human health, as well as, tetrachloroethylene, & 1,2,-dichloroethane 17 private wells con- taminated bad taste and odor from iron, manganese, zinc, phenol, chlorides, dichlorophenol, and chlorobenzene			$25,000 spent by EPA to ascertain "nature and extent" of contamination of groundwater
12. Saltville, Virginia (EPA/430/9-80/ 004) pp. 38-39; 246 wastes from alkali processing plant	chloride, mercury	1895-1972			The Holston River contaminat- ed by mercury lead to ban on fishing (1970)	$700-800,000 spent to date put- ting riprap on stream bank and to stabilize levees; remedial plan would cost $4 million; removal of mercury would cost $23-32 million

TABLE 1 (Continued)

Site	Toxic Substances	Time Period	Water Quality	Damage Results			
				Human Health	Environment	Clean-up Cost	
13. Gary, Indiana (EPA/430/9-80/ 004) pp. 99, 11 solvent recovery facility and stor- age of plating wastes, solvents, acids, and cyanide	phenols, chromium, cyanides, arsenic, lead	1975-		fire hazard potential exposure to solvent fumes	soil and water contamination	estimates: $1.8 million to remove and dispose of on- site wastes and monitoring; $3.1 million to minimize future pollution problems	
14. Hamilton, Ohio (EPA/430/9-80/ 004) p. 32 hazardous chemical waste disposal	arsenic, cyanides, 1-dichloroethane, benzene, phenol, acetone, xylene, toluene, hexane, 1,1, 1-trichlorethane, dichlorobenzene, napthalene	since 1976		periodic sickness of nearby workers and people visiting nearby recreational facilities	soil contamina- tion possibly groundwater contamination		

to reduce the probability of death or illness or use these resources for other purposes in the economy. It is also politically impossible for a government to satisfy all aspects of economic demand that it faces. The objective of a cost-benefit analysis is to compare the costs and benefits of a governmental program (environmental, health services, occupational health and safety, etc.) in monetary terms (a common denominator) in order to achieve more efficient use of resources.

No consensus exists as to how to express the dollar value of life and health, even recognizing that valuations should be made. In general, there are two approaches to placing a dollar value on life and death: the "willingness-to-pay" approach, and (2) the "human capital" approach. The "willingness-to-pay" approach considers the amount of money that an individual is willing to pay to avoid or reduce a given amount of his (her) probability of death or illness (mortality or morbidity). While conceptually this is a sound approach to obtaining the value of live and health, it is extremely difficult to obtain empirically meaningful estimates. The magnitude of these estimates depends upon the respondents' understanding of the question and their perception of their own health status. It is an approach comparable to the concept of demand for life and health, except with no price constraint on the part of the respondent. Few previous studies have been able to provide dollar estimates of life and health using the "willingness-to-pay" approach. Graham and Vaupel (1980) found that out of 24 benefit-cost analyses of government programs, only 7 used a willingness-to-pay value, with a median value of $625,000 per life.

The second method is the "human capital" approach which measures the present value of foregone earnings due to illness or premature death. This approach considers the production aspect of life, which includes the change in the value of services in the market and non-market (household work) due to the change in an individual's mortality or morbidity.

The values of human life based on the human capital approach are affected by one's age, education, race, sex, and occupational status. This production aspect of the value of life or health usually presents only a part of what an individual or society would be willing to pay for the reduction of mortality or morbidity. Thus, the estimates from the human capital approach are a lower limit on the "true" value of "willingness-to-pay." The human capital approach is easier to obtain. One can use published labor statistics and disease incidence reports to derive the value of life or health. Graham and Vaupel (1980) found that out of 24 benefit-cost analyses of governmental programs, 15 used a foregone-earnings value, with a median value of $217,000 per life.

The costs of illness can be considered from two aspects: (1) direct costs, and (2) indirect costs. Direct costs are the value of resources used to cure the illness and provide care during the illness and include such things as physician's services, hospital services, pharmaceutical services, and so on. Indirect costs include the value of lost time or productivity on the part of the patient or members of the patient's household. Numerous studies have estimated costs of illness by disease category (Hu and Sandifer, 1981), based on mortality and morbidity rates of various diseases as reported by the National Center for Health Statistics, and the wage and employment status of various population groups as reported by the Bureau of Labor Statistics. Direct cost information can be obtained from the statistics reported by the Health Care Financing Administration, Department of Health and Human Services.

B. *Valuation of Environmental and Water Quality*

The damages from hazardous waste disposal frequently affect animals (e.g., fish deaths from water contamination or cattle deaths from poisoning), trees, and vegetation. Unlike the case of human life and health, market values exist for fish, cattle, trees, and vegetation. These values can be obtained from *Fisheries of the United States* (U.S. Department of Commerce), and *Agricultural Statistics* (U.S. Department of Agriculture).

One type of environmental damage resulting from hazardous waste disposal is the deterioration of underground and ground water quality. The direct effect is the contamination of wells and streams. As reported in Table 1, some wells have had to be closed for health reasons. There are two alternatives for measuring the costs of damage to water quality. If the damage is to a well, the replacement cost for a well can be used as the damage cost. Several EPA studies (*Preliminary Assessment of Suspected Carcinogens in Drinking Water,* EPA Report to Congress, December 1975; *Economic Impact Analysis of the Promulgated Trihalomethane Regulation for Drinking Water,* EPA-529/9-79-022, September 1979) have provided estimates on water treatment or well replacement cost. If the damage is contamination of a stream, cleanup costs can be used as damage costs. Many of the damage cost reports have estimated these site-specific clean-up costs.

The value of recreation losses due to environmental damages should also be included in damage costs. Recreation costs have been measured in terms of the willingness-to-pay approach by using a demand function for recreation visits. The observation of changes in travel and associated costs due to environmental changes enables estimates to be made of the value of recreation. This study will not include the value of recreation losses. First, many hazardous waste disposal sites and/or their neighboring locations by definition are not used for recreational purposes at all. Second, even if such losses would be the result of a change in water quality, the monetary imputation of the change in water quality and the change in land value would implicitly cover the potential loss of recreation value.

Table 2 shows the assumed dollar values of life, costs of illness, and environmental and water quality in 1980 dollars. Table 3 shows the estimated damage costs for the 14 cases summarized in Table 1. It can be seen that damage

TABLE 2

Assumed Dollar Values of Life, Costs of Illness, and Environmental and Water Quality in 1980 Dollars

	Assumed Values
Value of a life	$625,000 (willingness-to-pay approach)
	$217,000 (human capital approach)
Costs per hospital day	$250
Costs per hospital day — Ancillary service	1/3 of hospital day, $84
Average length of stay per case	5 days
Costs per psychiatric session	$50
Costs per physician visit	$30
Costs per cow	$5,000
Opportunity costs of wages lost per day	$40
Value of farmland in Western mountain states, @ acre	$750
Damage costs of a well	$12,700

TABLE 3

Cost Estimates of Previous Damage Costs

Case Code[1]	Costs in $1,000[2]
1	33.6
2	422.5
3	28,250.0
4	378.4
5	38.1
6	50.0
7	1,250.0
8	15,018.0
9	20,000.0
10	15,000.0
11	215.0
12	27,500.0
13	4,900.0
14	100.0

[1]The code number is identical to the case number in Table 1.
[2]Based on damage costs reported in Tables 1 and 2.

costs vary from about $33,000 at one site to about $28 million at another. The $28 million case involves a reported $25 million in clean-up costs. It should be noted that not every damage case includes all human health, environment, and water quality damages. It depends upon the nature of the waste disposal. Therefore, some of these damage costs include only the relevant or reported costs. On the other hand, governmental fines are not included in the damage cost estimations. From the societal point of view, governmental fines are transfer payments between the private and public sectors; they do not change the value of resources in the entire society, and, therefore, no productivity loss is involved.

IV. *Recommendations for Future Empirical Estimation*

To estimate either the environmental or human health damage costs due to toxic waste disposal, it is necessary to have the following information:

1. The linkage between the *sources, amount,* and *type* of hazardous wastes and the extent of environmental damages over time.
2. The linkage between the *sources, amount,* and *type* of environmental damages and human exposure,
3. The linkage between the *duration, amount,* and *type* of human exposure and human health effects, and
4. The *monetary value* of the environmental damages and human health effects.

Without data on any of the four, the estimation of damage costs of wastes disposal is very difficult or unreliable. The present status of these four types of information indicates that although there are studies or reports on the linkage between sources and type of hazardous wastes and environmental damages, there is a lack of information on the amount of toxic substances involved at the various disposal sites. There are studies relating to sources, type and, and amount of environmental damages and human exposure. But these linkages are established by inference or deduction and often lack scientific documentation or experimentation. The information on the amount and extent of human exposure to toxic substances and health effects are not often quantified. Finally, the monetary valuation of environmental damages and human health effects has been estimated in numerous studies, but there is no consensus as to their accuracy. This paper suffers the same limitations, a function of the state of the availability of relevant data. This paper has omitted information on the *amount* of toxic wastes and *concentration* of toxic elements. It directly links the type of toxic substances to the extent of environmental and human health damages.

The key recommendations for future cost analysis are as follows:

—There is a need for a complete list of toxic or hazardous substances and the amount of each substance for each damage case.

—There is a need for documentation and quantification of the relationship between hazardous substances and the extent of their environmental and health impact. Numerical accounts of the impact and the population base are required to estimate the effect of an environmental regulation.

—There is a need for actual cost or standard health cost references (i.e., value of life or cost of illness) so that cost estimation can be made.

—Finally, there is a need for linkage of all of the above-stated information in each damage case. Only when all of this information is provided can a meaningful damage cost analysis be done.

ACKNOWLEDGEMENTS

Professor Richard Rosenberg at The Pennsylvania State University provided helpful suggestions and comments. This paper was funded by the U.S. Department of Energy. The contents of this paper do not necessarily reflect the official views or policies of the authors' affiliation or the funding agency.

REFERENCES

Action, J. P. *Evaluating Public Programs to Save Lives: The Case of Heart Attacks.* Rand Report, R-950-RC, Santa Monica, California, The Rand Corp., January, 1973.

Cooper, B. S. and D. P. Rice. "Economic cost of illness revisited." *Social Security Bulletin,* 39:2 (February 1976), pp. 21-36.

Institute of Medicine. *Costs of Environmental Related Health Effects.* Washington, D.C.: National Academy of Sciences, January 1981.

Jones-Lee, M. "Valuation of reduction in the probability of death by road accident," *Journal of Transport Economics and Policy* (January 1969), pp. 37-47.

Klarman, H. E. "Socioeconomic Impact of Heart Disease" in *The Heart and Circulation: 2nd National Conference on Cardiovascular Diseases* (ed. by E. Andrus and C. Maxwell). Bethesda, MD, Federation of American Societies for Experimental Biology, 1965, pp. 693-707.

Mishan, E. J. "Evaluation of life and limb: A theoretical approach," *Journal of Political Economy* 79:4 (July-August 1971), 687-705.

Mushkin, S. J. and D. W. Dunlop. *Health: What is it Worth? Measures of Health Benefits.* Elmsford, NY: Pergamon Press, 1979.

Rice, D. P. *Estimating the Cost of Illness.* Health Economics Series, No. 6, Washington, D.C.: U.S. Public Health Service, Division of Medical Care Administration, 1966.

Schelling, T. C. "The life you save may be your own" in *Problems in Public Expenditure Analysis* (ed. by Samuel B. Chase). Washington, D.C.: The Brookings Institution, 1969.

Thaler, R. and S. Rosen. "The value of saving a life: Evidence from the labor market" in *Household Production and Consumption* (ed. by N. Terlackyj). New York: National Bureau of Economic Research, Columbia University Press, 1975.

U.S. Environmental Protection Agency. *Hazardous Waste Disposal Damage Reports.* Document No. 1. June 1975. [EPA/530/SW-151].

_____. *Hazardous Waste Disposal Damage Reports.* Document No. 2. December 1975. [EPA/530/SW-151.2].

_____. *Hazardous Waste Disposal Damage Reports.* Document No. 3. June 1976. [EPA/530/SW-151.3].

_____. *Leachate Damage Assessment.* "Case Study of the Fox Valley Solid Waste Disposal Site in Aurora, Illinois." June 1976. [EPA/530/SW-514]. (Authored by Kenneth A. Shuster).

U.S. Environmental Protection Agency. *Leachate Damage Assessment.* "Case Study of the Peoples Avenue Solid Waste Disposal Site in Rockford, Illinois." June 1976. [EPA/530/SW-517]. (Authored by Kenneth A. Shuster).

_____. *Leachate Damage Assessment.* "Case Study of the Sayville Solid Waste Disposal Site in Islip (Long Island), New York." June 1976. [EPA/530/SW-509]. (Authored by Kenneth A. Shuster).

_____. *Damages and Threats Caused by Hazardous Material Sites.* January 1980. [EPA/430/9-80/004].

_____. *The Prevalence of Subsurface Migration of Hazardous Chemical Substances at Selected Industrial Waste Land Disposal Sites.* October 1977. [EPA/530/SW-634].

_____. *Background Document.* 40CFR Part 265, Subpart N. Interim Status Standards for Landfills. May 2, 1980.

_____. *Final Background Document.* 40CFR Part 265, Subpart P. Interim Status Standards for Hazardous Waste Facilities for Thermal Treatment Processes Other than Incineration and for Open Burning. April 1980.

_____. *Background Document.* 40CFR Parts 264 and 265, Subpart H. Financial Requirements. December 31, 1980.

_____. *Economic Impact Analysis of Hazardous Waste Management Regulations of Selected Generating Industries.* (SW-1820). Prepared by Energy Resources Co., Inc. Cambridge, Mass. under contract No. 68-01-4819, for EPA, Office of Solid Wastes, 1979.

Zeckhauser, R. "Procedures for valuing lives," *Public Policy,* 23:4 (Fall 1975), pp. 420-464.

Hazardous and Toxic Wastes: Technology, Management and Health Effects. Edited by S.K. Majumdar and E. Willard Miller. © 1984, The Pennsylvania Academy of Science.

Chapter Twenty-Three

Requirement Liabilities of Waste Generators

Emlyn Webber[1] and Rex A. Hunter[2]
[1]Vice President Sales
VFL Technology Corporation
42 Lloyd Avenue
Malvern, PA 19355

[2]RAH Consulting
1190 Charter Rd.
Warminster, PA 18974

A basis of most existing hazardous waste control legislation is that responsibility begins with and rests primarily with the generator of the waste. Storage, transport, treatment, and disposal are each covered by specific guidelines and control requirements with a chain of custody and responsibility initiating with the generator, and being transferred through each successive stage of the waste handling process.

Thus, it is extremely important to a generator to understand not only his responsibility while hazardous waste exists at his property, but also what liability (both legal and financial) the generator may retain after he has had the waste removed for ultimate disposal.

This chapter will summarize relevant national and local legislation affecting the generator of hazardous waste. As with any regulation of this type, it is always prudent to check with your local and state regulators to ascertain any specific local requirements, to determine current interpretation of statutes, and to maintain awareness of revisions in codes, and local and financial requirements.

ARE YOU A GENERATOR?

Prior to determining what responsibility you are assuming, it is imperative that you determine whether in fact you *are* a generator of hazardous waste. Most legislation places the onus on the individual to ascertain his status as a generator. Conversely while an individual has responsibility to determine his status, regulators have the authority to monitor this assessment and confirm a generator's legal status.

The basic criteria for defining a generator's status are type of material generated, quantity of material, and method of handling/storage practiced on your property.

Small Volume Generator

Individuals or companies who generate small volumes of materials are exempt from meeting most hazardous waste management regulations but under general liability law, any waste must be properly disposed.

Most states follow federal guidelines in defining small generator status. In summary, anyone who generates less than 1,000 kilograms per calendar month is a small generator. Some states, including New Jersey have lowered the limit for defining small generator states.

Small generator status relieves one from most administrative requirements but does not relieve the generator from assuring proper disposal for wastes generated.

Are You More Than A Generator?

As mentioned, ultimate responsibility for assuring correct disposal begins with the generator. However, a generator can directly assume additional liability if in fact, he is classified as a transporter, storer, disposer or treater of hazardous waste.

Rather than discuss these potential liabilities, this chapter will summarize what actions can be monitored to assure one does not become classified as more than a generator.

A transporter of hazardous waste is rather self explanatory, and as long as the waste does not leave your site you are not considered a transporter.

In general, if hazardous waste is stored for more than 90 days, the facility is defined as a hazardous waste storage facility. As long as you do not store waste for more than 90 days, you are not classified as a storage facility, even though approved storage procedures must be followed.

A disposal facility can briefly be described as a facility at which hazardous waste is intentionally placed into any land or water, at which waste will remain after closure.

The treatment of waste can be described as any process designed to change the physical, chemical, or biological characteristics of the waste so as to affect the degree of hazard of a waste.

If you, as a generator, also perform these additional operations, additional regulatory responsibilities and liabilities will be assumed.

Is Your Material Really a Hazardous Waste?

Whether a waste is defined as hazardous depends on administrative fiat, test results of the waste, and the physical condition of the waste. In general, a waste is considered hazardous if it exhibits ignitability, corrosivity, reactivity, or toxicity. For each of these characteristics federal and state regulations exist which describe, in detail, test procedures and results which will classify a waste as hazardous or non-hazardous.

Some tests are straight forward. For instance, a waste with pH less than or equal to 2, or greater than or equal to 12.5 is defined as corrosivily hazardous. Tests for toxicity on the other hand, are more complex with levels of contaminants which leach out of the waste defining classification.

As a further assistance, the Federal EPA, and most state agencies have listed a large number of chemical and elemental compounds as being defined as hazardous, and thus no specific testing is required.

To repeat a phrase, it is always the generator's responsibility to determine whether his waste is hazardous, and as always, ignorance is not a legitimate excuse.

WHAT ARE YOUR RESPONSIBILITIES?

Since the inception of RCRA, it has been known as the "cradle-to-grave" waste management system. RCRA was so dubbed because of the volume of required paperwork designed to track a waste material through the generation, transportation, and disposal phases. The person responsible for initiating the required paperwork, in addition to its maintenance is the Generator.

For a thorough review of the Generator's responsibilities under federal law, the reader is referred to the code of Federal Regulations (40 CFR 262, in particular). Rules and regulations for Pennsylvania based generators can be found in 25 PA Code Chapter 75. What will be presented below is a synopsis of these requirements.

Step 1

Let us assume that, from the information presented above, you have decided that you are a generator of hazardous waste. Now what? First and foremost you must notify the EPA of this and request an EPA Identification Number. There are good reason both legally and practically for doing this. Under the Act, Section 3008, a generator who knowingly conceals the fact that he is a generator by falsifying or destroying records or making false statements to the authorities is subject to criminal penalties of "not more than $25,000. . . for each day of violation, or to imprisonment not to exceed one year. . . or both."

From a practical standpoint, transporters and disposers operating under this system will be required to show your EPA ID Number on their paperwork (as well as their numbers). Since the above noted criminal penalties apply to transporters and treatment, storage, and disposal facilities as well as generators, the odds of finding a reputable firm who would be willing to help you around this point are not good. This only leave the alternative of using somewhat less than reputable firms (read: midnight dumpers). In light of the current emphasis placed on clean-up of abandoned sites, this alternative appears to be a good way of assuring that you will receive future publicity-maybe not all favorable-and you may even be a candidate for repaying the federal government three times the cost to clean up some abandoned site (Comprehensive Environmental Response, Compensation and Liability Act of 1980, P.L. 96-510 Section 107). Surely you would find it easier and simpler to file EPA Form 8700-12, would you not?

Temporary Storage

Now you are a generator operating within the requirements and you generate your first bit of hazardous waste. You have made a decision to not be a storage facility for any number of good reasons, but does this mean that you need a conveyor belt or some other device carrying the waste off-site the moment it is generated? Of course not, but some rules do apply. Hazardous waste may be stored on site for up to ninety (90) days without a separate permit as a storage facility. In order to be allowed this accumulation time without a permit, the generator must comply with certain rules. However these are more common sense than common law.

Storage drums must be non-leaking. They must be compatible with the material stored. They must be closed when not putting in or taking out waste. They must not be handled in any manner which would cause them to lead or rupture. Any they must be inspected on regular intervals for leaks and deterioration. These requirements are really nothing more than procedures you already have established to protect your expensive raw materials. Why treat chemnical waste with any less respect than you would virgin chemicals?

In addition to these storage procedures, the generator is required to have an accurate means of inventorying these wastes. Accumulation dates must be clearly marked and the containers must be labeled as "Hazardous Wastes." The development of employee training programs, spill prevention plans, and contingency plans is also a regulatory necessity. These plans and programs are just like those required for treatment, storage, and disposal facilities as well they should be. The potential for accidental discharge of hazardous wastes exists no matter how long or how short the storage time.

We have made the point before that a generator can store hazardous wastes for ninety days and not be considered a storage facility. Should it become necessary because of "unforeseen, temporary, and uncontrollable circumstances," a generator has the right to seek a 30 day extension of this time period. These extensions are granted at the discretion of EPA on a case-by-case basis. The generator needs to be aware of all of his options, but owes it to himself to exercise these carefully. The intent of this provision is not to give the generator more time to accumulate enough materials to minimize possible transportation or disposal surcharges, but to not penalize those generators making an honest effort to comply with regulations.

This is a good place to stress that a generator may be working under two sets of regulations, if his state has already promulgated their own regulations. The reason for bringing this out here is that Pennsylvania does not allow for the granting of this time extension. The implications of this fact are that, at the time of this writing at least, that a Pennsylvania generator could be in violation of PA Act 97 The Solid Waste Management Act, but not RCRA. Until Pennsylvania receives authority to administer the whole program, the generator must comply with the strictest regulation. The area of overlapping laws and regulations is very complex and areas of potential problems and conflicts will only be noted here.

OFF-SITE DISPOSAL

Now that you have accumulated some hazardous wastes, how do you properly dispose of them? Where will they go, how will they get there, what must they be shipped in? All of these are important questions.

Getting Ready for the Transporter

First, the generator is obligated to package, label and placard the waste according to U.S. Department of Transportation regulations. The reader is referred at this point to several sections of the Code of Federal Regulations Title 49 Transportation for more in-depth review. These sections are 49 CFR 172, 172 Subpart F, 172.304, 173, 178, and 179. These regulations specify the types of containers to be used, proper labelling procedures, and how hazardous wastes are to be placarded. In general, new and/or reconditioned 55-gallon steel drums will fit these specifications—both open head and closed head bung hole type.

The text of the identifying label to be affixed to these containers is presented in the regulations. This consists of a generic warning stating the contents are hazardous wastes, lists phone numbers to be used in emergency response situations and identifies the generator and drum contents. Any container of 110 gallons or less must be labelled in this fashion. Both the completion of this labelling information and the placarding of the initial transporter are the responsibility of the generator.

The selection of a transporter for your hazardous waste can be aided along by conducting a review of the transporters capabilities. This is, of course, in addition to asking him how much it will cost you. Here are some questions to ask your potential transporter to help assure you that he is complying with the regulations.

Of primary importance is his EPA Identification Number. Do not do business with anyone who cannot substantiate that he has this. This point cannot be overstressed. Once you operate any link in this chain of proper waste disposal without proper authorization, the legal penalties to be expected are severe as noted earlier. There are enough firms operating within the system that risking working outside of the system is highly unwarranted. Consider this Rule No. 1!

Other areas to explore are: Is the carrier capable of taking the material from your facility to the disposal facility? If not, what other carriers are involved?

How is the material transfer handled? How long will it take from the time the waste leaves your plant until it is accepted by the disposal facility? Will other wastes be added to your wastes' container? These matters are all important in complying with the manifest system.

The manifest system is a means of keeping track of how much hazardous waste is involved, where it goes, and who handles it. This system is the true crux of the "cradle-to-grave" hazardous waste management system. As such, it is somewhat cumbersome and confusing. The more times materials change hands or are altered the more cumbersome and confusing it becomes. Add to this the fact that, as of this writing, interstate shipments require additional manifesting for each state entered. One bright note is that by the time you read this, a uniform manifest system should be in effect.

Let us explore some of the ideal and worst case possibilities involving transporters from the standpoint of generator's liability. If the initial transporter can deliver the waste to its final destination, that is ideal. If not, each new transporter must be licensed and approved and more links in the paperwork chain have been added. If the transporter can deliver your materials without the use of temporary transfer or holding facility, that is ideal. If the material is to sit at a temporary facility for more than ten days, this facility must be permitted as a hazardous waste treatment, storage, and disposal facility. If the transporter can make the delivery in a timely fashion such that the generator receives a copy of the manifest signed by the owner or operator of the disposal facility within thirty-five (35) days of the date the waste was accepted by the initial transporter, that is ideal. If not, the generator must initiate action to either obtain this or explain to the regulatory authorities why not. If the transporter adds material of another DOT shipping description to your containers, he must now comply with the laws and regulations covering questions. If this addition were made and no report of it was ever filed, you are still the generator of record. Imagine all of the headaches and complications this could cause down the road if the added material turned up later. It is probably safe to assume that secret addition of materials to an already manifested wastes is not the technique which would be used to dispose of innocuous materials, but rather some of your less desirable wastes.

Accidental discharges of hazardous wastes in-transit and their clean-up is an area to discuss with your transporters. Under federal and Pennsylvania law, the transporter is responsible for clean-up of these incidents. However, it seems obvious that one man driving a truck loaded with eighty drums of waste cannot single handedly clean-up these approximately 500 pound drums. Discuss with your transporter what provisions he has made for these incidents. Pennsylvania law requires a contingency plan on the part of the transporter and safety training and equipment. The EPA has established national contracts for the provision of emergency response to these types of incidents. The generator's legal and financial liability is not clearly addressed in the regulations, but your name is on those drums floating in the stream or emitting toxic gases. The Boy Scouts said it best when they said "Be Prepared." Someone has to pay for the clean-up costs.

The Treatment, Storage, and Disposal Facility

Owners and operators of treatment, storage and disposal facilities (TSD's) are regulated just like the rest of us. In fact, the regulations surrounding these facilities make up the largest part, in volume at least, of the RCRA regulations.

Now you need to select this facility and an alternate. The listing of an alternate facility is required on the waste manifest. For openers, go back to Rule No. 1. No EPA ID No., no waste. But there are more things to consider than just that. Certain states have exclusions on all of their facilities.

Some examples of these are provided by special chemicals such as PCB's (polychlorinated biphenyls) and pesticides. The disposal of PCB's is strictly regulated by the Toxic Substances Control Act (ref: 40 CFR 761). The disposal of pesticides and pesticide by products such as 2,3,7,8-TCDD (dioxin) is highly regulated as to place and type of disposal. State-wide exclusions exist for certain broad classes of materials. In Pennsylvania, no landfills are permitted to accept materials containing greater than one percent oil and grease. Of course, the classic example of these broad exclusions is the on-again off-again disposal of liquids in landfills. At the heart of this are even more basic questions of what is a liquid and what is a solid.

Because the generator never losses liability for the safe disposal of hazardous wastes, but as evidenced by Superfund, only shares it, the best advice is to be as thorough and candid about the nature of your waste as possible when evaluating disposal options. Never is a long time. You are as responsible for the safe disposal of that material as the TSD. He is your hired contractor is this situation. In this broad sense, the generator's liability is to make sure that the TDS he is dealing with complies with the Standards for Owners and Operators of Hazardous Waste Treatment, Storage, and Disposal Facilities (40 CFR 264) and similar state statutes once the states take over the administration of the program.

How do you do this? Start off by determining his permit status. There are two types of permits—Part A and Part B. All TSD's have a Part A permit (or if not you are talking to a guy with no EPA ID No.). A Part B permit is a very in-depth engineering analysis of the facility design, operation, location, and emergency procedures. At this time there are very few facilities which have a Part B, but eventually all operating facilities will either have to obtain one or cease to exist. If you are talking to a facility who has a Part B permit, you know you are talking to one who has been put under the microscope and reviewed by the EPA, their state regulatory agency, and the public.

Also you can review his past performance and talk with other clients. You can make your own assessment of his technology. What is state-of-the-art today, may be illegal in ten years. Do whatever is necessary to be sure this supplier of services measures up to the standards you set for all of your other suppliers of goods and services. A few disgruntled customers can cost you profit dollars. A few disgruntled regulatory people can shut you down.

THE BIG FILE CABINET

Of course, no regulatory program would be complete without its fair share of paperwork. We have mentioned some of this before in other sections. Here is a more thorough look at this area.

As stated, all generators must file EPA form 8700-12 in order to obtain an EPA Identification Number.

The generator must retain a signed copy of each manifest. Initially, he will maintain the copy signed by himself and the transporter. When he receives a copy from the TSD, this will supplant the original. These must be kept for three years.

Generators who ship wastes off-site must file Annual Reports on EPA forms 8700-13 and 8700-13A. These represent a summary of manifested wastes sent off-site where, how much, and what type. Again copies will be maintained for three years.

Any tests you made to determine if a material was hazardous or to classify it must be maintained for three years after the last date this material was sent off-site.

Also, remember back when you did not get a completed manifest from the TSD? If you did not get it within forty-five days from the time the transporter signed for the waste, you must file an Exception Report which includes a copy of the manifest and an explanation of what you are doing to locate the waste.

In addition to these requirements, the regulators can require additional reports they deem necessary on the disposition of hazardous wastes. This is in addition to the right of the regulators to extend the period of record maintenance in cases of enforcement actions or any other reason they deem it necessary.

One small consolation is the fact that you decided not to be a TSD facility. Part B applications are measured in pounds not pages.

SUMMARY

This chapter presents a brief look at the liabilities and responsibilities of generators of hazardous wastes under the Resource Conservation and Recovery Act. Most states are moving to implement their own regulations and take over administration of the program. In order to do this the state law must be at least as stringent as the federal law.

As always in cases involving legal matters the generator is urged to become thoroughly familiar with the regulations, consult with competent professionals in the field, and lend their participation to the rulemaking process.

PART 5

Environmental and Health Effects

In our modern industrial society there is an increasing variety of hazardous and toxic materials that affect the quality of the environment. Many of these substances have been proven to have harmful health effects. The exposure of humans to toxic substances can occur in varied ways such as from polluted air and contaminated water.

The initial paper of this section presents some epidemiological methods for studying the potential health effects of toxic waste sites. Some of the difficulties of epidemiologic investigation include the long latent period for most chronic diseases to develop after exposure to a causative agent, the problem of selecting an appropriate control population, the difficulty of securing statistical measures that will not fluctuate markedly in a relatively small population and thus have little or no epidemiologic significance, and the problem of keeping the public informed of actual conditions.

Coal gasification, including underground coal conversion, coking, and the production of gas was developed in the nineteenth century. The second chapter of this section addresses the effects on health of coal tar wastes from the production of gas. The study focuses on the wastes from a contamination site at Stroudsburg, Pennsylvania which was the first plant in the nation to receive Emergency Superfund money for clean-up and containment action. The paper reviews the chemical and physical properties of coal tar wastes, as well as their toxicity, and the mechanisms by which they are distributed throughout the ground water environment.

The final two chapters consider the effects of hazardous wastes on health. Because of the expanded production and use of chemicals and radioactive materials today, there is an increasing potential for the individual to be exposed to these dangerous conditions. These chapters consider such aspects of this important problem such as the processes of risk assessment, estimating exposure and health effects, and public perceptions. Examples of environmental pollution that create potential health hazards include the eruption of volcanoes such as the explosion of Krakatoa, Love Canal and Three Mile Island. From these examples it is recognized that hazardous wastes originate from not only man-made sources but also from natural disasters.

The problems of a potential health danger from hazardous wastes in recent years are due to human errors in judgment. There is little or no evidence that physical health problems have resulted from these mistakes. However, fear has created conditions that have caused severe psychological problems. Because satisfactory answers are still not available continued vigilance is a necessity. It is essential that expanded research programs be established to determine health safety levels of toxicity from hazardous and toxic wastes.

Hazardous and Toxic Wastes: Technology, Management and Health Effects. Edited by S.K. Majumdar and E. Willard Miller. © 1984, The Pennsylvania Academy of Science.

Chapter Twenty-Four

Epidemiologic Methods for Investigating Potential Health Effects of Toxic Waste Sites: A Case of Asbestos

George K. Tokuhata, Dr. P.H., Ph.D.
Adjunct Professor of Epidemiology and Biostatistics
Graduate School of Public Health
University of Pittsburgh
Pittsburgh, PA 15261

Man is exposed to a rapidly increasing variety of toxic chemicals and other hazardous substances in the environment. Some of these substances are known to have mutagenic, teratogenic or carcinogenic capabilities while many others are suspected as having similar potentials including adverse consequences in reproduction. Since the consequences of exposing humans to such harmful agents usually occur long after an agent is introduced into the environment, early identification of potential health hazards is crucial in order to minimize exposure and resultant risks.

The task of assessing health risks associated with toxic agents, however, is extremely complex because of the multiplicity of ways in which they can interact with biologic systems. When human cells are exposed to such agents, the affected cells, particularly their DNA component, may respond differently depending upon the host susceptibility as well as the extent and nature of the exposure.

The consequences of such exposure also vary according to whether or not somatic or germ cells are affected or both somatic and germ cells are equally affected. When somatic cells are affected, only the index hosts may manifest adverse health consequences; affected germ cells may result in adverse health consequences in future generations through mutation.

Governmental agencies, particularly public health, are responsible for detecting environmental health hazards caused by toxic agents, natural or man-made, of exposed humans. Therefore, it is important that health scientists and administrators are aware of the applications and limitations of the available epidemiologic methods of investigating potential health hazards to local populations residing in the vicinity of any toxic waste sites.

It is the intent of this paper to describe, in general terms, how such health risks may be investigated epidemiologically together with the identification of some of the essential factors involved in such investigations, including inherent limitations and technical problems in conducting such studies.

To illustrate the application of such epidemiologic methods, one actual study of potential health effects associated with an asbestos waste dump in a small community in Pennsylvania is also presented.

I. *Epidemiologic Methods of Investigation:*

The basic question raised here is whether or not the health of local residents living near a given toxic waste site is adversely affected. To answer this question, epidemiologists usually formulate a null-hypothesis assuming that there are no health effects. They, first, design a study, collect and analyze the data, and then perform significance tests in order to determine if the original null-hypothesis can be rejected, i.e., whether or not the observed magnitude of health risks are significantly greater than would be expected in normal populations. If the observed risks are significantly greater than would be expected, they would conclude that the health of local residents living in the vicinity of toxic sites is adversely affected.

Having shown the basic framework or the sequence of statistical inference, some systematic discussion of various components and factors involving this type of epidemiologic investigation is in order.

A. *Identification of Toxic (Hazardous) Substances:*

Contrary to popular conception, the exact identity of toxic (hazardous) substances in the environment is not always known. In other words, local residents whose health has been allegedly affected, as well as local health authorities, are unaware as to exactly what kinds of toxic chemicals or substances are present in the waste sites under question.

It is important that such substances are properly identified prior to the initiation of any epidemiologic investigation. Otherwise, it would be difficult, if not impossible, for epidemiologists to know what health consequences or diseases to look for, as based on earlier related studies. It must be noted, however, that the existing scientific literature may not be complete enough to cover that particular substance which is being investigated. It is quite possible that only limited adverse health risks are reported in the literature at any given time of review. Therefore, it is prudent to consider additional diseases and conditions not previously recognized as also being associated with the substance; the selection of proper diseases and conditions under this circumstance should be guided by experienced epidemiologists.

One of the most difficult potential problems is that investigating scientists may not be fully cognizant of the complex interaction that might have taken place between the original agent or substance, identified in the industrial waste and the physical and chemical environment to which it came in contact, creating some new substances which could be more toxic and harmful than the original waste itself. If the circumstances indicate, complete chemical assay would be needed to identify all such substances, original and newly synthesized.

B. *Human Exposure to Toxic (Hazardous) Substances:*

Before discussing epidemiologic methods for detecting possible health effects of toxic waste sites, investigators should be aware of several important factors in human exposure to such toxic substances.

First, there are several different pathways or transport media through which humans come in contact with toxic agents or substances, thus affecting different anatomical sites or organ systems in the body. Specifically, humans may be exposed through *air* in the form of gas, fume, mist, or dust; the primary concern here is *inhalation* of such substances which may affect the respiratory system.

Humans may also be exposed through contaminated *water* containing toxic substances; the primary concern here is *ingestion* which may affect the *digestive system.*

Another transport medium is *soil* through which contaminated vegetation containing substances may be ingested directly or consumed indirectly by humans through yet another pathway, such as contaminated milk of cows raised on contaminated vegetation in the area.

In both air- and water-mediated exposures, there is always a possibility of immediate contact with humans which may affect *skin* and *other external organs,* such as eyes.

Second, initial recognition of the potential health problems associated with toxic waste sites emerges (comes to attention) differently with respect to the source and magnitude of contamination. The most common combination is that the source of potential hazards, such as a given waste site, is known but the extent and nature of human exposure is unknown. In other words, a certain waste site is already identified, but how much local residents are contaminated is usually undetermined.

A less common combination is that the extent of human exposure to a given contaminant is known, but the exact source of contamination is undetermined. For example, blood lead levels among school children may be significantly elevated, but the origin of lead pollution has not been determined. Does the lead pollution originate from auto exhaust fumes, a nearby battery smelter or domestic pica in old residential homes?

Third, in any epidemiologic investigation of this type, three important "exposure factors" need to be considered, namely (a) *duration* of exposure, (b) *mode* of exposure, and (c) *total (cumulative) dose* of contamination. While the total (cumulative) dose of contamination is usually the most important independent variable, it is the combined function of the *duration of exposure* and the *level of exposure* at any given time that may be even more important. For certain toxic agents, such as ionizing radiation, the mode of exposure is extremely important, i.e., investigators must consider whether or not a total dose is delivered in one brief exposure or smaller doses, on a continuous basis or intermittently for a long period of years. The concept of *latent period* in disease is not the major consideration in the case of teratogenic and/or mutagenic effects.

Fourth, investigators should also be aware of a number of important epidemiologic considerations before designing a study of this type. It is well to keep in mind that genetic-constitutional host susceptibility can affect the incidence of diseases, particularly chronic disorders, such as malignant neoplasms (cancer) among those who have been exposed to any toxic substances in the environment. In other words, the marked heterogeneity of humans contributes to the varied susceptibility to a given agent.

In addition to this basic biologic propensity, at least three basic demographic characteristics are considered to be of primary importance in evaluating potential health risks; these are (a) age, (b) sex, and (c) race or ethnicity. Many diseases, particularly chronic conditions, including cancer, are age-dependent, i.e., the risk increases with the passage of years except for childhood cancer. The age factor is also important in evaluating certain congenital anomalies, such as Down's Syndrome (Mongolism) in which maternal aging materially increases the probability of this particular birth defect in the offspring.

Apart from the obvious anatomic and endocrinologic differences between males and females, sex distinction is an important consideration. Specifically, an observed male predilection in any disease can often provide an important clue as to whether or not the disease under study is occupationally influenced. On the other hand, diseases predominantly observed in one sex against another can also give epidemiologists a helpful clue as to possible hormonal linkage in the etiology of such diseases.

With respect to race and ethnic background of the population under study, one may entertain the possible biologic differences on the one hand and the potential importance in life style, including eating and personal habits, on the other.

The hypothesis to be tested in any epidemiologic study of this type should have an acceptable or sound biologic basis in terms of a cause-and-effect relationship. However, such knowledge may not always exist beforehand; in this situation, investigators must approach the problem macroscopically with all potentially significant etiological avenues taken into account.

Equally important, methodologically, is the knowledge of other major factors which are known to be related to a given disease or anomaly, already identified as being caused by a particular waste substance. When analyzing the data, these factors must be considered in order to concretely establish a causal relationship from the observational data. For example, a direct link can not be established between lung cancer and a given toxic substance without considering the effect of smoking, a powerful ubiquitous bronchogenic carcinogen.

C. *Epidemiologic Methods for Detecting Health Effects:*

Having identified toxic substances in the waste sites and having discussed a number of important considerations in human exposure to such substances, a systematic presentation of epidemiologic methods of investigation is in order.

Delineation of the Study Population:

Before designing an epidemiologic study, the investigator must delineate the population (geographic boundary) allegedly affected as being exposed to a toxic waste dump. This seemingly simple task is actually one of the most difficult, yet most important, decisions to make. This difficulty rests upon several factors: (a) The size of the population to be studied is particularly important when a relatively small number of people are involved, such as in the connection with cancer. (b) If the study area has a mobile population, it is difficult to define the base population as well as to determine the extent of exposure to a given toxic substance. (c) If the area under study constitutes cities or boroughs, as opposed to townships, there is a usual biased over-reporting of both births and deaths to such areas.

Study Design:

There are several different ways in which epidemiologic studies can be designed in order to assess the possible health effects of any given toxic waste site. Some methods have certain advantages over others. However, the investigator is often dictated by the availability of the needed data, the time and resource constraints to conduct a study and the socio-political milieu surrounding the initial recognition of the alleged health hazards.

In general, two standard approaches are used in designing such a study: (a) In a *cross-sectional comparison,* suitable control populations(s) are selected and health measures (risk indicators) at selected times are compared between the study area and the control area(s). In all epidemiologic studies, control populations should be similar (if not identical) to the study population with respect to all major characteristics except for the the the criterion factor, i.e., exposed or not exposed to a given toxic substance. (b) In a *longitudinal (historical) comparison,* the health measures (risk indicators) are compared before exposure and after exposure to a given toxic waste substance within the same population. Another form of longitudinal study is a "cohort" study where a given group of people exposed to a given toxic substance are followed for years and the incidence of certain diseases or anomalies are determined. The incidence of diseases or anomalies in this study cohort is compared with those chosen as controls. Selection of appropriate controls is as important as the conclusions drawn and are directly influenced by the type of controls selected for comparison. However, the use of the cohort design is not usually practical in assessing possible health effects of toxic waste sites.

One of the methodological problems in longitudinal studies is that the incidence of certain diseases is strongly influenced by changes in diagnostic opportunity and technique over time, which often results in an apparent increase in the incidence of that particular disease in the population where the true incidence may remain unchanged. This type of methodological problem can also occur when the control population is selected where the standard of medical practice is substantially different in a cross-sectional study. It is always difficult to find comparable control data in cohort studies except when a comparable control cohort was selected at the same time when the study cohort was identified and both are followed in the same manner during subsequent years.

Data Sources:

After the type of study design is determined, the investigator must then consider the kind of data to be collected or identified from the existing source(s). The most widely used, but with certain limitations and disadvantages, is *mortality data* which are routinely compiled by every state in the U.S. and abroad. For longitudinal comparisons, it is important to recognize that approximately every ten years the International Classification of Diseases makes certain changes which, in turn, make historical comparison of certain diseases difficult. Furthermore, unless multiple causes of death are routinely coded, only *underlying causes of death* are readily available for research and all *contributory causes of death* may remain unidentified. Nevertheless, death certificate information is probably the most systematically compiled health data that can be used for this type of epidemiologic investigation. For comparative purposes, the mortality rates are computed against the base population. Usually, the overall (all cause) mortality rate is expressed per 1,000 population, whereas the cause-specific mortality rates are expressed per 100,000 population.

Morbidity data are more useful than mortality data but are usually not readily available for epidemiologic investigations in a given population. Isolated case statistics based on selected individual patients have not only limited value but, more importantly, could be misleading by providing an improper or erroneous impression of what is occurring in the population under study. Systematic ascertainment of morbidity data in any population is extremely involved and costly both in terms of time and expenditure; this type of data gathering, despite its recognized value and importance, is only infrequently reported. Use of hospital medical data should be treated with caution as they are not populated-based. Screening and diagnostic tests involving an entire population are prohibitive in terms of cost-benefit. Complete case ascertainment in the population and completeness in individual case data are equally important in any epidemiologic study.

Morbidity statistics are usually expressed in terms of *incidence* or *prevalence*. The *incidence* of a disease is computed as the number of newly diagnosed cases during a given period, such as one year, within a given population at risk; the rate is expressed per 100,000 population. In contrast, the *prevalence* of a disease is computed as the number of existing cases within a given population at risk as of a given time. This rate is also expressed per 100,000 population. For most acute diseases, the concept of incidence is more important; for most chronic diseases both incidence and prevalence are equally important. The pattern of disease prevalence of a community is determined jointly by the incidence of that disease and by the case fatality of that disease.

When necessary morbidity and/or mortality data have been collected according to the pre-determined study design, the investigator is now ready for analysis. For simple statistical analyses no major electronic data processing (EDP) equipment is necessary. In fact, it is faster and more efficient to use a small desk calculator when the data are not complex and no advanced statistical tests are needed. However, if the number of observations is large and multivariate analysis is required, application of EDP equipment is needed. Unlike most case-control studies based on retrospective data, many investigations of potential health effects upon local residents exposed to certain toxic waste products are usually conducted with only a limited number of confounding variables, such as age, sex and race, taken into account.

Under some unusual circumstances in terms of scientific or political pressure, or both, the investigator may be obliged or the government agency may decide to collect data beyond what is usally required for epidemiologic investigation. For example, the overall study protocol may include cytogenetic analysis designed to evaluate the extent of chromosome damage as a result of toxic exposure. In another instance, the study design may require a community-wide screening for blood lead among children as a demonstrated (documented) evidence of human exposure to lead contamination from lead waste products.

If human reproduction is questioned as being affected by certain toxic exposure, the study design may include systematic examination of human abortuses, which is not routinely done, in addition to the more conventionally conducted analyses of fetal and neonatal mortality, birth defects or postnatal maturity problems. The study of spontaneous abortions prior to the fifth week of gestation is extremely difficult because conception is often unrecognized at that early gestation by the pregnant women themselves. It is important to recognize that the fetus is extremely sensitive to environmental insult, be that chemical or physical toxicity, and that some systematic investigation of reproductive outcomes would be a profitable avenue in the investigation of potential health hazards of such environmental pollutants.

D. *Methodological and Technical Problems in Investigation:*

In the course of describing epidemiologic methods of investigation, a number of important considerations have been identified which the investigator should be aware. Since epidemiology is an applied science with limitations and qualifications, some systematic discussion of inherent methodological, as well as technical, problems associated with this type of epidemiologic investigation should be presented.

1. One should be aware that there is a long latent period for most chronic diseases to develop after exposure to a causative agent; this is particularly true in cancer. Furthermore, the latent period varies substantially from disease to disease, as well as from one individual to another for the same disease. Recognizing this important variability in natural history, the investigator must decide at which point in historical context observation should be made to assess potential health hazards. This can become a difficult problem to deal with when two or more diseases with substantially different latent periods are encountered.

2. The critical importance of selecting appropriate control population(s) has already been mentioned. This is true regardless of the type of study design. Since it may be impossible to find an identical control population, the investigator may select more than one control and/or comparable population as are feasible; some of the confounding variables can be adjusted statistically in the data analysis. However, certain variables cannot be adjusted statistically. For example, in a longitudinal (historical) study, changes in diagnostic methods and opportunities as well as general social conditions and medical care standards do affect the incidence and natural history of diseases in a community, but casual observations of end products alone cannot detect the effects of such social and medical changes.

3. Statistical measures, such as rates and ratios, fluctuate rather markedly in a relatively small population without any epidemiologic significance. Small populations cause methodological problems in detecting rare disease, such as certain specific forms of cancer. In many instances, the investigation of potential health hazards caused by toxic wastes involves such small populations. To overcome this problem, the investigator may expand the geographic boundary, thus increasing the population size by adding adjacent political subdivisions or municipalities. This methodological exercise is also beneficial in terms of minimizing potential residence-reporting bias. In Pennsylvania, both births and deaths are systematically over-reported to cities and boroughs while under-reported to townships (more rural areas); as a result, birth and death rates are unduly higher in many urban areas and lower in surrounding rural areas. The reason for this lies in the way the postal (mailing) address system (which is often different from the residential, actual, address system) was developed in Pennsylvania. Other states

with a similar address system are expected to pose similar methodological problems in this type of epidemiologic investigation.

4. Usually, waste dump sites produce "low risks" depending upon the characteristics of the toxic substances involved. Because of this, it is difficult to detect any subtle or small changes or differences in health status. It is usually not possible to establish clear dose-response relations under such circumstances.

5. Unfortunately, premature news media publicity usually occurs regarding the alleged health hazards in a community. This has a negative influence upon conducting an objective epidemiologic investigation of potential health hazards in the area. Under such circumstances the "blind technique" is not practical and furthermore there is bound to be biased data ascertainment.

6. Local migration into and out of the study area over time makes accurate assessment of time-dependent health effects of the exposed population very difficult. The validity of research findings without the migration factor taken into account depends upon the extent of migration that takes place over time.

7. There are technical difficulties in defining and measuring "exposure dose" (quantitative data) in certain investigations. For instance, the instruments employed may not be sensitive enough to register low-level exposure in the area under study.

8. Despite the strong community voice against the toxic waste dump that usually prevails, there is often less than adequate community support to conduct an epidemiologic study of the area. Poor (or lack of) cooperation from the medical community, in particular, often interferes with the investigator's efforts in ascertaining the needed data.

9. Cooperation of the accused industry responsible for production/disposal of the toxic substances is also usually negative. The investigating scientist is often kept from reviewing industrial medical records which would be of immeasurable help in assessing occupational risks associated with the particular product in question. Known occupational health risks would make community-wide risk analysis easier and more efficient.

10. In many situations, alleged health hazards to a local population emerge with the sole identification of the presence of the accused industry without accurate or complete listing of the toxic substances involved. Much less recognized at this stage of investigation is the specificity of diseases and anomalies presumably associated with such substances. This places the epidemiologist in a disadvantageous position in designing his investigative study. However, it is not usually the responsibility of any one scientist involved, but it rather reflects the current state of the art with respect to industrial toxicology and occupational/environmental epidemiology in general.

11. Finally, biologic interpretation of epidemiologic findings need to be addressed. Conclusions drawn from epidemiologic data are expressed in terms of statistical probabilities under a certain set of assumptions and constraints; they are *not* a statement of certainty on an individual basis. This important qualification rests on the fact that many confounding factors can and do contribute to the etiology of most chronic diseases and reproductive abnormalities in humans, and that not all of the important contributory factors may be taken into account in epidemiologic studies. This is particularly true in conducting field investigations to assess the potential health risks associated with toxic waste sites.

The investigator should be aware of all *major* and *known* causative factors in addition to the substance in the waste site being questioned and attempt to collect such data, although some of the data may not readily be available. By following this procedure, the investigator can then draw firmer conclusions regarding the potential health risks of the accused toxic substance in the environment. If some "new" findings regarding diseases and conditions are reported from an epidemiologic study, such findings may be subjected to an experimental study in animals so that the "new" findings may be substantiated. The reviewing of human clinical data in light of such "new" epidemiologic findings may also be needed. These additional efforts will contribute to the existing body of scientific knowledge in the field of environmental, industrial and occupational epidemiology.

II. *Asbestos Waste Piles and Health Effects:*
A. *Industrial Uses of Asbestos:*
Asbestos is a generic name for fibrous magnesium and calcium silicates which can be crushed and processed into a variety of flame-heat resistant materials. It exists in *four major forms,* namely, Chrysotile, Amosite, Crocidolite, and Anthophyllite, which may be distinguished by their relative concentrations of metals - calcium, magnesium, silicon, and especially iron - as well as by their gross physical properties[1]. Only Chrysotile, Crocidolite and Amosite are commercially important; Chrysotile accounts for 95% of the world's production. Most of Chrysotile is consumed by the asbestos cement industry for use in pipes, flat and corrugated sheets, shingles, etc. The second largest user of Chrysotile is the floor tile industry; the production of paint, roof coating, asbestos paper and electrical insulation, plastics and friction materials, such as brake linings, account for most of the remaining amounts used[1]. In 1977, the U.S. Consumer Product Safety Commission banned the use of respirable asbestos particles in patching compounds and artificial emberizing materials under the Consumer Product Safety Act.

B. *Epidemiology of Asbestos Fibers:*

In 1935, an association between exposure to airborn asbestos fibers and the development of cancer was suggested[2]; but it was not until 1947 when the relationship was demonstrated epidemiologically[3]. Many recent studies, such as those done in Holland and the United Kingdom, have reinforced the belief that occupationally-exposed workers are at significantly higher risk of developing neoplasia, particularly of the respiratory system (lung cancer), than their unexposed counterparts. One of the malignant tumors associated with asbestos is *mesothelioma.* Beginning in 1956, a clinician in South Africa observed a cluster of patients with mesothelioma, most of which were traced not to occupational exposures but to living near open pit asbestos mines as children[4].

Cases of mesothelioma frequently occur without coexisting asbestosis; evidence now suggests that ambient levels of asbestos fibers, considerably below those necessary to produce asbestosis, may be sufficient to produce mesothelimoa. Asbestosis is an irreversible and progressively disabling lung disease resulting from the retention of inspired asbestos particles in the lungs. The latent period (time span between the initial exposure and the clinical manifestation of the disease) for mesothelioma is approximately twenty to thirty years[5].

The marked excess of bronchogenic carcinoma found among asbestos workers has been linked to cigarette smoking[6].

Asbestos work standards in most industrial countries regulate worker exposure to protect against asbestosis and to reduce materially the riisk of asbestos-induced cancer. Since many governmental agencies and some occupational health scientists believe that there is no safe or "threshold" dose for a human carcinogen, it has been implied that only complete avoidance of exposure can assure protection against the carcinogenic effects of asbestosis[7].

C. *Asbestos Waste Piles: Ambler, Pennsylvania:*

In early 1975, the Pennsylvania Department of Health was requested by the Pennsylvania Department of Environmental Resources to investigate the possible health hazards of the asbestos waste piles located in the Borough of Ambler, Montgomery County, Pennsylvania. This request was prompted because of contradictory opinions and health concerns expressed by various persons at a public hearing. A survey conducted by the U.S. Environmental Protection Agency showed that the piles, as well as the factory, in Ambler served as emission sources for asbestos fibers[8].

The town of Ambler, where air samples were taken, had a population of 7,800 in 1970 and owed much of its growth and prosperity to the Keasby and Mattison (K & M) Company, located in Ambler, which has manufactured asbestos products since 1967. The company dumped the waste from its manufacturing process onto three waste piles, ranging from 16 to 60 feet in height over 16.9 acres. The principal ingredient of the piles was *calcium carbonate* (75%), which occurs in nature as limestone. The K & M Company was sold to the Nicolet Industries and Certain-Teed Company in 1972. This company continues to crush and dump faulty asbestos products onto the piles.

D. *Preliminary Epidemiologic Investigation:*

In view of the well-established epidemiologic evidence for carcinogenic effects of asbestos and, in particular, Wagner's report[9] that exposing children to asbestos-containing waste piles produced by asbestos mining operations was associated with a markedly increased incidence of mesothelioma after approximately forty years, the Pennsylvania Department of Health initiated a preliminary epidemiologic investigation of possible health hazards among Ambler residents. This investigation involved analysis of the already available mortality data for the Borough of Ambler and its surrounding communities where there are potential health hazards, which may be related to the asbestos waste piles.

Because Ambler Borough has a small population against which mortality rates were calculated, a 6-year *average death rate,* rather than a single year death rate, was used for eight different groups of cancers, respectively, for the 1968-1973 period. The 1970 census population was used as the denominator for these calculations. These average mortality rates computed for Ambler Borough were compared to those computed for the adjacent communities as well as for the State of Pennsylvania as a whole that served as controls. In all comparisons, mortality rates were adjusted for age.

TABLE 1

Observed and Expected Numbers of Average Annual Cancer Deaths
By Sites: Ambler Borough, 1968-1973

Cancer by Organ Site	Observed Number	Expected Number	Ratio: O/E	Significance Test (P)
Buccal Cavity and Pharynx	0.67	0.33	2.03	>0.05
Digestive Organs and Peritoneum	7.17	4.69	1.51	>0.05
Respiratory System	3.83	2.81	1.35	>0.05
Bone; Connect. Tissue; Skin; Breast	3.00	1.74	1.72	>0.05
Genital Organs	3.33	1.81	1.82	>0.05
Urinary System	0.67	0.74	0.91	>0.05
Others; Unspecified	0.83	1.62	0.49	>0.05
Lymphatics and Hematopoietic Tissues	1.50	1.39	1.08	>0.05

Note: The age-adjusted expected numbers of cancer deaths in Ambler by organ sites were derived from the 1970 cancer mortality experiences of Pennsylvania as a whole.

As shown in Table 1, no significant differences were found between Ambler and Pennsylvania as a whole with respect to any of the eight cancer sites under study, including cancer of the respiratory system which was thought to be of potential importance. Also, there were no significant differences in any of these cancer mortality rates computed for the six communities adjacent to the Borough of Ambler. However, it was not possible to single out *mesothelioma,* a rare form of respiratory cancer, partly because of the small population of Ambler. In addition, there was no indication that death rates due to other diseases of respiratory organs, such as bronchitis, emphysema and asthma were significantly elevated in Ambler.

In conclusion, in the absence of morbidity data for the study area, the existing mortality data were analyzed for the purpose of this preliminary investigation. The results from the mortality data provided no evidence that the residents of Ambler or those of adjacent communities have experienced increased risks of cancer, including those of respiratory organs. Although the latent period in asbestos-induced carcinoma is considered to be at least thirty years, these waste piles have existed for a longer period of time; thus, the data would have shown the carcinogenic effects, if any, among long-term residents.

In view of the negative findings from the present mortality study, no community-wide morbidity survey, which would be quite costly and time-consuming, was recommended or undertaken.

REFERENCES

1. Hendry, N. W. The Geology, Occurrences, and Major Uses of Asbestos. Annals N.Y. Acad. Sci. *32* (1):12-21, 1965.
2. Lynch, K. M. and Smith, W. A. Pulmonary Asbestosis III. Carcinoma of the Lung in Asbesto-Silicosis. Amer. J. Cancer *14*:56-64, 1935.
3. Merewether, E. R. A. Asbestosis and Carcinoma of the Lung. In - Annual Report of the Chief Inspector of Factories, 1947. London: H. M. Stationery Office, 1949.
4. Doll, R. Pott and the Prospects for Prevention. Brit. J. Cancer *32*:263-272, 1975.
5. Lieben, J. and Pistawka, H. Mesothelioma and Asbestos Exposure. Arch. Envir. Health *14*:559-563, 1967.
6. Selikoff, I. J. and Hammond, E. C. Multiple Risk Factors in Environmental Cancer. In: Fraumeni, J. F. Jr. (ed.): Persons at High Risk of Cancer: An Approach to Cancer Etiology and Control. New York: Academic Press, 1975.
7. Workplace Exposure to Asbestos - Review and Recommendations. DHHS (NIOSH) Pub. NO. 81-103, Washington, D.C. Govt. Printing Office, November, 1980.

8. Environmental Protection Agency Study, conducted October 22-24, 1973. Office of Air Quality Planning and Standards, Research Triangle Park, North Carolina.
9. Wagner, J. C., Sleggs, C. A., and Marchand, P. Diffuse Pleural Mesothelioma and Asbestos Exposure in the North Western Cape Province. Brit. J. Indust. Med. *17*:260-271, 1960.

Hazardous and Toxic Wastes: Technology, Management and Health Effects. Edited by S.K. Majumdar and E. Willard Miller. © 1984, The Pennsylvania Academy of Science.

Chapter Twenty-Five

Coal Tar Wastes: Their Environmental Fate and Effects

James F. Villaume, M.S., P.G.
Senior Project Scientist
Environmental Management Division
Pennsylvania Power & Light Company
Allentown, PA 18101

Coal gasification is involved in various commercial processes, including underground coal conversion, coking, and—in the past—the production of illuminating or "town" gas for lighting. These processes all produce a variety of liquid and semi-liquid wastes, consisting typically of aliphatic and aromatic hydrocarbons (referred to collectively as "coal tar"), cyanides, and ammoniacal compounds. This paper deals primarily with the coal tar wastes resulting from the production of town gas, with special emphasis on the contamination site at Stroudsburg, Pennsylvania, which has been studied extensively and which was the first site in the United States to receive Emergency Superfund money for clean-up and containment action (Villaume, 1982).

The paper reviews the chemical and physical properties of coal tar wastes, as well as their toxicity, present regulatory status, and the mechanisms by which they become distributed throughout the ground water environment. Also reviewed are several of the methods which have actually been used for or can be applied to the remediation of coal tar contamination sites.

STATEMENT OF THE PROBLEM

History of Town Gas Production. Following Great Britain's example, coal gasification plants for the production of illuminating gas (also referred to as manufactured, producer, or water gas) were installed in Baltimore in 1816, Boston in 1822, and New York in 1825 (Rhodes, 1979a). In these early plants, bituminous coal was "carbonized", or destructively distilled, in small cast iron retorts at temperatures of 600-800 °C (1100-1400 °F). By 1920 there were over 900 gasification plants in the United States (Rhodes, 1979b). With the availability of natural gas, most of these plants ceased operation by 1947.

Coal Tar in the Environment. Before its reuse potential was developed, the coal tar from town gasification plants was probably either burned on site as a supplemental fuel (Smith, 1979) or was disposed of near the plant either in open lagoons (Yazicigil and Sendlein, 1981) or in underground injection wells (Lafornara et al., 1982; Villaume et al., 1983). As a result of such practices, abandoned town gasification plant sites are now being recognized as potential environmental troublespots. Besides the site at Stroudsburg, where a town gasification plant was operated from the mid-1800s until about 1939, a second site, in Burlington, Vermont, also appears on EPA's list of 418 Priority Superfund Sites. Further, over fifty of these sites in Pennsylvania alone were reported to EPA under the notification provisions of Superfund.

Also, with the current interest in this country in alternative fuel supplies, attention has focused recently on the environmental effects of underground coal gasification (Campbell et al., 1979; Wang et al., 1982); and, while little information is available in the literature, similar environmental problems can be expected to exist as a result of the past disposal activities of the coking industry (Coates et al., 1982). Even the coal tar recycling industry has a site, in St. Louis Park, Minnesota, which appears on EPA's list of Priority Superfund Sites.

Regulatory Status of Coal Tar. Decanter tank tar sludge results from the spray cooling of coke oven gases during the byproduct recovery process. This sludge has been designated by EPA as a toxic Hazardous Waste subject to regulation under the Resource Conservation and Recovery Act of 1976 (RCRA). The basis for this listing are the toxicity and mobility of its two chief constituents, phenol and naphthalene.

Coal tar from town gasification plants is not specifically regulated; however, several of its principal constituents, such as anthracene, fluoranthene, napthalene, phenol, and phenanthrene, are listed under the Clean Water Act as Priority Pollutants, and so would subject it to regulation as a Hazardous Substance under Superfund. Further, the coal tar from the Stroudsburg site was found to have an arsenic concentration which could cause it to be classified as an "EP Toxic" Hazardous Waste under RCRA.

Toxicity of Coal Tar. While many commercial products containing coal tar are readily available for purchase by the general public in the form of medicinal skin preparations, wood preservatives, and asphaltic coatings, the National Institute for Occupational Safety and Health (1977) has concluded, based on a review of available toxicological and epidemiological evidence, that prolonged exposure to coal tar products can produce phototoxic effects and skin and lung cancer in humans; however, they point out that the composition of various coal tars, and therefore probably their carcinogenic potential, depends on the source of the tar and the methods of processing.

TABLE 1

Partial Chemical Analysis of The Stroudsburg Coal Tar

Parameter	Value	Units
Naphthalene	3.60	%
Fluoranthene	3.20	%
Phenanthrene	2.30	%
Anthracene	2.30	%
Dimethyl Naphthalenes	2.15	%
Trimethyl Naphthalenes	1.78	%
Methyl Phenanthrenes	1.50	%
Trimethyl Benzene	1.30	%
Fluorene	0.98	%
Acenaphthylene	0.74	%
Pyrene	0.56	%
Benzo(a)anthracene	0.31	%
Chrysene	0.31	%
Benzo(a)pyrene	0.10	%
Other	7.84	%
TOTAL	29.69	%
Acidity	0.62	mg KOH
pH	4.6	standard
Free Carbon (Carbon I)	<0.01	%
Ash	0.00	%
Total Carbon	90.77	%
Total Hydrogen	8.12	%
Total Nitrogen	0.17	%
Sulfur	0.65	%
Ammonia	0.26	ppm
Chloride	0.65	ppm
Cyanide	0.18	ppm
Aluminum	22.4	ppm
Arsenic	12.7	ppm
Barium	0.5	ppm
Cadmium	0.01	ppm
Copper	2.48	ppm
Iron	50.3	ppm
Lead	0.5	ppm
Manganese	2.11	ppm
Nickel	0.19	ppm
Vanadium	1.6	ppm
Zinc	0.13	ppm

At Stroudsburg, the toxic effects of coal tar seepage into the Brodhead Creek, a cold water fishery, were assessed by a variety of methods, including a macroinvertebrate and fish survey, tissue analysis, and in situ toxicity testing of caged trout. It was concluded from the macroinvertebrate and fish survey that the aquatic environmental effects of the seepage were indistinguishable from other environmental perturbations, such as torrential stream flows, industrial and municipal waste water discharges, and lack of a stable habitat caused, in part, by man-made changes to the natural stream channel configuration. The tissue analyses revealed no apparent biological accumulation of the measured coal tar constituents in the fish or macroinvertebrates tested and, when the caged trout were placed in similar habitats (upstream, or background, versus downstream stations), no mortality occurred which could be directly and conclusively attributed to the coal tar seepage. These findings are consistent with the non-detectable levels of coal tar contaminants measured in the mixed stream flow by gas chromatograph-mass spectrographic (GC-MS) analysis.

COAL TAR PROPERTIES

Chemical Properties. Coal tar is a complex mixture of free (uncombined) carbon and over 10,000 individual organic compounds (mostly in trace amounts), of which only about 300 have ever been positively identified (NIOSH, 1977). A partial chemical analysis of the coal tar from the Stroudsburg site is presented in Table 1. The principal organic constituents are the polynuclear aromatic hydrocarbons (PAHs), chiefly naphthalene, anthracene, and the methylated naphthalenes. Because over 60% of the coal tar distills off at temperatures above

TABLE 2

Physical Properties of the Stroudsburg Coal Tar

Parameter	Value	Units
Specific Gravity @ 45 °F	1.028	-
@ 60 °F	1.017	-
@ 100 °F	0.991	-
@ 140 °F	0.985	-
Distillation - 10%	420.	°F
- 50%	518.	°F
- 90%	662.	°F
Pour Point	< -30.	°F
Flash Point, Cleveland Open Cup	170.	°F
Heat Content	17,763.	btu/lb.
Surface Tension	28.8	dynes/cm.
Interfacial Tension w/ Water, Pt-Ir		
Ring Tensiometer, @ 72 °F	22.	dynes/cm.
Wetting Angle in Water Against Quartz,		
Advancing, @ 78 °F	125.	degrees
Absolute Viscosity @ 45 °F	18.98	centipoise
@ 122 °F	5.04	centipoise
@ 140 °F	3.89	centipoise
Viscosity Index, ASTM D2270	194.	-
Conductivity	< 0.1	umhos

250 °C (482 °F), identification of high molecular weight aromatics and aliphatics with greater than 300 atomic mass units is extremely difficult using standard GC-MS techniques. Of particular interest is the near absence of a significant acid fraction, most notably phenols and cresols (0.99 and 0.33 parts per million, respectively), which are generally considered typical coal tar constituents.

Of the other town gas coal tars reported in the recent literature, only that from Ames, Iowa, with its styrenes, thiophenes, indoles, aldehydes, and thiazoles, seems to differ significantly (Yazicigil and Sendlein, 1981). Even the contaminants from the coal tar recycling plant at St. Louis Park, Minnesota, which received wastes from a variety of different sources, do not differ significantly from those occurring at Stroudsburg (Hult and Schoenberg, 1981).

Physical Properties. The physical properties of the Stroudsburg coal tar are summarized in Table 2. Most significant is its specific gravity, which causes it to sink in water. This seems to be fairly typical of coal tars in general; however, the coal tar from Ames, Iowa, is reportedly less dense than water (Yazicigil and Sendlein, 1982). The viscosity of the Stroudsburg coal tar is similar to that of a medium-weight oil and changes little with temperature, as indicated by its viscosity index. It has a flash point of 82 °C (180 °F) and has a heat value in excess of 17,000 btu/pound, characteristic of a high-quality fuel oil. It is also practically non-conductive and is moderately acidic. Its other physical properties, such as interfacial tension and wetting angle, are discussed below in the section on environmental fate.

COAL TAR IN GROUND WATER

Coal tar contamination can occur in the ground water environment as either an aqueous (dissolved coal tar) or hydrocarbon (free or discrete coal tar) phase. The mechanisms responsible for the distribution of each of these phases is described below.

Aqueous Coal Tar. The principal control on the concentration of dissolved organic coal tar constituents in the ground water environment is their aqueous solubility, as shown in Table 3 for the Stroudsburg site. These contaminants move essentially as a solute with the ground water, subject to the attenuation mechanisms described below. Their distribution at a given site is therefore largely determined by the local ground water flow regime. While not enough data exist to determine whether there is a relationship between solubility and distance of transport in the case of Stroudsburg, there does appear to be a rapid fall off in concentration just beyond the free coal tar plume (Figure 1) in the down-gradient direction. The only contaminant detected at this point is naphthalene at the less than 10 parts per billion level.

While ground water contamination does not appear to be a critical problem at the Stroudsburg site, the coal tar wastes at Ames, Iowa, forced the abandonment

TABLE 3

Organic Contaminants in Shallow Ground Water at Stroudsburg

Contaminant	Molecular Weight	Aqueous Solubility (mg/l)	Conc. in Coal Tar (mg/l)	Max. Gr. Water Conc. (mg/l)
Base-Neutral Fraction				
Naphthalene, $C_{10}H_8$	128.16	31.7[1]	36,000.	3.525
Acenaphthylene, $C_{12}H_8$	152.21	-	7,400.	0.428
{Acenaphthene, $C_{12}H_{10}$	154.21	3.93[1]	7,200.	0.275
{Fluorene, $C_{13}H_{10}$	166.21	1.98[1]	9,800.	0.218
Anthracene, $C_{14}H_{10}$	178.22	0.073[1]	23,000.	0.085
Phenanthrene, $C_{14}H_{10}$	178.22	1.29[1]	23,000.	0.330
Fluoranthene, $C_{16}H_{10}$	202.26	0.26[1]	32,000.	0.038
Pyrene, $C_{16}H_{10}$	202.24	0.135[1]	5,600.	0.063
{1,2-Benzoanthracene, $C_{18}H_{12}$	228.28	0.014[1]	3,100.	0.023
{Chrysene, $C_{18}H_{12}$	228.28	0.002[1]	3,100.	0.031
3,4-Benzopyrene, $C_{20}H_{12}$	252.30	0.0038[1]	1,000.	0.013
3,4-Benzofluoranthene, $C_{20}H_{10}$	252.32	0.0015[2]	370.	0.015
Benzo(ghi)perylene, $C_{22}H_{12}$	276.34	0.00026[1]	<250.	<0.010
Indeno(1,2,3-cd)pyrene, $C_{22}H_{12}$	276.34	0.0002[2]	<250.	<0.010
Volatile Fraction				
Benzene, C_6H_6	78.11	1,780.[3]	-	0.241
Toluene, C_7H_8	92.13	538.[3]	-	0.960
Ethylbenzene, C_8H_{10}	106.16	159.[3]	-	1.193

NOTES: { Indicates isomers which are indistinguishable by GC-MS.

    [1] Data from Mackay and Shiu (1977).

    [2] Data from NBS Cert. of Analysis for Stand. Ref. Mat. 1647, 1981.

    [3] Data from McAuliffe (1966).

of five city wells and restricted the use of several others due to taste and odor problems (Yazicigil and Sendlein, 1982). At St. Louis Park, Minnesota, the creosote contaminated ground water has caused the closure of seven municipal wells and threatens the continued operation of others in the area (Hickok et al., 1982). At this site relatively high levels of PAHs were found in drift and bedrock wells over an area of several square miles and at depths as great as 700 feet.

Also of concern at coal tar contamination sites is the presence of heavy metals and cyanide. At Stroudsburg, aluminum, iron, manganese, and cyanide were detected at levels as high as 218, 460, 25.5, and 0.30 parts per million, respectively, in some of the shallow ground water wells sampled. By comparison, these contaminants were measured in the raw coal tar at levels of 22.4, 50.3, 2.11, and 0.184 parts per million, respectively. Sodium was also found in the ground water at 26.2 parts per million, but was not analyzed in the coal tar. Cyanide is a direct combustion product of the gasification process and was typically removed with an iron salt to form iron cyanide, or "blue billy" (Smith, 1979). Aluminum, iron, and manganese are principal constituents of the coal ash. The high sodium levels

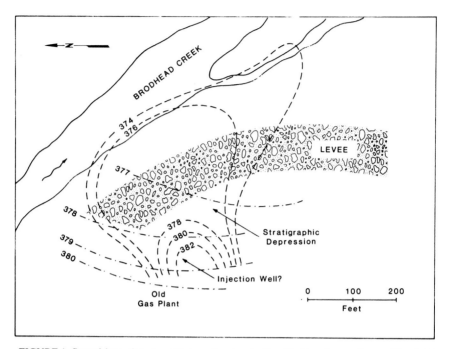

FIGURE 1. Stroudsburg site map with top of contamination (dash) and ground water (dot-dash) contours (in feet) shown. The ground water data is for June 12, 1981, prior to slurry wall construction. Almost no free coal tar occurs beyond the 374-foot contour.

may be the result of caustic usage at the plant, as suggested by Hult and Schoenberg (1981).

Free Coal Tar in the Unsaturated Zone. Movement of free coal tar through the unsaturated zone is only an important transport mechanism where the water table is deep or where the coal tar is less dense than water. In its movement toward the water table, the coal tar will travel downward under the influence of gravity. Some of the material will be adsorbed onto the soil particles and will be left behind as a residual phase. The remaining coal tar will continue to advance until it reaches the water table, where it will either float or sink, depending on its density. If it floats, it will travel in a direction consistent with the hydraulic gradient. If it sinks, then the complex mechanisms described below involving two-phase liquid-liquid flow will come into play. Of the coal tar which floats, some will rise upward under the influence of capillary pressure to form a small capillary fringe and some will be dissolved into the water below and be transported as a solute, as described above.

Free Coal Tar in the Saturated Zone. In the saturated zone coal tar which is denser than water will move as a separate liquid phase—sometimes contrary to the direction of ground water flow—according to the distribution of capillary pressure forces, as defined by the following equation:

$$P = \frac{2\gamma \cosine \theta}{R},$$

where γ is the interfacial tension between the coal tar and water, θ is the wetting (contact) angle formed by the coal tar against a solid surface in the presence of water, and R is the radius of the water filled pore which the coal tar is trying to enter. The interfacial tension may be viewed as an essentially liquid-liquid interaction between the coal tar and water and the wetting angle a liquid-solid interaction between the coal tar and the material of the pore wall. From the equation it follows that the higher the interfacial tension and the smaller the wetting angle and pore radius, the more difficult it will be for the coal tar to displace the pore water; or, once the interfacial tension and wetting angle have been established for a given system, the more difficult it will be for the coal tar to move into the smaller water filled pores. Thus, pore size distribution becomes all important in determining the coal tar distribution at a particular site (Villaume et al., 1983).

At Stroudsburg, for example, the coal tar was able to move through the clean (few fines), coarse gravels at the site but could not penetrate the underlying fine, silty sand because of the extreme capillary pressures involved, even though the porosities (but not the permeabilities) of the two materials are probably similar. For this reason the coal tar tended to collect in depressions in the top of the silty sand (Figure 2). It is from one such depression just below the old underground injection (disposal) well where approximately 8,000 gallons of nearly pure coal tar have so far been recovered through the in-ground pumping operation described below.

As with movement through the unsaturated zone, some of the coal tar will be immobilized by either capillary pressure forces, adsorption, or trapping in pore wall anfractuosities. Thus, a vertical coal tar distribution similar to that shown in Figure 3 will eventually develop if only gravity is active. Of course, ground water will cause some dispersion of the coal tar, as indicated by Figure 1, but the more than sixteen times greater viscosity of the coal tar and the capillary pressure forces set up in the porous media will hinder this movement.

Attenuation Mechanisms. Ehrlich et al. (1982) report a 150-fold decrease in the levels of phenolics and a ten-fold decrease in the level of naphthalene in the ground water 430 meters (1400 feet) down-gradient of the coal tar recycling facility at St. Louis Park, Missouri. Using sodium as a conservative tracer, they conclude that the phenolics and naphthalene are disappearing faster than expected if only dilution were occurring. They also note that the sorption of phenolics on aquifer sediments is negligible but that naphthalene is slightly sorbed. They conclude from this, and from the fact that methane and methane producing bacteria were found in water samples only from the contaminated zone, that anaerobic biodegradation of phenolics is primarily responsible for the observed attenuation. Similar evidence for the anaerobic degradation of naphthalene was not obtained.

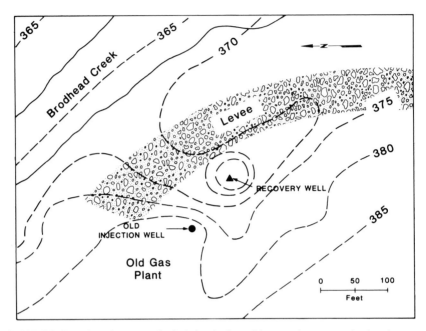

FIGURE 2. Top of sand contours (in feet) for the Stroudsburg coal tar contamination site.

Hickok et al. (1982), on the other hand, note that PAHs tend to adsorb strongly onto soil particles. Based on literature data, they estimate partition coefficients (the ratio of adsorbed to dissolved concentrations at equilibrium) of 10-100 liters/kilogram for the shallow sand near the St. Louis Park coal tar site. This implies retardation of PAH movement by a factor of approximately 60-600 relative to ground water, with retardation in the bedrock aquifers assumed to be much less.

REMEDIATION OF COAL TAR SITES

Because the environmental problems associated with abandoned town gasification plants and other facilities producing similar wastes have only recently been recognized, very little information is available in the literature concerning remediation of these sites; however, following is a summary of those remedial actions which either have been used successfully or would be appropriate for use at such sites.

Containment. Containment of the free coal tar contamination has been performed at the Stroudsburg site (Lafornara et al., 1982; McManus, 1982; Unites and Houseman, 1982; Villaume, 1982) and another abandoned town gasification plant site in Plattsburg, New York (Marean, 1981), using a bentonite slurry

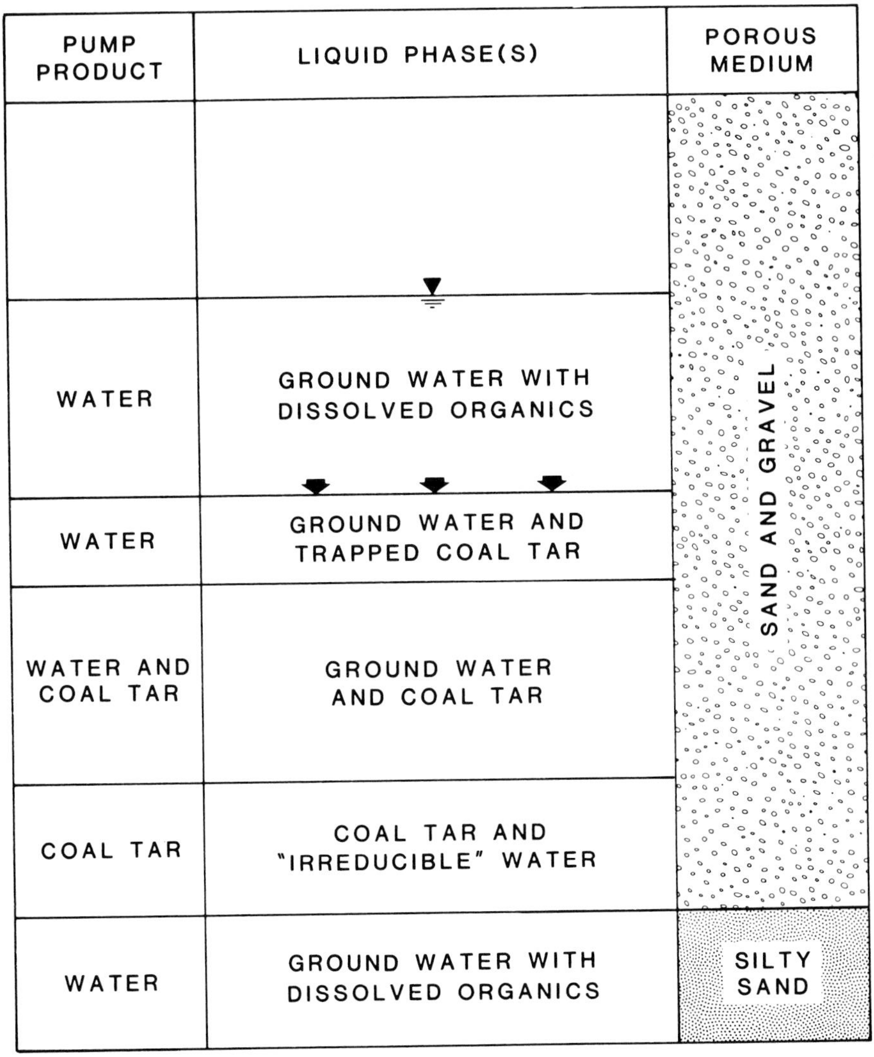

PUMP PRODUCT	LIQUID PHASE(S)	POROUS MEDIUM
WATER	GROUND WATER WITH DISSOLVED ORGANICS	SAND AND GRAVEL
WATER	GROUND WATER AND TRAPPED COAL TAR	
WATER AND COAL TAR	GROUND WATER AND COAL TAR	
COAL TAR	COAL TAR AND "IRREDUCIBLE" WATER	
WATER	GROUND WATER WITH DISSOLVED ORGANICS	SILTY SAND

FIGURE 3. Ideal distribution of coal tar in the porous materials at the Stroudsburg contamination site, as inferred from capillary pressure theory (after Arps, 1964). The arrows indicate the level of original coal tar injection.

wall. At Stroudsburg a 700-foot-long wall was constructed into the silty sand confining layer on the stream side of the western flood control dike along Brodhead Creek using money obtained under the emergency provisions of Superfund. The wall at each site cost approximately $350,000 (J. B. Marean, personal communication) and they have been in service for about one year.

Pumping. The removal of nearly pure coal tar has been performed successfully at Stroudsburg using a well and automatic pumping system (Villaume et al.,

1983). Over a period of nearly a year approximately 8,000 gallons of product with a water content of less than one per cent and virtually no ash have been recovered. The system operates by turning a pump on when the free coal tar surface reaches a preset level, which is determined by a conductance sensor placed in the well. A lower sensor turns the pump off again. Ground water is removed from the top of the well at a slow rate to cause the coal tar to upwell under the loss of hydrostatic pressure. This is necessary to overcome the very low transmissivity of the coal tar (50 gallons/day/foot compared to 4.2×10^4 gallons/day/foot for the ground water). The contaminated water is then discharged to an up-gradient leach field to provide an essentially closed ground water system. The recovered product is presently being sold as a fuel supplement, but it also has value as a chemical feedstock.

The cost of the Stroudsburg recovery system is about $35,000, which includes the pump controller and a 30-inch by 40-foot borehole and well assembly, as well as the associated piping. The average operating and maintenance costs are about $1,000 per month for pipe and product pump impeller replacement, tank and pump rental, and electrical service.

Stabilization. The stabilization of acid coal tar sludges has been successfully demonstrated in pilot studies using a twin-screw, co-rotating extruder-evaporator to mix in lime (for the neutralization of sulfuric acid), drive off volatile constituents, and blend in fuel oil (DiFillipo, 1983). The resulting product can be burned in a conventional boiler. The process can also be modified to produce a detoxified solid which can be landfilled.

Decanter tank tar sludge has also been successfully converted into a recycleable filter cake and oil. First, the sludge is mixed into a portable agitation tank with a precharged proprietary diluent. As a result, it loses its stickiness and can be pumped to a tank truck for transfer to a central treatment facility. Here the mixture is filtered to produce a dry, coal fines filter cake and a clean oil filtrate. The filter cake is returned to the coal pile and blended into the coke oven feed. The oil is suitable for use in conventional oil burning equipment. Alternatively, the sludge and diluent mixture can be ground in a ball mill and then added to the existing coal tar burning system (Yazicigil and Sendlein, 1980).

Hydraulic Isolation. Based on an analysis of local hydraulic gradients, Hickok et al. (1982) have recommended the use of 12-15 gradient control wells in five-six aquifers for dealing with the ground water contamination at the St. Louis Park site. Using this scheme, they estimate that the bedrock aquifers can be initially flushed in a number of decades, but that the drift will remain contaminated for several thousand years due to adsorption effects.

For the dissolved organics at Ames, Iowa, Yazicigil and Sendlein (1981) have likewise recommended a pumping trough barrier between the contamination source and city well field, a distance of about 800 meters (2,500 feet). Based on a digital computer model, they predict that a pumping rate of 800 gallons/minute would flush the contaminants out of the sand and gravel aquifer within two months, although it would also produce an undesirable water level decline in the vicinity of the well field. Under normal well pumping conditions, the taste and odor problems would not disappear for 2.5-3.0 years.

In Situ Water Flooding. Preliminary laboratory tests on the Stroudsburg coal tar have shown that the addition of an alkaline conditioning agent can significantly lower its interfacial tension with water. Thus, during a water flood type of operation similar to that used for secondary oil recovery by the petroleum industry, the tar should form smaller droplets and be able to flow much more easily through the porous materials at the site. These tests have also shown that a polymer can be used to increase the viscosity, and therefore the driving properties, of the flood water (Dr. P. Krumrine, personal communication). The resulting fluid could then be pumped to the surface for separation of the oily and water fractions. In this regard, other bench-scale tests performed by Pennsylvania Power & Light Company have shown that an 85/15% water/coal tar emulsion can be readily broken, without the need for heat or mixing, using readily available synthetic emulsion breakers. When applied, heat alone was found to be ineffective.

SUMMARY AND CONCLUSIONS

The current literature on the various types of commercial coal gasification facilities and their past disposal practices indicates that they may be responsible for many present-day environmental problems; and the number of these facilities in Pennsylvania suggests that the state may have a disproportionate share of those problems.

The organic tars from these sites, which are phytotoxic and carcinogenic to man, can migrate great distances through the ground water environment and can persist for many years. While various remediation techniques such as pumping, physical containment, hydraulic isolation, and waste stabilization are available to deal with this contamination, they are not always immediately and totally effective and can be very costly to apply.

REFERENCES

Arps, J. J. 1964. Engineering Concepts Useful in Oil Finding. *Bul. AAPG.* 48:159-165.

Campbell, J. H., F. T. Wang, S. W. Mead and J. F. Busby. 1979. Ground Water Quality near an Underground Coal Gasification Experiment. *J. Hydrol.* 44:241-266.

Coates, V. T., T. Fabian and M. McDonald. 1982. Nineteenth-Century Technology, Twentieth-Century Problems. *Mech. Engr.* 104:42-51.

DiFilippo, A. J. 1983. Resource Recovery of Heavy Coal Tar Based Sludge. Presented at 38th Annual Purdue Indus. Waste Conf.

Ehrlich, E. J. L. 1982. Degradation of Phenolic Contaminants in Ground Water by Anaerobic Bacteria: St. Louis Park, Minnesota. *Ground Water.* 20:703-715.

Hickok, E. A., J. B. Erdmann, M. J. Simonett, G. W. Boyer, and L. L. Johnson. 1982. Groundwater Contamination with Creosote Wates. Presented at Am. Soc. Civil Engr. Nat. Conf. on Environ. Engr., Minneapolis, MN.

Hult, M. F., and M. E. Schoenberg. 1981. Preliminary Evaluation of Ground Water Contamination by Coal Tar Derivatives, St. Louis Park, Minnesota. U.S.G.S. Open-File Report 81-72, 57 pages.

Lafornara, J. P., R. J. Nadeau, H. L. Allen and T. I. Massey. 1982. Coal Tar: Pollutants of the Past Threaten the Future. *In: Proc. 1982 Haz. Mat. Spill Conf., Milwaukee, WI,* Bur. Expl., Washington, DC, 37-42.

Mackay, D., and W. Y. Shiu. 1977. Aqueous Solubility of Polynuclear Aromatic Hydrocarbons. *J. Chem. and Eng. Data.* 22:399-402.

Marean, J. B. 1982. Coal Tar: One Utility's Approach to Dealing with a Widespread Problem. Presented at EEI Biologists Workshop, Albuquerque, NM.

McAuliffe, C. 1966. Solubility in Water of Paraffin, Cycloparaffin, Olefin, Acetylene, Cycloolefin, and Aromatic Hydrocarbons. *J. Phys. Chem.* 70:1267-1275.

McManus, T. J. 1982. Risk Assessment Case Studies: Go or No Go. Presented at Am. Soc. Civil Engr. Nat. Conf. on Environ. Engr., Minneapolis, MN.

Rhodes, E. O. 1979a. The History of Coal Tar and Light Oil. *In: Bituminous Materials: Asphalts, Tars, and Pitches, Volume III: Coal Tars and Pitches,* A. J. Hoiberg, Ed., Robert E. Krieger Publishing Company, Huntington, NY, 1-31.

Rhodes, E. O. 1979b. Water-Gas Tars and Oil-Gas Tars. *In: Bituminous Materials: Asphalts, Tars, and Pitches, Volume III: Coal Tars and Pitches,* A. J. Hoiberg, Ed., Robert E. Krieger Publishing Company, Huntington, NY, 33-55.

Smith, R. A. 1979. Redevelopment of Contaminated Land: Gas Works Sites. Interdept. Comm. on the Redevelop. of Contaminated Land, Dept. of Environ., U.K., 3rd ed., 15 pages.

Unites, D. F., and J. J. Houseman, Jr. 1982. Field Investigation and Remedial Action at Sites Contaminated with Coal Tars. *In: Proc. 5th Annual Madison Conf. of Applied Res. and Practice on Municipal and Indus. Waste,* Dept. of Engr. and Applied Sci., U. of Wis. Ext., Madison, WI, 344-355.

Villaume, J. F. 1982. The U.S.A.'s First Emergency Superfund Site. *In: Proc. 14th Mid-Atlantic Indus. Waste Conf.,* Alleman, J. E., and J. Kavanaugh, Eds., Ann Arbor Publishers, Ann Arbor, MI, 311-321.

Villaume, J. F., P. C. Lowe and D. F. Unites. 1983. Recovery of Coal Gasification Wastes: An Innovative Approach. *In: Proc. 3rd Nat. Sym. and Expo. on Aquifer Restor. and Ground Water Monitoring,* Nat. Water Well Assoc., Columbus, OH, in press.

Wang, F. T., S. W. Mead and D. H. Stuermer. 1982. Mechanisms for Groundwater Contamination by UCG: Preliminary Conclusions from the Hoe Creek Study. Presented at 8th Underground Coal Conv., Symp., Keystone, CO.

Yazicigil, H., and L.V.A. Sendlein. 1981. Management of Ground Water Contamination by Aromatic Hydrocarbons in the Aquifer Supplying Ames, Iowa. *Ground Water.* 19:648-665.

_____. 1977. Criteria for Recommended Standard: Occupational Exposure to Coal Tar Products, Nat. Inst. for Occupational Health and Safety, Washington, DC, 189 pages.

_____. 1980. Treatment Coverts Sludge into Dry Coal Fines Cake. *Iron Age,* November 3.

Hazardous and Toxic Wastes: Technology, Management and Health Effects. Edited by S.K. Majumdar and E. Willard Miller. © 1984, The Pennsylvania Academy of Science.

Chapter Twenty-Six

The Effects of Hazardous Waste on Public Health

Clark W. Heath, Jr., MD

Professor of Community Health
Master of Public Health Program
Department of Community Health
Emory University School of Medicine
Atlanta, Georgia 30322

Man has always lived in the presence of some degree of environmental exposure to toxic materials, in the form both of toxic chemicals and of background radiation. With the development of industrialized society, opportunities for exposure have increased so that control and prevention of exposure now represents a major task for modern public health.

The problem is of particular public health importance primarily because of increased population exposure to relatively low doses of toxic materials. Although prevention of high dose exposure is always a concern, opportunities for such exposure have now probably become less frequent than in earlier days when our awareness of chemical and radiation hazards was less acute. Today, although high individual exposures may be relatively well controlled, overall population dose has almost certainly increased due to wider production, distribution, and use of chemicals and radioactive materials and from concomitant and inevitable increases in toxic waste byproducts. Ionizing radiation has come to be widely used for medical and industrial purposes, and each year new chemicals and chemical products are developed for a myriad of applications.

In the face of this growing potential for toxic exposure, systems for preventing or limiting exposure have evolved in three major areas: 1) control of exposures in the workplace, 2) control of exposures from consumer products and uses, and 3) control of waste materials so that their disposal will not pose a hazard for human health. This chapter examines the public health aspects of hazardous waste disposal with emphasis on the process of risk assessment through use of toxicologic and epidemiologic data. Although the focus here is on hazardous chemicals, it should be recognized that the same issues of risk assessment and exposure control exist in relation to ionizing radiation from radionuclide wastes. For both chemicals and radiation, ultimate concern is for long-term health risks associated with low-dose exposure, risks which are principally expressed in terms of cancer, reproductive effects, and genetic alterations.

The Process of Risk Assessment

Assessment of health risks is required both for establishing universal safety standards for specific chemicals under specific exposure conditions and for evaluating potential hazards in particular local situations. In either setting, assessment involves evaluation of two kinds of data: exposure data and health outcome data. Success in assessing risk ultimately depends on the accuracy and precision with which measurements both of exposure levels and of health outcome frequencies can be made and the extent to which the two forms of data can be related to each other. All too often, particularly in individual toxic waste settings, it proves difficult to make accurate measurements of past exposure levels, to measure individual health outcomes precisely, or to relate exposure levels to specific outcomes in potentially exposed populations.

Data for risk assessment come both from human observations and from laboratory experimental studies. In both categories, practical considerations tend to limit useful data to relatively high doses of chemical exposure. As a result, it is inevitably necessary to make extrapolations from high to low dose exposures. Such extrapolations are based on theoretical models of dose-response relationships, the most common and simplistic of which is the linear non-threshold model which assumes that no dose level exists at which no biologic effect occurs and that degree of effect is directly proportional to dose. This model, widely used in setting safety standards, is generally considered to be a conservative approach in light of biologic repair mechanisms which may counteract damage to tissues and cells more efficiently at low dose levels than at high levels.

To the extent that risk assessment relies on non-human observations, extrapolations of a different sort are required to equate non-human biologic effects with human risk. At present this process of inter-species or in vitro/in vivo extrapolation depends upon various rather arbitrary assumptions and safety factors. In the end, however, it is hoped that risk assessment can be entirely based upon data which can directly relate non-human experimental findings and human subclinical observations to eventual human illness outcomes.

Estimating Exposure

The process of assessing extent of chemical exposure requires knowledge of 1) the physical properties of particular chemicals and of possible products resulting from their interaction with other environmental chemicals, 2) the environmental pathways by which human exposure may occur and the possible vehicles involved in those pathways, 3) the concentrations of chemicals in various environmental compartments, and 4) the extent to which particular chemicals may be absorbed into human tissue and then have active toxic effects.

Although such information may be developed with relative ease for certain chemicals for which existing toxicologic information is fairly complete, often the chemicals of concern in toxic waste settings, and their chemical interaction products, are not well characterized. Analytic laboratory techniques may also not be fully developed, making it difficult to measure levels in soil, air, or water with suitable precision. Adequate laboratory techniques for measuring environmental levels are in fact of critical importance for toxic chemicals which may be rapidly metabolized or cleared after absorption into tissue. Under such conditions, risk assessment will inevitably depend almost entirely on accurate measurements of environmental levels, coupled with information regarding biologic activity extrapolated from animal or in vitro experimentation.

Since chemical measurements in environmental samples are often expensive and time-consuming, it is important to select environmental sampling sites with care and to minimize laboratory variability so that maximum information can be derived from a limited selection of samples. Close attention must be given not only to possible air, water, and soil sources of environmental contamination directly adjoining the waste site but to paths of migration through surface water run-off and ground water aquifers by which human populations some distance away may receive significant exposure through contaminated water sources. Although most chemicals migrate rather slowly in aquifer channels, the potential for distant exposure can be substantial if contamination has existed for some length of time.

Plans for environmental chemical sampling are also greatly influenced by knowledge of how particular chemicals enter human tissue, to what degree they are absorbed, metabolized, and excreted, what particular tissues they affect, and at what dose levels they or their metabolic products exert toxic action. For chemicals which are efficiently absorbed into tissue, are only slowly excreted, and are highly toxic, environmental testing must be particularly thorough and make use of particularly sensitive and reliable laboratory methodology. For materials which are poorly absorbed or have relatively low biologic activity, somewhat less emphasis need be given to extensive environmental testing, particularly at low chemical levels.

Finally, occupational exposures incurred during work at a toxic waste site or in the process of cleaning up an abandoned site need to be evaluated. Such exposure may involve relatively high dose levels and may require special equipment and techniques for adequate protection of workers.

Estimating Health Effects

Risk assessment is ultimately a matter of measuring clinical illness in exposed population and relating those measurements to levels of exposure. Although this process is easy to fulfill for acute toxic effects at high doses, it has proved to be very difficult for low-dose exposures. At high dose levels, health effects are seen soon after exposure and with relatively high frequency, making it an easy task to link health effects to their toxic exposure causes. Depending upon the nature of the toxic material, the mode of exposure, and the organs principally affected, acute health effects can vary widely, commonly affecting skin, lungs, liver, gastrointestinal tract, kidneys, and nervous tissue. Detailed observation regarding such acute toxic illnesses form the principal basis for current knowledge regarding how toxic organic and inorganic chemicals affect human health.

Experimental animal studies, together with less direct human observations, however, make it clear that toxic chemical exposures, perhaps even at quite low doses, may also induce various forms of delayed illness such as cancer, reproductive effects, and chronic degenerative disorders at different tissue sites. For such delayed health effects, linkage of toxic chemical exposure to particular cases of illness is a difficult task because of several general features of such illness (Table 1): 1) long latency between toxic exposure and illness development, 2) the relative rareness of such illness on a population basis, and 3) the clinical non-specificity of such illness whereby it is usually impossible to identify the specific cause of a particular case of illness from the clinical features of that case.

Latency periods of 15 to 30 years are believed to characterize most cases of human cancer which may arise from toxic chemical exposure (1). Since the temporal coincidence of acute illness and toxic exposure is probably the major reason why we can easily identify the causes of acute toxic disease, it is not surprising, on grounds of latency alone, to find it difficult if not impossible to identify causes of cancer cases. Only for reproductive abnormalities (spontaneous abortions, low birth weight infants, birth defects) may it be possible to relate clinical outcomes to exposure, since latency, assuming that outcomes result from teratogenic effects of exposure during pregnancy, is necessarily limited to at most the nine month duration of pregnancy.

Linkage of outcome to exposure, in any case, must usually be done on a population basis using epidemiologic approaches by which rates of illness in exposed populations are compared to rates in unexposed or reference populations. The necessity of such methodology emphasizes the second difficulty encountered in such studies: the need to study relatively large populations since the disorders

TABLE 1

Features of Biologic Effects to Environmental Exposures Which May Limit Ability to Make Risk Assessments By Studying Health Outcomes

---

1. *Long and Variable Latency* - Inability to link cause and effect through closeness in times of exposure and illness onset.
2. *Infrequent Occurrence* - Epidemiologic investigation of rare diseases requires large populations for study, often over long periods of time.
3. *Clinical Non-specificity* - Inability to identify the cause of the case of disease purely from its clinical features.
4. *Multiple Causation* - Capacity for many different specific factors to contribute, alone or in combination, to the development of particular kinds of illness.
5. *Imprecise Information on Exposure* - Inability to develop objective and quantifiable estimates of exposure in individuals because of non-persistence of toxic chemicals in tissue after exposure.

---

being studied are relatively rare. Since most local studies of toxic waste sites involve rather small populations close to sites, population size is a crucial limiting factor for success in epidemiologic analysis.

Lastly, however, the task of linking clinical illness to toxic exposure is made even more difficult by the clinical non-specificity of cancer and other chronic illnesses. From the clinical features of particular cases it is almost always impossible to know what caused illness, whether a particular case was induced by ionizing radiation or by exposure to a specific chemical, for example. This requires the epidemiologist to pay close attention in data collection to possible causes other than toxic waste exposure (for instance, cigarette smoking in the instance of lung cancer). The epidemiologist must at the same time consider other confounding factors such as socioeconomic status which may be related to causative factors and which hence may distort comparisons of disease rates. As a result of these study design requirements, arising in large part from the clinical non-specificity of cases, epidemiologic associations between illness and toxic waste exposure are difficult to establish with certainty unless exposures have been extraordinarily high and unless acute illness, uniquely associated with the particular exposure, has been documented.

Some help for the epidemiologist may conceivably come 1) from parallel studies done regarding illness in animals, wild or domestic, living near toxic waste sites, or 2) by focusing on subclinical human biologic effects, such as frequency of chromosome breakage, slowed nerve conductive velocity, sperm alterations, and abnormal results of mutagenicity testing of human urine. Although most such approaches are still in the developmental phase, they are promising in the sense that they may enable the epidemiologist to make somewhat more powerful observations with respect to human populations of limited size in which direct measurement of clinical illness is hampered by small numbers of cases. However, none of these several non-human or sub-clinical approaches has yet been developed to the stage where we are able to predict with any certainty that finding a particular subclinical abnormality (increased chromosome breakage, for instance) indicates increased future risk of particular clinical illness in the population studied.

As a result of these various difficulties inherent in measuring and interpreting health effects, whether clinical or subclinical, assessment of risks associated with toxic exposures must rely heavily on measurement of environmental levels of toxic materials with interpretation of those levels in terms of human illness risk. Although data regarding human health effects may provide indirect supportive evidence, in the last analysis risk assessment depends upon documentation of environmental chemical levels. In the absence of reliable and objective exposure measurements, not only is risk assessment extrapolation seriously compromised, but the limited ability of epidemiologists to relate exposure to biologic effect is reduced even more. In studies of health effects related to past exposures, virtually the only avenue open to the epidemiologist is when the chemicals of concern are biologically persistent, that is, are stored for long periods of time in fat or other tissues, so that past exposures can be estimated in individual subjects by means of blood or tissue assays. Lacking such an objective approach to individual exposure, estimate of exposure remains a matter of imprecise exposure surrogates such as distance of residence from the waste site, duration of residence in the area, or subjective recollections of exposure incidents.

Public Perceptions of Toxic Waste Health Risks

Public understanding of toxic chemicals, how they can affect human health, and how risk assessments are made is an important aspect of waste management, and it often has a direct influence on the process of risk assessment. Waste control is of obvious public interest both because of its implications for health and because it often has economic implications by affecting local property values and giving rise to legal disputes involving sizable economic settlements. In contrast to public health concerns such as natural disasters and infectious disease exposure which mostly arise from causes not of man's making, toxic waste chemical exposures result from human action and hence are frequently open to legal argument.

In the face of such legal, emotional, and socioeconomic pressures, public perceptions of health risks easily become oversimplified, distorted, and misguided. There are two main forms of misconception, one centering on the principle of dose-response, the other on clinical specificity. Often, when people learn that a particular chemical can cause cancer, they fail to appreciate the further fact that causation is dose-dependent, that low doses are less likely to cause cancer than high doses, even though no threshold for effect may exist. The idea that risks are relative is much less easily understood and accepted than the idea of risk itself. This difficulty is reflected historically in the so-called Delaney clause which was incorporated into food safety law in the United States in the 1950's and which prohibits the marketing of any food containing any amount of chemical shown to have oncogenic potential, however small that amount might be.

The second common misconception is to assume that, since a particular chemical can cause cancer, all cancers in persons so exposed are caused by that exposure. This neglects the fact that cancer, like many if not most diseases, is clinically non-specific, that many different etiologic factors may contribute to case occurrence, and that, when we examine any individual case, we are nearly always unable to know what particular factor or set of factors caused that particular case. Again, the simple fact of causal association is much more easily grasped than the complex idea of multiple competing causes without clinical markers.

These two areas of misconception and oversimplification need constant attention when toxic chemical risks are discussed with the public. If health professionals can convey to the public the relative nature of exposure risks and the degree to which causes of particular cases or illness can and cannot be deciphered, the difficulties caused by this kind of public misunderstanding of scientific issues can be greatly reduced.

Measuring Exposure and Health Risk: Love Canal as an Example

The risk assessment issues described above can be illustrated by reviewing the process by which levels of toxic waste exposure and human health risk have been evaluated at the Love Canal chemical waste dump. That process, which is not yet completed, began in 1978 with public recognition of the problem and action to prevent further exposure. Risk assessment was approached both through measurement of levels of chemicals in environmental media and through studies of biologic effects and disease frequencies in the surrounding population. This work involved efforts by many different public and private groups and was conducted in a setting of great public anxiety and intense legal dispute.

The Love Canal dump site is located in Niagara Falls, New York, where it was used in the 1940's and 1950's for disposal principally of waste chemical byproducts from the manufacture of organic pesticides. In 1953 the site was sealed and became the property of the city school board. An elementary school was built adjoining the midsection of the site, streets and utility lines were installed across it, and residential neighborhoods were constructed to the east, west, and north, including homes built directly along the edges of the canal. By the mid 1970's it became increasingly apparent that the site was no longer sealed and that canal chemicals were appearing at the surface of the site and migrating through water and soil into adjoining residential properties. Starting in the spring of 1978, the New York State Department of Health (NYSDH) began chemical analysis of soil, air, and water from the dump site and its vicinity, and, at the same time, initiated a questionnaire survey of residents living close to the site (2). These studies were expanded over the next two years and were supplemented by other environmental health studies conducted by local organizations and federal agencies.

Environmental Testing. Environmental testing sought to define levels of exposure from contact with soil, surface water, ground water, outdoor air, and indoor air. Sampling was particular intense on the canal itself and in the two rings of homes closest to the canal, with later extensions to areas where draining from the canal might have carried chemical contamination (the tracks of natural surface swales and the outfall of storm sewers into local streams). Testing was performed both by the NYSDH before reconstruction drainage work on the site was begun in 1978 and by the U.S. Environmental Protection Agency in 1980 after drainage work was complete. Emphasis was placed on detection of the kinds of organic chemicals dumped at the site (chlorobenzene, for instance) and at particular locations where human contact might have been most likely (soil in yards, air in homes, basement sump pump water). On the whole, most chemicals tested were of the kind which are not stored in human tissue but are promptly cleared or metabolized. As a result, no useful direct information could be developed regarding past levels of exposure through testing human serum or fat.

In general, results of environmental testing, both before and after drainage reconstruction, showed only very low levels of chemicals, probably not increased significantly above background levels, with the exception of samples from the canal itself, from storm sewers draining it, and from certain homes located very close to the site (3,4). Final decisions regarding rehabilitation of the area, with special concern for future human residence, depend greatly on the results of such environmental testing. The process of decision-making, however, is extraordinarily difficult since no fixed criteria exist for making absolute judgments regarding the adequacy of particular environmental sampling procedures, frequencies, and results.

Health Studies. The initial health survey at Love Canal sought interview information for a wide range of acute and chronic symptoms and illness among persons living close to the canal site, compared with persons living north of the area. No clear patterns of increased risk were seen, although the subjective nature of much of the data made interpretation difficult. Since information regarding pregnancy outcomes could not rule out the possibility that abortion and birth defect frequencies might be increased near the canal, records of physicians and hospitals were sought to confirm diagnoses. This study of reproductive outcomes was then expanded to include families living on or near natural drainage swales and later to include all women of childbearing age ever to have lived in the

TABLE 2

Past Abnormal Pregnancy Outcomes of Women
Living as of June 1978 in the Love Canal Area (2,5)

Place of Residence in the Canal Area	Total Pregnancies	Total Live Births	Spontaneous Abortions		Congenital Malformations		Low Birth Weight (Under 2500 Grams)	
			No.	%*	No.	%**	No.	%**
Within One Block of the East and West Edges of the Canal Site	79	65	15	19.0	4	6.2	1	1.5
Further East in the Canal Area Not Near Drainage Swales	108	85	25	23.1	10	12.0	13	15.7
Not near Drainage Swales	164	144	21	12.8	7	4.9	10	6.9
North in the Canal Area	125	110	11	8.8	8	7.3	3	2.7

* Percent of total pregnancies
** Percent of total live births

canal area. Pregnancy outcome measures were also extended to include frequency of low birth weight births.

Particular emphasis was placed on pregnancy outcomes since it was reasoned that unborn embryos and fetuses might be particularly sensitive to effects of chemicals, would provide a known and relatively short latency period, and hence might be useful indirect human predictors of eventual oncogenic and mutagenic risks. Although analysis of this complex set of pregnancy outcome data has yet to be finished, no clear pattern of increased risk has yet been seen in relation to closeness to the canal or to its potential drainage channels (Table 2) (5). Interpretation of the data is limited both by the relatively small numbers of births in the exposed cohort and by the absence of any means for directly measuring the cumulative exposure of individual persons to canal chemicals.

These same limitations apply in varying degrees to subsequent studies carried out in the Love Canal area regarding cancer incidence and frequency of chromosome abnormalities. Cancer incidence for the 12 year period 1966-77 was examined in data from the NYSDH state-wide cancer registry, comparing rates in the census tract containing the Love Canal with rates for other Niagara Falls census tracts and with rates for the state as a whole (6). No patterns of increased incidence were seen except for a small excess of lung cancers in women (Table 3). Special attention was given to cancers arising in liver or hematopoietic tissues because of the known toxicologic patterns of particular chemicals dumped at the canal site.

Interpretation of these data is limited by each of the several factors discussed earlier: small numbers, lack of precise and individualized exposure information, lack of data regarding competing causes of cancer (especially levels of cigarette smoking), and perhaps inadequate allowance for elapsed latency. Use of an entire census tract to define a population "exposed" to the Love Canal had the effect of diluting the potential degree of exposure in the study population and hence diminishing the power of the study to detect exposed-unexposed differences in disease frequency. Although the census tract base provided adequate numbers from which to derive estimates of cancer incidence, the tract encompasses an area considerably larger than the canal and its immediate vicinity.

Cytogenetic studies were performed in two stages, a preliminary stage in 1980 when analyses were made in 36 volunteer subjects from the Love Canal area (7) and a later phase in 1981-2 when findings in 46 canal area residents were compared with findings in 44 persons from a nearby control area (8). The second study included persons who had lived in homes adjoining the canal and in whose homes increased levels of canal-related chemicals had been found in 1978. Although the earlier study had initially been interpreted as showing increased levels of chromosomal breakage, this conclusion was later shown to be probably incorrect. The second study produced no evidence of increased chromosome damage when Love Canal and control subjects were compared.

As with other health-related studies in the Love Canal area, interpretation of

TABLE 3

Observed and Age-Standardized Expected Numbers of Cancer Cases,
By Sex, Love Canal Census Tract, 1966-1977 (6)

Cancer Site	Male Cases			Female Cases		
	Observed Number	Expected Number	Expected 95 Percent Confidence Interval	Observed Number	Expected Number	Expected 95 Percent Confidence Interval
Digestive	13	15.7	8-24	15	13.8	7-22
Respiratory	25	15.0	8-23	9	4.6	1- 8
Genital	13	7.9	3-14	8	12.5	6-20
Urinary	7	6.0	2-11	1	2.4	0- 5
Leukemia, Lymphoma & Liver Cancer	3	6.3	2-12	6	4.7	1- 9
All sites	71	60.8	46-77	71	65.4	50-82

these results is hampered by a number of considerations. Although chromosome studies have the theoretical advantage of measuring subclinical effects which may represent sensitive early indicators of later clinical disease, it has yet to be established that such a subclinical-clinical link exists or, if it does, what its dimensions are. Beyond that fundamental difficulty, however, are limitations in sample size, lack of precise exposure data for individual subjects, and the fact that chromosome tests were performed on biologic specimens drawn some time after unrestrained exposure to canal chemicals had ceased and after some subjects had moved away from the area.

Despite the shortcomings of these various kinds of health effect studies (pregnancy outcomes, cancer, chromosomal abnormalities), their negative or equivocal results may have some limited usefulness, in conjunction with environmental testing, for assessing health risk in situations like the Love Canal. Still, in the end, risk assessment must rely primarily on interpretations of environmental toxic chemical measurements. Since such measurements in the Love Canal area have on the whole yielded very low values (chemical concentrations in the low parts per billion range), it is not surprising that evidence has not been found to link presumed exposure with subsequent clinical or subclinical health effects. Additional evidence has been sought by studying characteristics of local wild animal populations (meadow voles) (9). Although the results of those studies suggest possible chemical effects on growth and reproduction in such animals in the Love Canal area, many uncertainties exist both in comparing wild animal populations in different settings and in extrapolating such results to the human situation.

Summary

Assessment of health risks associated with hazardous waste material involves evaluation of 1) levels of environmental toxin exposure and 2) extent of possible health effects. Such information provides the basis for general safety standards connected with exposure to hazardous waste and for judging potential for health hazards at particular waste sites. Assessment of acute health effects in high dose exposure situations is a relatively simple matter in contrast to chronic effects resulting from low dose exposures. The relationship of clinical or subclinical effects to low doses of chemicals is difficult to establish because of long latency periods, imprecise exposure information, rare disease outcomes, and clinical non-specificity. Health studies have given particular attention to cancer, abnormal reproductive outcomes, and chromosomal abnormalities as potential sequelae to low dose hazardous waste exposures. Such biologic effects are difficult to evaluate in the absence of high risk ratios or large study populations and because of uncertainty in predicting clinical outcomes from subclinical findings. Most risk evaluations, therefore, must rely primarily on interpretation of environmental chemical levels. Such interpretation is often uncertain, because of need to extrapolate from high dose information to the human setting. Public perceptions of risk present further difficulties since they are often influenced by emotional and economic fears and by tendencies to overlook 1) the relationship of illness risk to chemical dose and 2) physicians' inability to identify specific causes of illness by observing specific clinical features. For optimal management of hazardous waste problems, public health officials must address public concerns and misconceptions at the same time as they are assessing levels of exposure and potential health effects.

REFERENCES

1. Armenian, H. K., and Lilienfeld, A. M. 1983. Incubation period of disease. *Epidemiol. Rev.* 5:1-15.
2. Heath, C. W., Jr. 1983. Field epidemiologic studies of populations exposed to waste dumps. *Envir. Health Perspect.* 48:3-7.
3. Kim, C. S., Narang, R., Richards, A., et al. 1980. Love Canal: Chemical contamination and migration. Proceedings, National Conference on Management of Uncontrolled Hazardous Waste Sites. Environmental Protection Agency, pp. 212-219.
4. Environmental Protection Agency. 1982. Environmental monitoring at Love Canal. 3 Volumes. Washington, D.C., U.S. Government Printing Office. Document No. EPA-600/4-82-030a.
5. Vianna, N. J. 1980. Adverse pregnancy outcomes - potential endpoints of human toxicity in the Love Canal. preliminary results. In: *Human Embryonic and Fetal Death* (Proceedings, 10th Annual New York State Department of Health Birth Defects Symposium). I. H. Porter and E. B. Hook, Eds., Albany, N.Y., pp. 165-168.

6. Janerich, D. T., Burnett, W. S., Feck, G., et al. 1981. Cancer incidence in the Love Canal area. *Science* 212:1404-1407.

7. Picciano, D. 1980. Pilot cytogenetic study of the residents living near Love Canal, a hazardous waste site. *Mammalian Chromosome Newsletter* 2186-2193.

8. Heath, C. W., Jr., Nadel, M. R., Zack, M. M., Jr., et al. Cytogenetic findings in persons living near the Love Canal.*J. Amer. Med. Assoc.,* in press.

9. Rowley, M. H., Christian, J. J., Basu, D. K., et al. 1983. Use of small mammals (voles) to assess a hazardous waste site at Love Canal, Niagara Falls, New York. *Arch. Envir. Contam. Toxicol.* 12:383-397.

Hazardous and Toxic Wastes: Technology, Management and Health Effects. Edited by S.K. Majumdar and E. Willard Miller. © 1984, The Pennsylvania Academy of Science.

Chapter Twenty-Seven

Health Effects of Hazardous Wastes

George Lumb, M.D., M.R.C.P., F.R.C.Path.
Professor of Pathology
Director of Environmental Pathology
Hahnemann University
Broad & Vine
Philadelphia, PA 19102

The effect on health of various environmental factors is of great importance. The reaction to this problem has been largely emotional and much more needs to be done to assemble facts on actual sickness rates related to potentially injurious substances. These include those resulting from natural phenomena and those derived principally from industrial sources.

The safe disposal of hazardous wastes has become a matter of urgency in recent years. In Pennsylvania, where the fourth largest quantity of hazardous wastes in the United States is generated, considerable attention is paid to the problem and guidelines for disposal methods have been and are now being defined.

Reaction to the health problem tends to be extreme. At one end of the spectrum are those who are willing to follow a prescribed law, providing it does not seriously interfere with production and profitability. At the opposite pole are those who advocate closing all plants and factories suspected of generating any potentially dangerous products or by-products.

It is important to define a rational attitude and it is particularly important for physicians to understand the problem so that they can adequately advise their patients.

Most physicians, have been taught to think in terms of crisis medicine. When a patient feels ill, an attempt is made to define the problem and to treat it. Most study of disease is related to pathogenesis, or the natural history and spectrum of an established process. When confronted with a variety of possible etiologic agents, physicians are not accustomed to assess the capacity for disease production in the absence of prior knowledge of established associations with illness. There is a lack of full understanding that the establishment of disease is multifactorial. Thus, even the clinical state of tuberculosis depends not only on the tubercle bacillus as an etiologic agent, without which the disease will not develop, but, also on other factors, such as climate, nutrition, social background, and poverty.

Examples of Hazard

Three examples of occurrences giving rise to environmental pollution and potential health hazard follow. They will introduce the problems of aerial pollution; mixed hazardous waste ground pollution; and the specific potential problem arising from the use of nuclear fuel in power plants.

1. Krakatoa

In August 1833 the volcano on the island of Krakatoa near Java in the South Pacific erupted violently changing the island into a rocky waste land in a few days. In addition, there were noted all over the world, and particularly in London, England, changes in the sunset, with extraordinary color manifestations. It was considered by many scientists at the time, that the changes resulted from dust thrown up by the volcano which had surrounded the world. This was almost certainly true. The color changes, however, may well have been confused with those caused by smog and aerial pollution as this was the period of the development of the industrial revolution in England when a tremendous amount of soft coal was being burned. The English artist Turner had painted wild sky colours earlier in the century. Turner had been regarded by many of his critics as being virtually mad, but we can look at his paintings now and realize that what he was seeing and interpreting, with a little artistic license, was in fact aerial pollution or smog as seen in the early 19th century.

This story introduces the topic of aerial pollution, a form of hazardous waste which can be produced by natural causes, such as occurred most recently with the volcanic eruption at Mt. St. Helen. This eruption, although it introduced large volumes of potentially toxic gases, including sulphur dioxide, into the atmosphere, does not appear to have caused or exacerbated human disease in any significant way other than the obvious results of immediate trauma. This does not mean that the incident can be ignored and monitoring of toxic substances in the air and in acid rain must continue. Other forms of so called naturally occurring aerial pollution include blowing dust from overplowed farm land, which may or may not contain hazardous chemicals associated with pesticides and insectisides. Included in this category are certain substances which appear to cause their injurious effects in the lungs as a result of individual hypersensitivity. Farmers lung associated with a mold growing in stored hay and sugar cane fiber inhalation are examples of this type of problem, along with such conditions as mushroom pickers' and pigeon fanciers' pneumonitis. Aerial pollutants of significance are asbestos, coal dust, silica, and beryllium, all associated with various forms of pneumoconiosis and these are well recorded elsewhere.

A serious form of hazardous waste which has increased in importance in recent years occurs as a result of gaseous effluents, such as sulfur dioxide, nitrogen oxide and dioxide, carbon monoxide and excesses of ozone. These come mainly from power plants and other manufacturing sites fired by fossil fuels. Some of these gases are also produced by automobiles and are particularly dangerous in times of high humidity in places such as parking garages and tunnels. Recent alarming occurrences amounting to the extent of epidemics of respiratory deaths in Dinora, Pennsylvania, and London, England, have been related to pollutants of this type occurring in smoggy atmospheres.

The aerial pollutants can spill into rivers and lakes as a result of precipitation of so called acid rain, and when this is mixed with particulate material, including chemicals such as lead, cadmium, mercury and nickel, can be particularly dangerous in causation of disease. Nickel and cadmium oxides and salts in particular, are components of the "soot" generated in power plant smoke stacks. Reference will be made later to the fact that although considerable advances have been made in documenting levels of pollutants, very inadequate information is available on "effects" of pollutants and health hazards. This particularly relates to "safe levels" with regard to many of these polluting substances.

The principal forms of disease which are produced by the above mentioned hazardous substances are various types of chronic obstructive lung conditions and cancer. Those who smoke tobacco are at greater risk than non-smokers.

2. Love Canal

In 1892, William T. Love arrived in Niagara Falls. He wanted to build an industrial city with convenient access to inexpensive water power and major markets. He realized that the water power would provide the cheapest electric power for it was essential in 1892 that industry be located near its power source because it was impossible to transmit electricity over distances.

In 1894 work on the canal began. In that year, however, a full-scale economic depression was beginning and people began to cut off financial support. Worse than this, Louis Telsa discoverd the means to provide alternating current, and thus to transmit electrity over distances. The purpose of Mr. Love's proposal disappeared and the project was abandoned. All that remained was a piece of the canal which children who lived in the area used as a swimming hole until the 1920's.

In the middle of the 1920s, this swimming hole became a disposal site used by several chemical companies in the area and also by the City of Niagara Falls. Chemicals of unknown kind and quantity were buried there until 1953 when the ditch was full of waste material and the site was covered with earth.

At the end of the 1950's the political fathers of the area decided to put the land to use. They proposed a housing development which would include an elementary school. The Chemical Company, which owned the land, agreed to sell the area to the city authorities.

The development proceeded; a school was built; and two straight streets, 99th Street and 97th Street, were laid out. Between these two streets was the

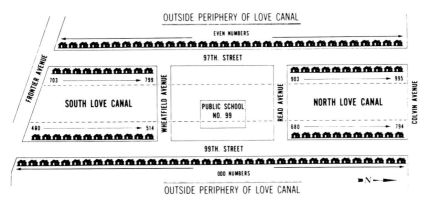

FIGURE 1. Plan of Love Canal area.

former Love Canal with its buried dump of chemicals.

Almost in the middle of this area and directly over the former canal site, the elementary school was built (Fig. 1).

The Niagara Falls area experienced some extreme rains and one of its worst blizzards ever in the early 1970s. Soon after the blizzard, dark fluid was found seeping into the basements of the houses close to the canal. Large lumps of blackish material began to appear in the yard of the school and the backyards of the houses bordering the chemical dump. It was quickly realized that these were part of the chemical dump which was beginning to surface. The story of Love Canal, which has been termed a horror story and a disaster, became the focus of attention and mounting hysteria on the part of the people living in the area, and, finally took state and federal intervention for the evacuation of families.

In the spring of 1978, an extensive air, soil, and ground water sampling analysis program began[1]. It included identifying a number of compounds in the basements of 11 homes adjacent to Love Canal. It was shown that the houses on South 97th and 99th streets had the highest levels of contamination (Fig. 1).

The ten most significant chemicals identified at the site are shown in Figure 2. The highest value obtained for benzene appears to be considerable. The present acceptable industrial level for benzene is 10 parts per million and OSHA is seeking to lower the safe limit of benzene to one part per million. This request has been denied on the basis that not enough evidence is available to support such a reduction. The highest level observed in the Love Canal area for benzene, however, was 270 mcg/m^3 or 0.00027 mgm/liter. This represents an equivalency of 0.085 parts/million. This is based on the known fact that one part/million for benzene is 0.00319 mg/liter.

An investigation of miscarriages in the area revealed that more than twice

COMPOUNDS	NO. OF TIMES FOUND IN HOUSES	PERCENT OF TOTAL HOUSES SAMPLED	HIGHEST VALUE OBSERVED
Chloroform	23	26	24 ug/m^3
Benzene	20	23	270 ug/m^3
Trichloroethene	74	84	73 ug/m^3
Toluene	54	61	570 ug/m^3
Tetrachloroethene	82	93	1140 ug/m^3
Chlorobenzene	6	7	240 ug/m^3
Chlorotoluene	32	36	6700 ug/m^3
m+p xylene	35	40	140 ug/m^3
o-xylene	17	19	73 ug/m^3
Trichlorobenzene	11	13	74 ug/m^3

FIGURE 2. Ten most significant toxic substances found in basements.

the anticipated number of miscarriages (2.08) occurred among the women living in the 99th Street South section. There were no significant differences between the observed and expected distributions for the other sections.

The women's background histories lacked evidence of abnormal pregnancies before coming to the Love Canal area. A significant number of pregnancies were studied as most of the families were young.

The study of 99th Street South was interesting because the oldest houses were there and the average duration of residence of women was longer. The women with miscarriages had resided on the canal for an average of 18.58 years versus an average length of residence of 11.52 years for those without miscarriages. This difference was statistically significant representing a chance-occurrence probability of 4 in 1,000. The data also indicated that this occurrence was not due to differences in age or number of pregnancies reported by the women.

Although the numbers are small, there do seem to be some important correlations.

A chromosome damage study was performed on certain residents of the Love Canal area. Although chromosomal changes in a population may have meaning, at the present state of our knowledge, they cannot be interpreted in an individual.

The study performed on the Love Canal population led to problems of interpretations and differences of opinion between various experts who examined the resuls.

Unfortunately some of the results were released to the press. The news distressed and led to hysterical outbursts on the part of people living near Love Canal.

At the present time and with the data available, it appears that the tests performed on people in the region of Love Canal have shown no specific abnormalities except for a possible increase in incidence of miscarriages in the women living in one particular area immediately adjacent to the canal dump. This is the area where the highest levels of poisonous materials, including benzene, were found but these levels were extremely low.

Love Canal has become a classic example of thoughtless ground pollution, leading to problems which were handled in an extremely inept manner. The results in terms of effects on physical health appear to be minimal at the present time, although no prognosis can be given for the development of disease in the long term. Psychologic upset was considerable and expense, as yet not compensated by those responsible for dumping the hazardous material, has been excessive.

Various members of the New York State Department of Health have been most cooperative in supplying information relating to Love Canal, and the latest information is as follows[5]. The school and all the houses surrounding the dump with the exception of two in the so called outer ring, have been demolished. The owners of the two remaining houses refused demolition. The chemical dump area had a special drain system put in around it and a cap of impervious clay put over the top of it and around it. This was the original plan and the Health Authorities feel that this is satisfactory. The water draining away from the site passes through a special treatment plant and is carefully tested before it is returned to the general water system. The air is being tested from time to time. In the opinion of the Health Department, the containment of the chemical dump is now satisfactory. This shows that if such a site is constructed properly, it can, at very high cost, be safe.

No medical report has been issued, but an ongoing epidemiologic study continues. No significant findings have been recorded to date. Future plans for the area of contamination are under discussion, but no building is considered in the immediate future[15].

Many other examples of hazardous waste dumping have been investigated, but none, in the United States, have revealed significant health effects other than mental stress. Greater attention must be paid to long term surveillance of populations at risk, in order to uncover variations in disease patterns. This will require greatly improved data reporting, which in turn will require considerable physician participation.

3. Three Mile Island

At 4 a.m. on the morning of March 28, 1979, which is not the best time to have problems, alarm bells sounded in the facility indicating that something had gone wrong. Immediate steps were taken to identify the problem. A vital valve in the reactor cooling system had stuck in an open position when it should have returned to a closed position after correcting a minor problem. This could be described as a mechanical failure. It took several hours to determine that this event had actually taken place, partly due to the design of the equipment, which made it difficult to identify the exact location of the problem, and partly due to the fact that the operators were not trained to deal with this particular problem. This combination could be described as design failure and human failure. All three, mechanical, design and human failure, taken by themselves were relatively simple. Together, they led to a situation which came very close to the problem which had been considered an impossible accident. This is core melt down. Needless to say, this did not occur. For various reasons and after a great deal of frustrating indecision, the various valves were closed and the problem which had occurred at 4 a.m., was corrected by approximately 18 minutes after 6 a.m., but, after this, the principal damage to the core began to take place because of the heat which had developed. The media coverage showing scientists, managers and politicians in various attitudes of apparent indecision and incompetence is well known.

When considering the hazards related to nuclear energy it must be recognized that there is a curious prejudice in making any considered judgements.

In the minds of many people, nuclear energy immediately raises the spectre of nuclear arms and nuclear explosions. The big yellow mushroom, rising to the sky, and spreading death and destruction around it.

The development of nuclear energy, however, involves extremely small quantities of radioactive material as compared with those used for the detonation of destructive weapons. Even smaller quantities of material are radioactive at any one time. The processes in the various manufacturing procedures involved are restricted to well-defined and well-protected areas. The nuclear material is used to produce high pressure steam which then powers turbines and generators to produce electricity. In other words relatively small amounts of uranium isotopes are used instead of gas or oil. The rest of a nuclear generating plant is exactly the same as any other electrical power generating plant.

So far, most scientific energy has been directed to the areas of nuclear fuel fabrication and energy production. In the future much more attention must be paid to the disposal of the low-level radioactive waste material.

Although mistakes have been made, as they always will be at the time of development of any new process, the safety rate for nuclear energy plants has been extremely good when compared with the coal mining and petrochemical industries.

Since the accident at Three Mile Island, careful estimations have been made of what might have happened if melt down had occurred. The three independent reports, one to the President, one to the Governor, and one to the National Radiation Commission (NRC) indicate that containment would have been achieved. A personal visit to the site and interviews with Management and N.R.C. personnel confirmed this conclusion to the satisfaction of this author[6-9]. Containment means that any radioactive material resulting from the melting of the reaction chamber would have remained within the protected area. It must be remembered that there was no question of an explosion.

What Did Happen?

Approximately 2.5 million curies of radioactive noble gases and 15 curies of radioiodines were released. These resulted in an average dose of 1.4 millirems (m rem) to the approximately 2,000,000 people in the site area.

In order to understand these numbers, the following figures are important. The average dose of 1.4 m rems which was received by the people around TMI, is less than 1% of the annual dose to any normal human being from both natural background and medical practice. There is a natural background of radiation in the air at any time. For further comparisons, the 1.4 m rem dose has been compared to the differences in annual doses in background radiation from living in a brick house versus living in a wooden frame house. Brick would produce an additional 14 m rems per year. Another comparison would be someone living at the high altitude of Denver compared with somebody living at sea level. The person living in Denver would get an additional 80 m rems per year.

Another comparison which was made by the independent reviewing group for the N.R.C. was to estimate the maximum probability dose which could have been received by any one person located off site. They used an extreme example by assuming an individual standing where maximum exposure was most likely to occur for 24 hours a day for 6 days with no clothes on, in the open. It was calculated that such a person would have received a dose below 100 m rems. Such an extreme exposure would provide that individual with an additional life-time fatal cancer risk of 1 in 100,000. The additional life-time fatal cancer risk to the individual receiving the actual average off-site dose of 1.4 m rems mentioned above, is about 1 in 5,000,000.

In summary, therefore, taking reasonable and even unreasonable examples, it has to be accepted that the risk to the population around Three Mile Island as a result of the accident was virtually zero. Taking another and simpler example, the risk of the accident compared with the risk of skin cancer from sunbathing around a swimming pool for an entire summer, is so small, that it cannot be compared.

Those few workers inside the plant who, had to perform jobs at the time of the accident which entailed exposure, received a dose of approximately 4 rems which is in excess of the quarterly limit of 3 rems allowed by the Nuclear Regulatory Commission (NRC). (NRC allows 5 rem/yr with no more than 3 rems in any single quarter).

The independent commission reviewing the accident did conclude that during the first several days after March 28, there was a potential for severe overexposure even though it did not occur. This, they felt, was principally due to the inadequacy of the radiation protection and the overall safety protection program at Three Mile Island. They further concluded that this was probably due to the fact that both the owners and operators of the plant and also NRC considered the safety program of secondary importance compared to making the plant work. Unfortunately this is a universal problem and it is not only related to nuclear energy plants. It is imperative that pressure be brought to insist on adequate safety precautions throughout industry. Death occurs just as easily from a beam falling on the head or by an explosion in a coal mine as by a nuclear reaction. In fact, the beam and the explosion are much more certain and dramatic than the long range potential of cancer risk.

A health effect research program was set up following the Three Mile Island accident headed by the Director of Epidemiologic Research for the Pennsylvania Department of Health.[19-21] It is long-term and somewhat exhaustive in its concept. It represents a mini-version of the continuing surveillance of the health hazards following the Hiroshimo and Nagasaki bomb incidents. It involves the following features:

1. A census of all the people working in the plant and living within five miles of TMI which will be kept up on a long-term basis.
2. A pregnancy outcome study.
3. A special study of congenital or neo-natal hypothyroidism, because of the very small amount of radioactive iodine released at the time of the accident.
4. A health behavioral stress study is underway.
5. A cardiac mortality study will be performed. It was suggested after the accident that the stress and worry might lead to an increased number of cardiac deaths among people already suffering from heart disease.
6. A long-term disease surveillance including mental health will be followed over an indefinite period.

So far no excess of abnormal findings has been discovered.

In summary, and perhaps oversimplifying, nobody was quite sure what would happen following the accident at Three Mile Island. In fact, nothing serious did happen, but, once again, as at Love Canal, enormous emotional upset and expense was generated.

EFFECTS OF CHEMICAL CONTAMINATION

There are reasonably well-developed data relating to the extent of chemical contamination. Estimating the effects of chemical contamination, however, is very difficult.

In New York State, reasonably accurate figures are available for the pollution levels of many areas such as Lake Ontario, the other major lakes and the Hudson River. The compounds involved in this pollution include dioxins, Mirex, polychlorinated biphenyls (P.C.B.'s). "acid rain", and, in the lower Hudson River, cadmium. Information is available about the enormous quantity of gasoline spilled from tractor trailers on the roads, and the discharge of mercury and nickel which seep and leach into water supplies. Land disposal sites for suspected synthetic chemical contamination have been mapped out, although illegal and unknown dumps are everywhere.

The question arises whether there is evidence of diseases either appearing de novo or increasing in number as a result of this chemical contamination. Most of the emphasis on potential hazards has been centered on the demonstration of carcinogenesis. Latent periods between exposure and emergence of significant numbers of new tumors, or increases in known tumors, will probably be long. Examples such as the vaginal tumor related to DES, or the angio-sarcomas related to vinyl chloride, demonstrate this.

Vinyl Chloride is a highly toxic substance and must be very carefully controlled. The toxic effects are reasonably well documented and should be recognized by physicians. The most important damaging effects are the potential for carcinogenicity and, in particular, the production of the interesting and otherwise unusual tumor, the angiosarcoma, particularly in the liver.

It is interesting and important that this is one of the few examples where experimental animal tumor production predicted the site of tumor occurrence in man.

Exposure to vinyl chloride is associated with multiple systemic disorders including a sclerotic syndrome, acro-osteolysis (sometimes associated with a Raynaud-like symptomatology), thrombocytopenia and liver damage, including parenchymal cell damage. Fibrosis of the capsule, perioportal fibrosis associated with hepatomegaly and splenomegaly having all been described. Although skin and bone changes may disappear when the patient is removed from contact with vinyl chloride, the thrombocytopenia persists after termination of exposure.

Reduced pulmonary function has been observed in workers exposed to vinyl chloride and the prevalence of this impairment was similar in smokers and non-smokers suggesting that occupational or other environmental factors were operative[23-26].

New incidence of tumors may be masked among the commonly occurring carcinomas. For example, it is apparent that cigarette smoking causes the vast majority of lung cancers. It is also clear that the cigarette effect could be masking smaller numbers of lung cancers caused by other factors. Enstrom[27] has examined lung cancer death rates among non-smokers. He observed that lung cancer death rates doubled among male non-smokers during the 10-year interval of 1958-68, while female non-smoker death rates were unchanged. The death rate of non-smoker male lung cancer in 1966-68 was roughly equivalent to the death rate for stomach or pancreatic cancers. The figures since 1968 are not very well recorded, but a recent New York State cancer registry survey[8] seems to confirm the trend to an increasing incidence in male non-smokers. Thus, there seems to be reason to suspect another unidentified causative factor in lung cancer. Albert et al.[29] suggested that an alert system might be based on cancer registry monitoring of young adult non-smokers who have squamous or small cell undifferentiated lung cancer. This suggestion is based on the fact that exposure to bischloromethyl ether in the work place is associated with squamous and undifferentiated small cell carcinoma. While very few lung cancer patients are reported that meet these conditions, attention to the non-smoking lung cancer patients as a take-off point for field investigations could be of value.

Any investigations of this type require that cigarette smoking statistics must be added to cancer registry data in order to be useful for environmental monitoring of lung cancer. It also points out the need for careful and uniform histologic analysis of tumors in order to identify sub-sets of this type.

In New York State recently[30] an attempt was made to see whether, from the cancer registry data, it could be determined in retrospect that there was a relationship between beta-naphthylamine or benzidine, both well-known chemical carcinogens, and bladder cancer. It was found that there was so much "noise" in the system that it would have been impossible to identify this known fact based on the statistical evidence of the registry. This clearly suggests that present data systems have to be modified and improved. Similarly with regard to mesothelioma, it is interesting to note that when the records in New York State were reviewed in retrospect, no change in incidence of mesotheliomas could be identified despite the tremendous increase in asbestos usage[31].

This finding may mean that there had been no effect, or that there is a long latency which had not yet appeared. It seems much more likely, however, that the information was not detectable or was lost in a larger group. It is significant that there were 2275 cases of pleural and peritoneal tumors recorded of unidentified type during the same period, compared with the very small number of mesotheliomas.

POLLUTION RISKS AND THEIR ASSOCIATION WITH LOW LEVELS OF POLLUTANTS

Remarkably little is known of the effects of very low doses of toxic substances.

There are two possible relationships between carcinogenesis incidence and pollutant concentration. A very simple graph identifies these[32]. (Fig. 3).

The lower curve (A) shows a threshold theory where it is frequently assumed that the cancer incidence varies as the square of the dose at low levels, leading to an approximate threshold level.

The second curve (B) is the so-called linear theory in which incidence is considered to be proportional to the dose and, therefore, the incidence passes back to the zero point.

The importance of determining which relationship obtains in a given situation is illustrated by the following. If, in a population of 200 million people, 200,000 deaths a year would be expected due to a pollutant at a given level, the question can then be posed: suppose the exposure to this 200 million people was reduced by a factor of 100? According to curve (A), the threshold theory, very few people

ILLNESS MEASURE

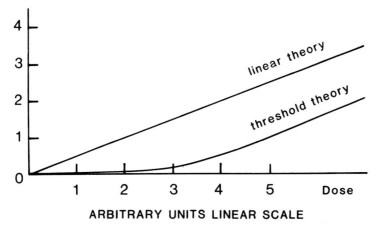

FIGURE 3. Linear and threshold theory.

or no one would die. According to curve (B), the linear theory, 2,000 people would die per year.

Most of our knowledge about pollutant effects comes from small populations, low dosages, and low incidences. In fact, we have no valid evidence to support a distinction between these two curves. Even animal experiments have not helped to resolve this matter. Where valid data are available for part of the curve, extrapolation back cannot be made. For example, a linear response has been seen in a study of cigarette smoking. It was shown that 40 cigarettes per day give approximately 10 times the hazard of 4 cigarettes per day[33]. However, there are no data at very low levels of, for example, 1 cigarette a week.

The fact that something can be done to modify the effect of an environmental factor is suggested by the work of Auerbach et al[34] who examined large numbers of bronchial biopsies from men dying of causes other than lung cancer between the years 1955-60 and also between the years 1970-77. The objective was to see whether the atypical changes which occur in the bronchial mucosa of smokers had diminished during the years 1970-77 following a period when low tar cigarettes and filters had been used. Their findings indicate, as would be expected, that the atypicalities were significantly higher among the smokers than the nonsmokers and that the changes could be dose related to the amount of cigarettes smoked. They were also able to show in the biopsies between 1970 and 1977, that, although the difference between smokers and nonsmokers still existed, the changes in the smokers had significantly diminished.

There are also data of linearity with chemical cancers induced in animals. The induction of angiosarcomas in rats by vinyl chloride at levels less than 1000 parts/million is linear with the concentration. It also appears that the slope of the curve is consistent with the slope obtained from 68 known human cases of liver angiosarcoma in workers exposed to high levels of vinyl chloride[35]. What is not known is whether liver angiosarcomas occur at concentrations of 20 parts/million or below, so the possibility of a threshold remains.

A further implication of this is as follows. The permitted exposure level for vinyl chloride for workers in factories is one part/million. No liver angiosarcomas have been observed in animals or man at 20 parts/million or below because no experimental data are available. If 20 parts/million is the real threshold, then exposure at one part/million would lead to no hazard. On the other hand, if there is no threshold, and one extrapolates the linear response curve of high concentrations, one cancer every year might occur among 10,000 workers, or, 14,000 cases a year if the entire population of the USA were exposed.

By setting the levels at one part/million in the factory, the average exposure of the 5 million Americans who live near plants is kept to considerably less than one part/billion of vinyl chloride. This should lead to one cancer in two years even on a linear theory. Most people would agree that this is as close to zero as it is reasonable to come. It is also true that if the linear theory was used to enforce total removal of the toxic product, it would bring the industry to a standstill. Peto[36] has indicated that, in those cases where a particular carcinogen causes a high incidence of tumors, an incremental dose of the substance will lead to a linear response.

In summary, a linear theory gives an opportunity to work out figures and statistical evaluations which are useful, but these figures must be used sensibly including a cost-risk-benefit analysis. As a result, the Delaney clause will probably have to be modified.

SIGNAL LESIONS

A method which may be useful in providing information related to environmental hazards is the identification of so-called "signal lesions" which are lesions, either common or uncommon, that appear in unusually large numbers in any population group. Accurate recording of "normal" incidence is necessary in order to identify "abnormal" incidence accurately.

We have examples of how this is effective, and can be made more effective. For instance, rare lesions, such as mesothelioma, or, angiosarcoma of the liver, can be carefully recorded and correlated with the exact exposure to the environmental hazards of the patients involved.

One way that signal lesions could be used would be to examine a condition of common occurrence. As an example, pulmonary fibrosis could be studied and the lesions and mechanism of death carefully identified and tabulated. This could be followed by identification of occupational and associated exposures of the patients involved. Such a study of bronchitis in 7 Japanese cities[37] showed its incidence to be proportional to the sulphate level. Similar findings have been reported for the mortality of patients with bronchitis in the United States[38,39], and England.[40,41]. Studies would require a collaborative national effort on the part of pathologists and a standardization of reporting techniques. It would cost money, but would be likely to produce better results than the very costly long-term animal studies that have so far provided such a small amount of useful information.

DISCUSSION

In most recent examples of problems associated with hazardous wastes in the United States, one finds a very similar sequence of events. A mixture of negligence, human and mechanical error and considerable expense incurred in correction processes. The major health problems appear to have been psychological although continued surveillance is required to rule out long term physical disabilities. This does not mean that an attitude of complacency can be adopted. On the contrary, increased vigilance is needed in order to examine minor changes in disease patterns more carefully. It is easy to quote isolated incidents of major problems, such as explosions in chemical factories, extreme water contamination by chemicals, such as have occurred with Mercury in Japan, or the results of Nuclear explosion testing with inadequate safety precautions. These, however, are the type of major accidents and disasters which will inevitably continue to occur and unfortunately it is true, that we learn from our mistakes.

It is in the areas of discovery of new hazardous agents; establishment of safety levels if they exist for particular substances; and the effects of low levels of pollutants on general health; that our efforts must be concentrated.

Not enough is known of the relationship of chemicals to the many diseases classified as idiopathic. Thus, chronic cardiovascular disease and behavioral disorders, or impaired intellectual development may be linked to exposure to heavy metal contaminants, including mercury and lead in childhood.

In recent years legislation has been introduced in many states, including Pennsylvania, which requires industry to make known to its work people, any hazardous substances with which they may come in contact in the course of their work. This is a valuable legal action, but such action necessitates much greater knowledge of the effects of toxic substances, toxic dose levels, and methods of treatment. Industry must be held responsible for adequate disposal of hazardous waste and extreme penalties must be imposed on those who dump waste in an inadequate manner, irrespective of whether or not human or animal ill effects result.

It is essential that while research continues to establish safe levels and thresholds of toxicity of hazardous materials, equal efforts should be exercised by physicians, researchers and legislators working in collaboration, to reduce obvious industrial pollutants to the lowest possible level compatible with economic stability.

At the present time it appears that "cleaning up the mess" is more expensive than paying greater attention to "preventing the mess".

Certainly Love Canal and Three Mile Island seem to support this contention.

REFERENCES

1. Love Canal: Public Health Time Bomb, Office of Public Health, State of New York, Special Publication, 1978.
2. Kolata, G. B. Love Canal: False alarm caused by botched study. Science 208:1239-1242, 1980.
3. Greenwald, Director, Division of Epidemiology, State of New York, Dept. of Health, Office of Public Health. (Personal Communication) 1981.
4. N. J. Vianna, State of New York, Dept. of Health. (Personal Communication) 1982.
5. E. Anders Carlson, State of New York, Dept. of Health Bureau of Toxic Substances Assessment (Personal Communication) 1983.
6. Rogovin, M. Frampton, G. T., Jr. Three Mile Island, A Report to the Commissioners and to the public, Vol. 1; 1979; Available from: GPO Sales Program, U.S. Nuclear Regulatory Commission, Washington, D.C.
7. Ad Hoc Population Dose Assessment Group. Summary and Discussion of Findings from: Population Dose and Health Impact of the Accident at the Three Mile Island Nuclear Station. A preliminary assessment for the period 3/28-4/7/79. Available from: HFX-25, Bureau of Radiological Health, Rockville, MD.
8. U.S. Nuclear Regulatory Commission, TMI-1 Restart Hearings (Docket 50-289), Testimony of Bruce Molholt, Ph.D., In Support of Off-Site Contentions of the Environmental Coalition of Nuclear Power; Submitted March 16, 1981.
9. Ad Hoc Population Dose Assessment Group. The Three Mile Island Population Registry, Report One: A general Description, October 1981. Available from: Government Printing Office, Washington, D.C.
10. Tokuhata, G. K. Three Mile Island Health Effects Research Program: proceedings of a symposium; Pennsylvania Academy of Science 54; 19-21, 1980.
11. Tokuhata, G. K., Smith, M. W. History of Health Studies around Nuclear Facilities: a Methodological Consideration, Environ. Res. 25, 75-85, 1981.
12. Tokuhata, G. K. Health Studies in the Three Mile Island Area. Paper presented at the annual meeting of the American Nuclear Society, Miami

Beach, Florida, June 7-12, 1981.
13. Tokuhata, G. K., Kim, J., Newlin, V., Staub, S., Bratz, J., Digon, E. Maternal Body Weight, Weight Gain During Pregnancy, Height and Pregnancy Outcome. Paper presented at the 109th annual meeting of the American Public Health Association, Los Angeles, California, Nov. 4, 1981.
14. Tokuhata, G. K. Impact of TMI Nuclear Accident Upon Pregnancy Outcome, Congenital Hypothyroidism and Infant Mortality, *Energy, Environment and the Economy,* pub., Pennsylvania Academy of Sciences, 1982.
15. Tokuhata, G. K. Fetal and Infant Mortality and Congenital Hypothyroidism Around TMI. Paper presented at the International Symposium on Health Impacts of Different Sources of Energy, Nashville, Tenn., June 22-26, 1981.
16. Tokuhata, G. K. Pregnancy Outcome Around Three Mile Island. Paper presented at a conference, Linking Public Health Social Worker and Public Social Services for Comprehensive Care for Mothers and Children, March 29-April 2, 1981.
17. Bromet, E. Three Mile Island: Mental Health Findings. Pittsburgh: Western Psychiatric Institute and Clinic, October, 1980.
18. Houts, P. S., Miller, R. W., Tokuhata, G. K., Ham, K. S. Health-Related Behavioral Impact of the Three Mile Island Nuclear Incident Part I. Report submitted to the TMI Advisory Panel on Health Research Studies, PA. Dept. of Health, April 8, 1980.
19. Houts, P. S., Miller, R. W., Tokuhata, G. K., Ham, K. S. Health Related Behavioral Impact of the Three Mile Island Nuclear Incident Part II. Report submitted to the TMI Advisory Panel on Health Related Studies, PA. Dept. of Health, November 21, 1980.
20. Houts, P. S., DiSabella, R. M., Goldhaber, M. D. Health-Related Behavioral Impact of Three Mile Island Nuclear Incident Part III. Report submitted to the TMI Advisory Panel on health Research Studies, PA. Dept. of Health, May 12, 1981.
21. Hu, Teh-wei, Slaysman, K. S. Health-Related Economic Costs of the Three-Mile Island Accident. A final report submitted to the Division of Epidemiological Research, PA. Dept. of Health, July 1981.
22. New York State Departments of Environmental Conservation and Health: Toxic Substances in New York's Environment - an interium report. DEC Technical Report, Albany, May 1979, pp. 1-48.
23. Creech, J. L., Jr. and Johnson, M. N. Angiosarcoma of liver in the manufacture of polyvinyl chloride. Jour. Occ. Med. *16*:150-151, 1974.
24. Nicholson, W. J., Hammond, E. C., Seidman, H., and Selikoff, I. J. Mortality experience of a cohort of vinyl chloride - polyvinyl chloride workers. Ann. N.Y. Acad. Sci. *246*:225-230, 1975.
25. Thomas, L. B., and Popper, H. Pathology of angiosarcoma of the liver

among vinyl chloride-polyvinyl chloride workers. Ann. N.Y. Acad. Sci. *246*:268-277, 1975.

26. Thomas, L. B., Popper, H., Burke, P. D., and Selikoff, I. J. Vinyl-chloride induced liver disease. From idiopathic portal hypertension (Banti's Syndrome) to angiosarcomas. New Engl. Jour. Med. *292*:17-22, 1975.

27. Enstrom, J. Rising lung cancer mortality among nonsmokers. Jour. Natl. Cancer Inst. *62*:755-760, 1979.

28. Greenwald, P. Epidemiological Monitoring of Health Effects from Environmental Contamination. Preventive Medicine (in press).

29. Albert, R. E., Pasternak, B. S., Shore, R. E. and Nelson, N. Identification of occupational settings with very high risks of lung cancer. Jour. Natl. Cancer Inst. *63:*1289-1290, 1979.

30. Greenwald, P. and Laurence, C. E. The state of records for environmental research. Public health conference on records and statistics. National Center for Health Statiostics, DHEW Pub. No. (PAS) 79-1214, Hyattsville, Aug., 1979, pp. 296-299.

31. United States Bureau of Mines: Mineral facts and problems, U.S. Government Printing Office, Washington, D.C. 1956 and 1976.

32. Wilson, R. "Risk caused by low levels of pollution". Yale Jour. Biol. Med. *51*:37-51, 1978.

33. Doll, R. and Hill, A. B. Mortality in relation to smoking: ten years' observations. Brit. Med. Jour. *1*:1399-1410, 1964.

34. Auerbach, O., Hammond, E., Garfinkel, L., New Eng. Jour. Med., *300*:381, 1979.

35. Maltoni, C., Lefamine L: Experimental Carcinogenesis, Vinyl Chloride Research Plan and Early Results. Env. Res. 7:387, 1974 plus later data. These data are used by AM Kuzmack and RE McGaughy (EPA report, Dec 1975) in a risk analysis for vinyl chloride. This is also included in Wilson R: Risk Benefit Analysis for Toxic Chemicals: Vinyl Chloride. American Chemical Society Meeting, San Francisco, CA, Sept 1977.

36. Peto R: The Carcinogenic Effects of Chronic Exposure to Very Low Levels of Toxic Substances. National Institute for Environmental and Health Sciences (NIEHS) Extrapolation Conferences, Pinehurst, 1976. See also New Scientist (letter) Oct 11, 1977, p. 111.

37. Nishiwaki, Y. et al, Atmospheric Contamination of Industrial Areas Including Fossil Fuel Stations and the Method of Evaluating Possible Effects on Inhabitants, Report to the Conference on Environmental Effects of Nuclear Power Stations IAE-SM-145/16, International Atomic Energy Agency, Vienna.

38. Lave, L. B., Seskin, E. P.: Air Pollution and Human Health. Science, 169:723-733 (1970).

39. Report from Biomedical and Environmental Assessment Group, BNL 20582, 30 July 1974, Hamilton LD, ed, Brookhaven National Laboratory, Upton, NY.
40. W. W. Holland and D. D. Reid, Lancet, 1965-I, 445 (1965).
41. D. D. Reid, Lancet, 1958-I, 1289 (1958).

Appendix

Hazardous • National
Waste Sites: • Priorities List[1]

- ## Current NPL
  in order of ranking
- ## Proposed Update to NPL
  in order of ranking
- ## NPL and Proposed
  Update Sites, by State

The National Priorities List (NPL) identifies the targets for action under the Comprehensive Environmental Response, Compensation, and Liability Act of 1980 (CERCLA or "Superfund"). The law sets up a Trust Fund (expected to total $1.6 billion) to help pay for cleaning up hazardous waste sites that threaten public health or the environment. The fund is administered by the U.S. Environmental Protection Agency (EPA).

To date, EPA has inventoried almost 16,000 uncontrolled hazardous waste sites. To be eligible for a remedial action under Superfund, however, a site must be included in the NPL. In October 1981, EPA published an interim priority list of 115 sites and followed with an expanded eligibility list of 45 additional sites in July 1982, for a total of 160 sites. EPA published a list of 418 sites as a proposed rule in December 1982, including 153 of the 160 sites previously published. Times Beach, Missouri, was proposed in March 1983, bringing the total proposed list to 419. After a period of public comment, EPA analyzed the comments and is now publishing the NPL, also referred to as the current NPL, as a final rule. At the same time, EPA is proposing new sites in its first NPL update, which CERCLA requires at least annually.

[1]Reprinted from *United States Environmental Protection Agency,* August 1983, HW-7.1, Office of Solid Waste and Emergency Response.

This publication includes three lists:

- The 406 sites that met the criteria for inclusion in the current NPL, as ranked by EPA.
- The 133 sites on the first proposed NPL update, as ranked by EPA.
- NPL and proposed updates sites arranged by State. NPL sites are listed first alphabetically, followed by update sites listed alphabetically.

CURRENT NATIONAL PRIORITIES LIST (NPL)

Each entry on the NPL of 406 sites contains the name of the site, the State and city or county in which it is located, and the corresponding EPA Region. The entries are arranged according to their scores on the Hazard Ranking System (HRS). HRS scores are designed to take into account a standard set of factors related to risks from migration of substances through ground water, surface water, and air. the list serves both as an information and management tool, allowing EPA to decide which sites warrant detailed investigation to determine what, if any, response is needed.

The 406 sites are in nine groups, 50 to a group (except for the last group). The scores on the NPL (except for some sites designated by States as their top priority) range from 75.6 to 28.5, the cut-off point selected for the December 1982 proposal to identify at least 400 sites, the minimum number specified by CERCLA. The range within most of the groups of 50 is only 3 to 5 points. EPA considers sites within these groups to have approximately the same priority.

CERCLA allows each State or territory to designate a top priority site; 36 did so. These sites must be listed in the first 100, regardless of score. Of the 36 top priority sites, 9 are included in the first 100 on the basis of their scores. The remaining 27 have scores that would not place them in the top 100. They are included at the bottom of the second group of 50. Of the 27, 7 have scores below 28.5. Top priority sites are designated by asterisks.

Of the 419 proposed sites, 14 are not included in the final rule, and one has been split into two sites, making a total of 406 sites on the current NPL published in August 1983. The 15 sites and their disposition:

- Five sites are not included because of score changes resulting from EPA's review of comments and other information. A total of 143 scores changed, but in only these five cases did the new score drop below the cut-off point of 28.5. These five sites will not be eligible for future remedial funding. They are:
    Crittenden County Landfill, Marion, Arkansas
    Flynn Lumber Co., Caldwell, Idaho
    Parrot Road Dump, New Haven, Indiana
    Phillips Chemical Co., Beatrice, Nebraska
    Van Dale Junkyard, Marietta, Ohio

- One site, Gratiot County Golf Course, St. Louis, Michigan, has been cleaned up by the responsible party and so is not included in the final list.
- One site, Plastifax, Inc., Gulfport, Mississippi, is not included because it is no longer designated the State's top priority, and its score is below the cut-off point.
- On seven sites, rulemaking is pending because EPA is still reviewing the technical comments and resolving final scores. These seven sites will continue to be eligible for remedial funding. EPA will gather data needed to calculate final scores. The seven sites are:
  Kingman Airport Industrial Area, Kingman, Arizona
  Airco, Calvert City, Kentucky
  Bayou Sorrel, Bayou Sorrel, Louisiana
  Clare Water Supply, Clare, Michigan
  Electrovoice, Buchanan, Michigan
  Littlefield Township Dump, Oden, Michigan
  Whitehall Well Field, Whitehall, Michigan
- One site, Vestal Water Supply, Vestal, New York, has been split into two sites—Vestal Water Supply Well 1-1 and Vestal Water Supply Well 4-2—because EPA discovered that there are two separate plumes of contaminated ground water from two separate sources. Both remain on the current NPL.

Status: Each entry is accompanied by one or more notations describing the current status of response and enforcement activities at the site. The notations are:

  V Conditions are currently being addressed through EPA-sanctioned voluntary actions by parties responsible for wastes at the site.
  R Conditions are currently being addressed through response actions funded by CERCLA.
  E Conditions are currently being addressed through a State or Federal enforcement action against parties responsible for wastes at the site.
  D EPA is considering various alternatives for this site. This includes all sites not listed in any other category.

Site Distribution: Of the 50 States, the District of Columbia, and six territories, 48 are represented. New Jersey has the largest number of sites with 65, followed by Michigan (41), Pennsylvania (30), New York (27), and Florida (25). Alaska, the District of Columbia, Georgia, Hawaii, Mississippi, Nebraska, Nevada, the Virgin Islands, and Wisconsin have no sites on the current NPL.

Group 1

EPA REG	ST	SITE NAME *	CITY/COUNTY	RESPONSE STATUS #			
02	NJ	Lipari Landfill	Pitman	V	R	E	
03	DE	Tybouts Corner Landfill *	New Castle County		R	E	
03	PA	Bruin Lagoon	Bruin Borough		R		
02	NJ	Helen Kramer Landfill	Mantua Township		R		
01	MA	Industri-Flex	Woburn	V	R	E	
02	NJ	Price Landfill *	Pleasantville		R	E	
02	NY	Pollution Abatement Services *	Oswego		R	E	
07	IA	Labounty Site	Charles City	V		E	
03	DE	Army Creek Landfill	New Castle County			E	
02	NJ	CPS/Madison Industries	Old Bridge Township			E	
01	MA	Nyanza Chemical Waste Dump	Ashland		R		
02	NJ	Gems Landfill	Gloucester Township		R	E	
05	MI	Berlin & Farro	Swartz Creek		R		
01	MA	Baird & McGuire	Holbrook		R	E	
02	NJ	Lone Pine Landfill	Freehold Township		R	E	
01	NH	Somersworth Sanitary Landfill	Somersworth				D
05	MN	FMC Corp.	Fridley	V		E	
06	AR	Vertac, Inc.	Jacksonville	V		E	
01	NH	Kes - Epping	Epping		R	E	
08	SD	Whitewood Creek *	Whitewood	V			
08	MT	Silver Bow/Deer Lodge	Silver Bow Creek				D
06	TX	French, Ltd.	Crosby		R		
01	NH	Sylvester *	Nashua		R	E	
05	MI	Liquid Disposal Inc.	Utica		R		
03	PA	McAdoo Associates *	McAdoo Borough		R	E	
06	TX	Motco *	La Marque		R		
05	OH	Arcanum Iron & Metal	Darke County *			E	
06	TX	Sikes Disposal Pits	Crosby		R		
04	AL	Triana Tennessee River	Limestone/Morgan			E	
09	CA	Stringfellow *	Glen Avon Heights		R	E	
01	ME	McKin Co.	Gray		R		
06	TX	Crystal Chemical Co.	Houston		R	E	
02	NJ	Bridgeport Rental & Oil	Bridgeport	V	R	E	
08	CO	Sand Creek	Commerce City				D
01	MA	W. R. Grace Co. (Anton Plant)	Acton	V		E	
05	MN	Reilly Tar *	St. Louis Park		R	E	
02	NJ	Burnt Fly Bog	Marlboro Township		R	E	
04	FL	Schuylkill Metal Corp.	Plant City				D
05	MN	New Brighton/Arden Hills	New Brighton		R	E	
02	NY	Old Bethpage Landfill	Oyster Bay			E	
04	FL	Reeves Se Galvanizing Corp.	Tampa				D
08	MT	Anaconda Smelter - Anaconda	Anaconda	V			
10	WA	Western Processing Co., Inc.	Kent			E	
04	FL	American Creosote Works	Pensacola				D
02	NJ	Caldwell Trucking Co.	Fairfield			E	
02	NY	GE Moreau	South Glens Falls			E	
05	IN	Seymour Recycling Corp. *	Seymour	V	R	E	
06	OK	Tar Creek	Ottawa County		R		
07	KS	Cherokee County	Cherokee County				D
02	NJ	Brick Township Landfill	Brick			E	

#: V = Voluntary or Negotiated Response; R = Federal and State Response;
E = Federal and State Enforcement; D = Actions to be Determined.
* = States' Designated Top Priority Sites.

Group 2

EPA REG	ST	SITE NAME *	CITY/COUNTY	V	R	E	D
05	MI	Northernaire Plating	Cadillac		R		
10	WA	Frontier Hard Chrome	Vancouver		R	E	
04	FL	Davie Landfill	Davie				D
04	FL	Gold Coast Oil Corp.	Miami	V			
09	AZ	Tucson Int'l Airport	Tucson	V	R		
02	NY	Wide Beach Development	Brant				D
09	CA	Iron Mountain Mine	Redding		R		
02	NJ	Scientific Chemical Processing	Carlstadt			E	
08	CO	California Gulch	Leadville		R		
02	NJ	D'Imperio Property	Hamilton Township		R		
05	MN	Oakdale Dump	Oakdale	V		E	
05	IL	A & F Materials	Greenup		R	E	
03	PA	Douglassville Disposal	Douglassville				D
02	NJ	Krysowaty Farm	Hillsborough		R		
05	MN	Koppers Coke	St. Paul				D
01	MA	Plymouth Harbor/Cannon Eng	Plymouth		R	E	
10	ID	Bunker Hill Mining	Smelterville			E	
02	NJ	Universal Oil Products (Chem Div.)	East Rutherford			E	
09	CA	Aerojet General Corp.	Rancho Cordova			E	
10	WA	Com. Bay, S. Tacoma Channel	Tacoma		R	E	
03	PA	Osborne Landfill	Grove City				D
02	NY	Syosset Landfill	Osyter Bay				D
09	AZ	Nineteenth Avenue Landfill	Phoenix			E	
10	OR	Teledyne Wah Chang	Albany				D
05	MI	Gratiot County Landfill *	St. Louis	V		E	
01	RI	Picillo Farm*	Coventry		R	E	
01	MA	New Bedford*	New Bedford	V	R	E	
06	LA	Old Inger Oil Refinery*	Darrow		R		
05	OH	Chem-Dyne *	Hamilton	V	R	E	
04	SC	Scrdi Bluff Road*	Columbia	V	R	E	
01	CT	Laurel Park, Inc*	Naugatuck Borough			E	
08	CO	Marshall Landfill *	Boulder County				D
05	IL	Outboard Marine Corp. *	Waukegan		R	E	
06	NM	South Valley *	Albuquerque				D
01	VT	Pine Street Canal *	Burlington				D
03	WV	West Virginia Ordnance *	Point Pleasant		R		
07	MO	Ellisville Site *	Ellisville		R		
08	ND	Arsenic Trioxide Site *	Southeastern		R		
09	TT	PCB Wastes *	Pacific Trust Terr.		R		
03	VA	Matthews Electroplating *	Roanoke County		R		
07	IA	Aidex Corp. *	Council Bluffs		R	E	
09	AZ	Mountain View Mobile Homes *	Globe		R	E	
09	AS	Taputimu Farm *	American Samoa				D
04	TN	North Hollywood Dump *	Memphis		R		
04	KY	A. L. Taylor (Valley of the Drums)	Brooks		R	E	
04	NC	PCB Spills *	210 Miles of Roads		R	E	
09	GU	Ordot Landfill *	Guam		R		
08	UT	Rose Park Sludge Pit *	Salt Lake City	V			
07	KS	Arkansas City Dump *	Arkansas City		R		
09	CM	PCB Warehouse *	North Marianas		R		

#: V = Voluntary or Negotiated Response; R = Federal and State Response;
   E = Federal and State Enforcement; D = Actions to be Determined.
* = States' Designated Top Priority Sites.

Group 3

EPA REG ST		SITE NAME *	CITY/COUNTY	RESPONSE STATUS #	
02	NY	Sinclair Refinery	Wellsville	R	
04	AL	Mowbray Engineering Co.	Greenville		D
05	MI	Spiegelberg Landfill	Green Oak Township	R	
04	FL	Miami Drum Services	Miami	R E	
02	NJ	Reich Farms	Pleasant Plains	E	
02	NJ	South Brunswick Landfill	South Brunswick	V	
04	FL	Kassauf-Kimerling Battery Disp.	Tampa	E	
05	IL	Wauconda Sand & Gravel	Wauconda	R	
01	NH	Ottati & Gross /Kingston Steel Drum	Kingston	R E	
05	MI	Ott/Story/Cordova	Dalton Township	R	
02	NJ	NL Industries	Pedricktown	E	
02	NJ	Ringwood Mines/Landfill	Ringwood Borough		D
04	FL	Whitehouse Oil Pits	Whitehouse	R	
05	MI	Velsicol Michigan	St. Louis	V E	
05	OH	Summit National	Deerfield Township	V E	
02	NY	Love Canal	Niagara Falls	R E	
05	IN	Fisher Calo	La Porte	V E	
04	FL	Pioneer Sand Co.	Warrington	E	
05	MI	Springfield Township Dump	Davisburg	R	
03	PA	Hranica Landfill	Buffalo Township		D
04	NC	Martin Marietta, Sodyeco	Charlotte		D
04	FL	Zellwood Groundwater Contam	Zellwood		D
05	MI	Packaging Corp. of America	Filer City		D
02	NY	Hooker - S Area	Niagara Falls	E	
03	PA	Lindane Dump	Harrison Township	E	
08	CO	Central City, Clear Creek	Idaho Springs	R	
04	FL	Taylor Road Landfill	Sefner	E	
01	RI	Western Sand & Gravel	Burrillville	R E	
02	NJ	Maywood Chemical Co.	Maywood/Rochelle Pk.	E	
06	OK	Hardage/Criner	Criner	R E	
05	MI	Rose Township Dump	Rose Township	R	
05	MN	Waste Disposal Engineering	Andover		D
02	NJ	Kin-Buc Landfill	Edison Township	V R E	
05	OH	Bowers Landfill	Circleville		D
02	NJ	Toms River Chemical	Toms River		D
05	MI	Butterworth #2 Landfill	Grand Rapids	E	
02	NJ	American Cyanamid Co.	Bound Brook	E	
03	PA	Heleva Landfill	North Whitehall Twp.		D
02	NY	Batavia Landfill	Batavia		D
01	RI	L&RR, Inc.	North Smithfield	E	
04	FL	NW 58th Street Landfill	Hialeah	E	
04	FL	Sixty-Second Street Dump	Tampa		D
05	MI	G & H Landfill	Utica	R	
02	NJ	Metaltec/Aerosystems	Franklin Borough	E	
02	NJ	Lang Property	Pemberton Township		D
02	NJ	Sharkey Landfill	Parsippany, Troy Hls.		D
09	CA	Selma Treating Co.	Selma	E	
06	LA	Cleve Reber	Sorrento		D
05	IL	Velsicol Illinois	Marshall	V	
05	MI	Tar Lake	Mancelona Township	R	

#: V = Voluntary or Negotiated Response;   R = Federal and State Response;
   E = Federal and State Enforcement;   D = Actions to be Determined.
* = States' Designated Top Priority Sites.

Group 4

EPA REG	ST	SITE NAME *	CITY/COUNTY	RESPONSE STATUS #		
02	NJ	Combe Fill North Landfill	Mount Olive Twp.			D
01	MA	Re-Solve, Inc.	Dartmouth		R E	
02	NJ	Goose Farm	Plumstead Township		R	
04	TN	Velsicol (Hardeman County)	Toone	V		
02	NY	York Oil Co.	Moira		R	
04	FL	Sapp Battery Salvage	Cottondale		R	
07	KS	Doepke Disposal, Holliday	Johnson County			D
01	RI	Davis Liquid Waste	Smithfield		R E	
01	MA	Charles-George Reclamation	Tyngsborough		E	
02	NJ	King of Prussia	Winslow Township			D
03	VA	Chisman Creek	York County			D
05	OH	Nease Chemical	Salem			D
02	NJ	Chemical Control	Elizabeth		R E	
05	OH	Allied Chemical & Ironton Coke	Ironton			D
05	MI	Verona Well Field	Battle Creek		R	
01	CT	Beacon Heights Landfill	Beacon Falls			D
05	MN	Burlington Northern	Brained/Baxter			D
03	PA	Malvern TCE	Malvern		R	
02	NY	Facet Enterprises, Inc.	Elmira	V		
03	DE	Delaware Sand & Gravel Landfill	New Castle County			D
04	TN	Murray Ohio Dump	Lawrenceburg			D
05	IN	Envirochem	Zionsville		R	
05	IN	Midco I	Gary		R E	
04	FL	Coleman Evans Wood Preserving Co.	Whitehouse		E	
04	FL	Florida Steel Corp.	Indiantown			D
09	AZ	Litchfield Airport Area	Goodyear/Avondale			D
02	NJ	Spence Farm	Plumstead Township		R	
06	AR	Mid-South Wood Products	Mena		E	
04	FL	Brown Wood Preserving	Live Oak			D
02	NY	Port Washington Landfill	Port Washington		E	
02	NJ	Combe Fill South Landfill	Chester Township			D
02	NJ	Jis Landfill	Jamesburg/S. Brunswick		E	
03	PA	Centre County Kepone	State College Boroug		E	
05	OH	Fields Brook	Ashtabula		R	
01	CT	Solvents Recovery Service	Southington	V	E	
08	CO	Woodbury Chemical Co.	Commerce City		R	
01	MA	Hocomonco Pond	Westborough		R	
04	KY	Distler Brickyard	West Point		R	
02	NY	Ramapo Landfill	Ramapo		E	
09	CA	Coast Wood Preserving	Ukiah		E	
02	NY	Mercury Refining, Inc.	Colonie		E	
04	FL	Hollingsworth Solderless Terminal	Fort Lauderdale			D
02	NY	Olean Well Field	Olean		R	
04	FL	Varsol Spill	Miami	V	R	
08	CO	Denver Radium Site	Denver		R	
04	FL	Tower Chemical Co.	Clermont		R E	
07	MO	Syntex Facility	Verona	V	E	
08	MT	Milltown Reservoir Sediments	Milltown		R	
02	NJ	Pijak Farm	Plumstead Township		R	
02	NJ	Syncon Resins	South Kearny	V	E	

#: V = Voluntary or Negotiated Response; R = Federal and State Response;
   E = Federal and State Enforcement; D = Actions to be Determined.
* = States' Designated Top Priority Sites.

Group 5

EPA REG	ST	SITE NAME *	CITY/COUNTY	RESPONSE STATUS #	
09	CA	Liquid Gold Oil Corp.	Richmond	E	
09	CA	Purity Oil Sales, Inc.	Malaga		D
01	NH	Tinkham Garage	Londonderry	R	
04	FL	Alpha Chemical Corp.	Galloway		D
02	NJ	Bog Creek Farm	Howell Township	R	
01	ME	Saco Tannery Waste Pits	Saco	R	
04	FL	Pickettville Road Landfill	Jacksonville		D
03	PA	Palmerton Zinc Pile	Palmerton		D
05	IN	Neal's Landfill	Bloomington	E	
01	MA	Silresim Chemical Corp.	Lowell	R E	
01	MA	Wells G&H	Woburn	E	
02	NJ	Chemsol, Inc.	Piscataway		D
05	MI	Petoskey Municipal Well Field	Petoskey	R	
02	NJ	Fair Lawn Well Field	Fair Lawn		D
05	IN	Main Street Well Field	Elkhart		D
05	MN	Lehillier/Mankato	Lehillier	R	
10	WA	Lakewood	Lakewood	E	
02	NJ	Monroe Township Landfill	Monroe Township	E	
02	NJ	Rockawayt Borough Well Field	Rockaway Township		D
05	IN	Wayne Waste Oil	Columbia City	R E	
07	IA	Des Moines TCE	Des Moines		D
02	NJ	Beachwood/Berkley Wells	Berkley Township		D
02	NY	Vestal Water Supply Well 4-2	Vestel	E	
09	AZ	Indian Bend Wash Area	Scottsdale		D
10	WA	Com. Bay, Near Shore/Tide Flat	Pierce County	R E	
05	IL	Lasalle Electric Utilities	La Salle	R	
05	IL	Cross Bros/Pembroke	Pembroke Township	R	
09	CA	McColl	Fullerton		D
10	WA	Colbert Landfill	Spokane	R	
02	PR	Frontera Creek	Rio Abajo		D
02	PR	Barceloneta Landfill	Florida Afuera		D
03	MD	Sand, Gravel and Stone	Elkton	E	
05	MI	Spartan Chemical Co.	Wyoming	E	
02	NJ	Roebling Steel Co.	Florence		D
04	TN	Amnicola Dump	Chattanooga		D
02	NJ	Vineland State School	Vineland		D
03	PA	Enterprise Avenue	Philadelphia		D
01	MA	Groveland Wells	Groveland	R	
04	SC	Scrdi Dixiana	Cayce	E	
07	MO	Fulbright Landfill	Springfield		D
03	PA	Presque Isle	Erie		D
02	NJ	Williams Property	Swainton	R	
02	NJ	Renora, Inc.	Edison Township		D
02	NJ	Denzer & Schafer X-Ray Co.	Bayville	E	
02	NJ	Hercules, Inc. (Gibbstown)	Gibbstown		D
05	IN	Ninth Ave. Dump	Gary	V E	
06	AR	Gurley Pit	Edmondsen		D
01	RI	Peterson/Puritan, Inc.	Lincoln/Cumberland	V	
07	MO	Times Beach	Times Beach	R	
05	MI	Wash King Laundry	Pleasant Plains Twp.		D

#: V = Voluntary or Negotiated Response;   R = Federal and State Response;
   E = Federal and State Enforcement;   D = Actions to be Determined.
* = States' Designated Top Priority Sites.

Group 6

EPA REG	ST	SITE NAME *	CITY/COUNTY	V	R	E	D
05	MN	NL Industries/Taracorp/Golden	St. Louis Park	V			
01	MA	Cannon Engineering Corp. (CEC)	Bridgewater		R	E	
02	NY	Niagara County Refuse	Wheatfield				D
04	FL	Sherwood Medical Industries	Deland				D
05	MI	Southwest Ottawa Landfill	Park Township			E	
02	NY	Kentucky Ave. Well Field	Horseheads				D
02	NJ	Asbestos Dump	Millington				D
04	KY	Lee's Lane Landfill	Louisville				D
06	AR	Frit Industries	Walnut Ridge	V		E	
05	OH	Fultz Landfill	Jackson Township				D
05	OH	Coshocton Landfill	Franklin Township				D
03	PA	Lord-Shope Landfill	Girard Township			E	
10	WA	FMC Corp. (Yakima)	Yakima	V			
01	MA	PSC Resources	Palmer	V			
05	MI	Forest Waste Products	Otisville		R		
03	PA	Drake Chemical	Lock Haven		R		
03	PA	Havertown PCP	Haverford			E	
03	DE	New Castle Spill	New Castle County				D
05	IN	Lake Sandy Jo (M&M Landfill)	Gary				D
05	IL	Johns-Manville Corp.	Waukegan				D
05	MI	Chem Central	Wyoming Township				D
05	MI	Novaco Industries	Temperance				D
02	NJ	Jackson Township Landfill	Jackson Township			E	
05	MI	K & L Avenue Landfill	Oshtemo Township		R		
10	WA	Kaiser Mead	Mead				D
05	MI	Charlevoix Municipal Well	Charlevoix				D
02	NJ	Montgomery Township Housing Dev.	Montgomery Township				D
02	NJ	Rocky Hill Municipal Well	Rocky Hill Borough				D
02	NY	Brewster Well Field	Putnam County				D
02	NY	Vestal Water Supply Well 1-1	Vestal			E	
02	NJ	U.S. Radium Corp.	Orange				D
06	TX	Highlands Acid Pit	Highlands		R		
03	PA	Resin Disposal	Jefferson Borough			E	
08	MT	Libby Ground Water Contamination	Libby				D
04	KY	Newport Dump	Newport			E	
03	PA	Moyers Landfill	Eagleville			E	
04	FL	Parramore Surplus	Mount Pleasant	V			
05	MI	Hedblum Industries	Oscoda				D
08	WY	Baxter/Union Pacific Tie Treating	Laramie				D
02	NJ	Sayreville Landfill	Sayreville				D
01	NH	Dover Municipal Landfill	Dover				D
02	NY	Ludlow Sand & Gravel	Clayville				D
07	MO	Minker/Stout/Romaine Creek	Imperial		R		
01	CT	Yaworski Waste Lagoon	Canterbury			E	
03	WV	Leetown Pesticide	Leetown				D
02	NJ	Evor Phillips Leasing	Old Bridge Township				D
03	PA	Wade (ABM)	Chester		R	E	
03	PA	Lackawanna Refuse	Old Forge Borough			E	
02	NJ	Manheim Avenue Dump	Galloway Township				D
02	NY	Fulton Terminals	Fulton	V			

#: V = Voluntary or Negotiated Response;   R = Federal and State Response;
E = Federal and State Enforcement;   D = Actions to be Determined.
* = States' Designated Top Priority Sites.

Group 7

EPA REG	ST	SITE NAME *	CITY/COUNTY	V	R	E	D
01	NH	Auburn Road Landfill	Londonderry			E	
03	WV	Fike Chemical, Inc.	Nitro	V			
05	OH	Laskin/Poplar Oil Co.	Jefferson Township		R	E	
05	OH	Old Mill	Rock Creek		R		
07	KS	Johns' Sludge Pond	Wichita	V	R	E	
02	NJ	Swope Oil & Chemical Co.	Pennsauken				D
01	ME	Winthrop Landfill	Winthrop		R		
06	AR	Cecil Lindsey	Newport				D
05	OH	Zanesville Well Field	Zanesville				D
05	MI	Grand Traverse Overall Supply Co.	Greilickville				D
05	MN	South Andover Site	Andover				D
05	MI	Kentwood Landfill	Kentwood		R		
05	IN	Marion (Bragg) Dump	Marion				D
05	OH	Pristine, Inc.	Reading			E	
05	OH	Buckeye Reclamation	St. Clairsville				D
06	TX	Bio-Ecology Systems, Inc.	Grand Prairie		R		
01	VT	Old Springfield Landfill	Springfield				D
02	NY	Solvent Savers	Lincklaen			E	
03	VA	U.S. Titanium	Piney River			E	
05	IL	Galesburg/Koppers	Galesburg				D
02	NY	Hooker - Hyde Park	Niagara Falls	V		E	
05	MI	SCA Independent Landfill	Muskegon Heights			E	
09	CA	MGM Brakes	Cloverdale			E	
05	MI	Duell & Gardner Landfill	Dalton Township		R		
02	NJ	Ellis Property	Evesham Township		R		
04	KY	Distler Farm	Jefferson County		R		
10	WA	Harbor Island Lead	Seattle				D
05	OH	E.H. Schilling Landfill	Hamilton Township				D
05	MI	Cliff/Dow Dump	Marquette			E	D
06	MN	Homestake Mining Co.	Milan	V		E	
05	MI	Mason County Landfill	Pere Marquette Twp.			E	
05	MI	Cemetery Dump	Rose Center		R		
01	RI	Stamina Mills, Inc.	North Smithfield		R	E	
01	ME	Pinette's Salvage Yard	Washburn				D
06	TX	Harris (Farley St)	Houston	V		E	
03	PA	Old City of York Landfill	Seven Valleys			E	
05	IL	Byron Salvage Yard	Byron		R		
03	PA	Stanley Kessler	King of Prussia			E	
02	NJ	Friedman Property	Upper Freehold Twp		R		
02	NJ	Imperial Oil/Champion Chemicals	Morganville			E	
02	NJ	Myers Property	Franklin Township				D
02	NJ	Pepe Field	Boonton				D
05	MI	Ossineke Ground Water Contam	Ossineke		R		
03	WV	Follansbee	Follansbee				D
05	MI	U.S. Aviex	Howard Township			E	
06	NM	AT & SF / Clovis	Clovis				D
02	NY	American Thermostat Co.	South Cairo			E	
04	TN	Lewisburg Dump	Lewisburg				D
05	MI	McGraw Edison Corp.	Albion			E	
03	PA	Metal Banks	Philadelphia			E	

#: V = Voluntary or Negotiated Response; R = Federal and State Response;
E = Federal and State Enforcement; D = Actions to be Determined.
* = States' Designated Top Priority Sites.

Group 8

EPA REG ST	SITE NAME *	CITY/COUNTY	RESPONSE STATUS #
04 KY	B.F. Goodrich	Calvert City	D
05 MI	Organic Chemcials, Inc.	Grandville	E
02 PR	Juncos Landfill	Juncos	D
04 FL	Munisport Landfill	North Miami	D
02 NJ	M&T Delisa Landfill	Asbury Park	D
10 OR	Gould, Inc.	Portland	E
05 MI	Auto Ion Chemicals, Inc.	Kalamazoo	R
04 SC	Carolawn, Inc.	Fort Lawn	R E
05 MI	Sparta Landfill	Sparta Township	E
05 IL	Acme Solvent/Morristown	Morristown	R
01 ME	O'Connor	Augusta	D
05 MI	Rasmussen's Dump	Brighton	R
03 PA	Westline	Westline	D
05 MI	Ionia City Landfill	Ionia	R
05 IN	Wedzeb Inc.	Lebanon	E
02 PR	GE Wiring Devices	Juana Diaz	D
05 OH	New Lyme Landfill	New Lyme	D
02 PR	RCA Del Caribe	Barceloneta	D
03 PA	Brodhead Creek	Stroudsburg	R
05 MI	Anderson Development Co.	Adrian	E
05 MI	Shiawassee River	Howell	R
03 DE	Harvey & Knott Drum, Inc.	Kirkwood	R
04 TN	Gallaway Pits	Gallaway	E
05 OH	Big D Campground	Kingsville	D
03 DE	Wildcat Landfill	Dover	D
03 PA	Blosenski Landfill	West Caln Township	E
03 DE	Delaware City PVC Plant	Delaware City	D
03 MD	Limestone Road	Cumberland	E
02 NY	Hooker - 102nd Street	Niagara Falls	E
03 DE	New Castle Steel	New Castle County	D
06 NM	United Nuclear Corp.	Church Rock	D
06 AR	Industrial Waste Control	Ft. Smith	D
09 CA	Celtor Chemical Works	Hoopa	R
04 AL	Perdido Ground Water Contam	Perdido	D
02 NY	Marathon Battery Corp.	Cold Springs	D
03 PA	Lehigh Electric & Eng. Co.	Old Forge Borough	R E
05 OH	Skinner Landfill	West Chester	D
04 NC	Chemtronics, Inc.	Swannanoa	D
07 MO	Shenandoah Stables	Moscow Mills	E
06 LA	Bayou Bonfouca	Slidell	D
03 VA	Saltville Waste Disposal Ponds	Saltville	D
03 PA	Kimberton	Kimberton Borough	D
03 MD	Middletown Road Dump	Annapolis	E
10 WA	Pesticide Lab	Yakima	D
05 IN	Lemon Lane Landfill	Bloomington	D
10 ID	Arrcom (Drexler Enterprises)	Rathdrum	D
03 PA	Fischer & Porter Co.	Warminster	E
09 CA	Jibboom Junkyard	Sacramento	D
02 NJ	A. O. Polymer	Sparta Township	R
02 NJ	Dover Municipal Well 4	Dover	D

#: V = Voluntary or Negotiated Response;   R = Federal and State Response;
    E = Federal and State Enforcement;   D = Actions to be Determined.
* = States' Designated Top Priority Sites.

Group 9

EPA REG ST		SITE NAME *	CITY/COUNTY	RESPONSE STATUS #
02	NJ	Rockaway Township Wells	Rockaway	D
06	TX	Triangle Chemical Co.	Bridge City	R E
02	NJ	PJP Landfill	Jersey City	D
03	PA	Craig Farm Drum	Parker	D
03	PA	Voortman Farm	Upper Saucon Twp.	D
05	IL	Belvidere Municipal Landfill	Belvidere	D

#: V = Voluntary or Negotiated Response; R = Federal and State Response;
 E = Federal and State Enforcement; D = Actions to be Determined.
* = States' Designated Top Priority Sites.

PROPOSED UPDATE TO NPL

The 133 sites on the first proposed NPL update are also listed in order of their HRS scores, which range from 63.1 to 28.5, with two exceptions: Flowood, in Flowood, Mississippi, the State's top priority site, and Quail Run Mobile Manor in Gray Summit, Missouri, which EPA believes involves special circumstances. In the winter of 1982-1983, EPA detected widespread *dioxin contamination* at Quail Run. The Centers for Disease Control subsequently issued a public health advisory warning that residents were exposed to a significant risk from direct contact with dioxin. Because the risk of direct contact is not presently used in calculating total HRS scores, Quail Run's score is below the cut-off point. By being on the proposed update, Quail Run is eligible for remedial action funded by CERCLA. Later, EPA intends to propose a limited numbers of alternative criteria for listing.

Each proposed update site is placed on the basis of its score in one of nine groups corresponding to the nine groups in the current NPL—for example, the eight sites in Group I of the update have scores comparable to the scores of the first 50 sites in the current NPL.

Status: Each entry is accompanied by one or more of the notations describing the current status.

Site Distribution: Of the 50 States, the District of Columbia, and six territories, 28 proposed sites. Wisconsin is included for the first time. Alaska, the District of Columbia, Hawaii, Nebraska, Nevada, and the Virgin Islands continue to have no sites on either the current NPL or proposed update. The sites are distributed as follows:

- New Jersey and Wisconsin, 20 each
- Minnesota, 13
- Pennsylvania, 9
- California, 8
- South Carolina, 7
- Indiana, 5
- Alabama, Florida, Georgia, Ohio, and Washington, 4 each
- Colorado, Michigan, New Hampshire, Puerto Rico, and Texas, 3 each
- Connecticut, Idaho, Massachusetts, New York, and Oklahoma, 2 each
- Delaware, Louisiana, Missouri, Mississippi, Montana, and Oregon, 1 each

Group 1

EPA REG	ST	Site Name *	City/County	Response Status #		
03	PA	Tysons Dump	Upper Merion Twp.	R		
08	MT	East Helena Smelter	East Helena			D
06	TX	Geneva Industries (Fuhrmann)	Houston	R	E	
02	NJ	Vineland Chemical Co.	Vineland	V	E	
02	NJ	Florence Land Recontouring Lf.	Florence Township	V	E	
02	NJ	Shieldalloy Corp.	Newfield Borough		E	
05	WI	Omega Hills North Landfill	Germantown	V	E	
05	OH	United Scrap Lead Co., Inc.	Troy			D

Group 2

EPA REG	ST	Site Name *	City/County	Response Status #		
05	WI	Janesville Old Landfill	Janesville			D
04	SC	Independent Nail Co.	Beaufort			D
04	SC	Kalama Specialty Chemicals	Beaufort		E	
05	WI	Janesville Ash Beds	Janesville			D
05	OH	Miami County Incinerator	Troy			D
05	WI	Wheeler Pit	La Prairie Township			D
02	NY	Hudson River PCBS	Hudson River			D
01	CT	Old Southington Landfill	Southington	V	E	
04	MS	Flowood *	Flowood			D

Group 3

EPA REG	ST	Site Name *	City/County	Response Status #		
10	ID	Union Pacific Railroad Co.	Pocatello		E	
04	AL	Ciba-Geigy Corp. (McIntosh Plant)	McIntosh			D
05	MN	St. Regis Paper Co.	Cass Lake	V		
04	GA	Hercules 009 Landfill	Brunswick			D
05	MN	Macgillis & Gibbs/Bell & Pole	New Brighton			D
05	WI	Muskego Sanitary Landfill	Muskego			D
02	NJ	Ventron/Velsicol	Woodridge Borough		E	
04	SC	Koppers Co., Inc. (Florence Plant)	Florence		E	
02	NJ	Nascolite Corp.	Millville		E	
05	MN	Boise Cascade/Onan/Medtronics	Fridley			D
02	NJ	Delilah Road	Egg Harbor Township		E	
03	PA	Mill Creek Dump	Erie	R		
05	WI	Schmalz Dump	Harrison			D
08	CO	Lowry Landfill	Arapahoe County		E	

#: V = Voluntary or Negotiated Response: R = Federal and State Response;
   E = Federal and State Enforcement;   D = Actions to be Determined.
* = States Designated Top Priority Sites;
Note: Group Refers to the NPL Group with Similar HRS Scores;

Group 4

EPA REG	ST	Site Name *	City/County	Response Status #
04	SC	Wamchem, Inc.	Burton	D
02	NJ	Chemical Leaman Tank Liners, Inc.	Bridgeport	E
05	WI	Master Disposal Service Landfill	Brookfield	E
02	NJ	W. R. Grace Co. (Wayne Plant)	Wayne Township	D
04	SC	Leonard Chemical Co., Inc.	Rock Hill	V
04	AL	Stauffer Chem. (Cold Creek Plant)	Bucks	D
04	GA	Olin Corp. (Areas 1,2 & 4)	Augusta	V
05	OH	South Point Plant	South Plant	D
03	PA	Dorney Road Landfill	Upper Macungie Twp.	D
05	IN	Northside Sanitary Landfill	Zionsville	E
09	CA	Atlas Asbestos Mine	Fresno County	E
09	CA	Coalinga Asbestos Mine	Coalinga	D
02	NJ	Ewan Property	Shamong Township	D
10	ID	Pacific Hide & Fur Recycling Co.	Pocatello	R E
05	MN	Joslyn Mfg. & Supply Co.	Brooklyn Center	D
05	MN	Arrowhead Refinery Co.	Hermantown	D
05	WI	Moss-American (Kerr-McGee Oil Co.)	Milwaukee	D

Group 5

EPA REG	ST	Site Name *	City/County	Response Status #
01	MA	Iron Horse Park	Billerica	D
05	WI	Kohler Co. Landfill	Sheboygan	D
05	IN	Reilly Tar & Chemical Corp.	Indianapolis	D
05	WI	Lauer I Sanitary Landfill	Menomonee Falls	E
05	MN	Union Scrap	Minneapolis	D
02	NJ	Radiation Technology, Inc.	Rockaway Township	E
05	WI	Onalaska Muncipal Landfill	Onalaska	D
05	MN	Nutting Truck & Caster Co.	Faribault	D
02	PR	Vega Alta Public Supply Wells	Vega Alta	V R E
05	MI	Sturgis Municipal Wells	Sturgis	D
05	MN	Washington County Landfill	Lake Elmo	R
09	CA	San Gabriel Area 1	El Monte	D
09	CA	San Gabriel Area 2	Baldwin Park Area	D
06	TX	Pig Road	New Waverly	D
02	PR	Upjohn Facility	Barceloneta	V
03	PA	Henderson Road	Upper Merion Twp.	D
06	LA	Petro-Processors	Scotlandville	E
03	PA	Industrial Lane Landfill	Williams Township	D
03	PA	East Mount Zion	Springettsbury Twp.	D
02	NY	General Motors-Cent. Foundry Div.	Massena	D
03	DE	Old Brine Sludge Landfill	Delaware City	D
05	MN	Whittaker Corp.	Minneapolis	D

#: V = Voluntary or Negotiated Response: R = Federal and State Response;
   E = Federal and State Enforcement;   D = Actions to be Determined.
* = States Designated Top Priority Sites;
Note: Group Refers to the NPL Group with Similar HRS Scores;

Group 6

EPA REG	ST	Site Name *	City/County	Response Status #		
01	CT	Kellogg-Deering Well Field	Norwalk	V	E	
04	AL	Olin Corp. (McIntosh Plant)	McIntosh	V		
04	FL	Tri-City Oil Conservationist, Inc.	Temple Terrace			D
05	WI	Northern Engraving Co.	Sparta			D
01	NH	Kearsage Metallurgical Corp.	Conway	V	E	
04	SC	Palmetto Wood Preserving	Dixianna		E	
05	MN	Morris Arsenic Dump	Morris			D
05	MN	Perham Arsenic	Perham			D
01	NH	Savage Municipal Water Supply	Milford			D
05	IN	Poer Farm	Hancock County	R		
06	TX	United Creosoting Co.	Conroe			D
05	WI	City Disposal Corp. Landfill	Dunn			D
02	NJ	Tabernacle Drum Dump	Tabernacle Twp.			D
02	NJ	Cooper Road	Voorhees Township			D
04	FL	Cabot-Koppers	Gainesville			D

Group 7

EPA REG	ST	Site Name *	City/County	Response Status #		
05	MN	General Mills/Henkel Corp.	Minneapolis	R		
09	CA	Del'Norte Pesticide Storage	Crescent City			D
02	NJ	De Rewal Chemical Co.	Kingwood Township			D
04	GA	Monsanto Corp. (Augusta Plant)	Augusta			D
01	NH	South Municipal Water Supply Well	Petersborough			D
05	WI	Eau Claire Municipal Well Field	Eau Claire City			D
04	GA	Powersville	Peach County			D
05	MI	Metamora Landfill	Metamora			D
02	NJ	Diamond Alkali Co.	Newark		E	
02	PR	Fibers Public Supply Wells	Jobos			D
05	WI	Mid-State Disposal, Inc., Landfill	Cleveland Township		E	
08	CO	Broderick Wood Products	Denver			D
02	NJ	Woodland Route 532 Dump	Woodland Township			D
05	IN	American Chemical Service	Griffith			D
05	WI	Lemberger Transport & Recycling	Franklin Township		E	
10	WA	Queen City Farms	Maple Valley			D
05	WI	Scrap Processing Co., Inc.	Medford			D
02	NJ	Hopkins Farm	Plumstead Township			D
02	NJ	Wilson Farm	Plumstead Township	R		
06	OK	Compass Industries	Tulsa	R		
09	CA	Koppers Co., Inc. (Oroville Plant)	Oroville		E	
03	PA	Walsh Landfill	Honeybrook Twp.			D
02	NJ	Upper Deerfield Township Slf.	Upper Deerfield Twp.		E	

#: V = Voluntary or Negotiated Response:　R = Federal and State Response;
　E = Federal and State Enforcement;　D = Actions to be Determined.
* = States Designated Top Priority Sites;
Note: Group Refers to the NPL Group with Similar HRS Scores;

Group 8

EPA REG	ST	Site name*City/County		Status	Response #		
01	MA	Sullivan's Ledge	New Bedford				D
05	IN	Bennett Stone Quarry	Bloomington		R		
04	SC	Geiger (C&M Oil)	Rantoules				D
05	WI	Waste Research & Reclamation Co.	Eau Claire		V	E	
04	FL	Pepper Steel & Alloys, Inc.	Medley		V	R	E
05	MN	St. Louis River	St. Louis County				D
03	PA	Berks Sand Pit	Longswamp Township				D
04	FL	Hipps Road Landfill	Duval County		R		
05	WI	Oconomowoc Electroplating Co.	Ashippin			E	
08	CO	Lincoln Park	Canon City				D
02	NJ	Woodland Route 72 Dump	Woodland Township				D
10	OR	United Chrome Products, Inc.	Corvallis				D
02	NJ	Landfill & Development Co.	Mount Holly		V	E	
03	PA	Taylor Borough Dump	Taylor Borough				D
05	OH	Powell Road Landfill	Dayton				D
05	MI	Burrows Sanitation	Hartford		R		
10	WA	Rosch Property	ROY				D
04	AL	Stauffer Chem. (Le Moyne Plant)	Axis				D

Group 9

EPA REG	ST	Site name*City/County	Status	Response #
05	WI	Delavan Municipal Well #4	Delavan	D
09	CA	San Gabriel Area 3	Alhambra	D
09	CA	San Gabriel Area 4	La Puente	D
10	WA	American Lake Gardens	Tacoma	R
10	WA	Greenacres Landfill	Spokane County	D
06	OK	Sand Springs Petrochemical	Sand Springs	R
07	MO	Quail Run Mobile Manor	Gray Summit	R

#: V = Voluntary or Negotiated Response: R = Federal and State Response;
   E = Federal and State Enforcement;   D = Actions to be Determined.
* = States Designated Top Priority Sites;
Note: Group Refers to the NPL Group with Similar HRS Scores;

NPL AND PROPOSED UPDATE SITES, BY STATE

A total of 546 sites appear on the list arranged by State: 406 on the NPL, 7 on which rulemaking is pending, and 133 on the proposed update.

For each State, the NPL sites (including the seven pending) are listed first alphabetically, followed by the proposed update sites, also alphabetically. After the period of public comment and revisions, the pending and proposed sites that continue to score above the cut-off will be merged into the current NPL.

When all states are combined, New Jersey has the largest number with 85, followed by Michigan (48), Pennsylvania (39), Florida (29), New York, (29), Minnesota (23), Ohio (22), and Wisconsin (20).

Landfilling of hazardous substances was the major type of operation, taking place at 216 sites, or about 40 percent of the total. Other major activities were surface impoundments (30 percent), drums (22 percent), open dump (17 percent), above-ground tanks (11 percent), piles (9 percent), and transportation (9 percent). Also represented were: below-ground tanks, recycling/recovery, incineration, midnight dumping, chemical/physical treatment, waste oil processing, solvent recovery, landfarming, underground injection, and biological treatment.

Most of the site had problems involving ground water. AT 410 sites (75 percent), hazardous substances were detected in ground water. About 56 percent of the sites had problems with surface water, and 20 percent with air.

NATIONAL PRIORITIES LIST

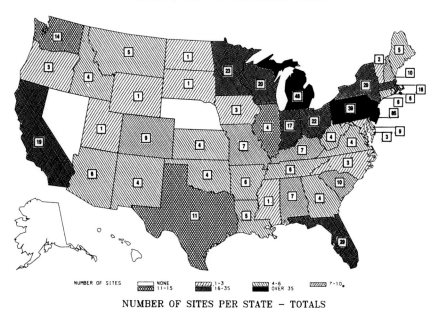

NUMBER OF SITES PER STATE – TOTALS

TYPES OF SITES

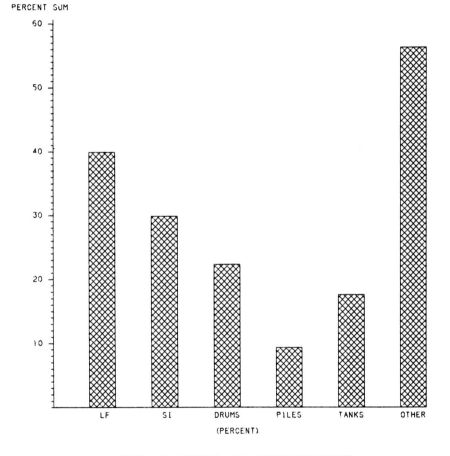

LEGEND: LF — LANDFILLS   SI — SURFACE IMPOUNDMENTS

NATIONAL PRIORITIES LIST

KINDS OF PROBLEMS

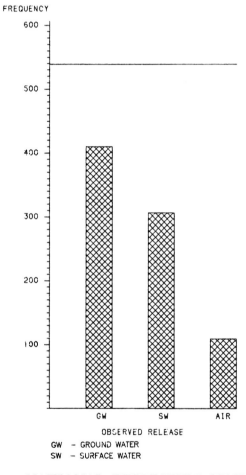

FREQUENCY

NATIONAL PRIORITIES LIST

National Priorities List (NPL) and Proposed Update List (PUL) of Hazardous Sites, by State (August, 1983).

ALASKA: None

ALABAMA: Mowbray Engineering Co., *Greenville;* Perdido Ground Water Contamination, *Peridido;* Triana-Tennessee River (once listed as Triana (Redstone) Arsenal), *Limestone/Morgan Counties;* Ciba-Geigy Corp. (McIntosh Plant), *McIntosh;* Olin Corp. (McIntosh Plant), *McIntosh;* Stauffer Chemical Co. (Cold Creek Plant), *Bucks;* and Stauffer Chemical Co. (Lemoyne Plant), *Axis.* (NPL 3, PUL 4, Total 7).

ARKANSAS: Cecil Lindsey, *Newport;* Crittenden County Landfills, *Marion;* Frit Industries, *Walnut Ridge;* Gurley Pit, *Edmondson;* Industrial Waste Control, *Fort Smith;* Mid-South Wood Products, *Mena;* and Vertac, Inc., *Jacksonville. (NPL 6, Total 6).*

AMERICAN SAMOA: Taputimu Farm*, *Island of Tutuila.* (NPL 1, Total 1).

ARIZONA: Indian Bend Wash Area, *Scottsdale-Tempe-Phoenix;* Kingman Airport Industrial Area[4], *Kingman;* Litchfield Airport Area, *Goodyear/Avondale;* Mountain View Mobile Home Estates* (once listed as Globe), *Globe;* 19th Avenue Landfill, *Phoenix; and* Tucson Airport Area, *Tucson.* (NPL 5, Total 5).

CALIFORNIA: Aerojet General Corp., *Rancho Cordova;* Celtor Chemical Works, *Hoopa;* Coast Wood Preserving, *Ukiah;* Iron Mountain Mine, *Redding;* Jibboom Junkyard, *Sacramento;* Liquid Gold Oil Corp., *Richmond;* MGM Brakes, *Cloverdale;* McColl, *Fullerton;* Purity Oil Sales, Inc., *Malaga;* Selma Treating Co., *Selma;* Stringfellow*, *Glen Avon Heights,* Atlas Asbestos Mine, *Fresno County;* Coalinga Asbestos Mine, *Coalinga;* Del Norte County Pesticide Storage Area, *Crescent City;* Koppers Co., Inc. (Oroville Plant), *Oroville;* San Gabriel Valley (Area 1), *El Monte;* San Gabriel Valley (Area 2), *Baldwin Park Area;* San Gabriel Valley (Area 3), *Alhambra;* and San Gabriel Valley (Area 4), *La Puente.* (NPL 11, PUL 8, Total 19).

NORTHERN MARIANA ISLANDS*: PCB Warehouse, *Saipan. (NPL 1, Total 1).*

COLORADO: California Gulch, *Leadville;* Central City-Clear Creek, *Idaho Springs;* Denver Radium, *Denver;* Marshall Landfill*, *Boulder County;* Sand Creek Industrial, *Commerce City;* Woodbury Chemical Co., *Commerce City;*

---

[1] Includes seven sites proposed in December 1982 on which rulemaking is pending.
[2] Designated as State's top priority.
[3] Not included because site was cleaned up.
[4] Rulemaking pending.
[5] Not included because proposed score was revised.
[6] Not included because no longer designated as State's top priority.

Broderick Wood Products, *Denver;* Lincoln Park, *Canon City; and* Lowry Landfill, *Arapahoe County.* (NPL 6, PUL 3, Total 9).

CONNECTICUT: Beacon Heights Landfill, *Beacon Falls;* Laurel Park, Inc. (once listed as Laurel Park Landfill), *Naugatuck Borough);* Solvents Recovery Service of New England, *Southington;* Yaworski Waste Lagoon, *Canterbury;* Kellogg-Deering Well Field, *Norwalk;* and Old Southington Landfill, *Southington. (NPL 4, PUL 2, Total 6).*

DISTRICT OF COLUMBIA: None

DELAWARE: Army Creek Landfill (once listed as Delaware Sand & Gravel - Llangollen Army Creek Landfills), *New Castle County;* Delaware City PVC Plant (once listed as Stauffer Chemical Co.), *Delaware City;* Delaware Sand & Gravel Landfill (once listed as Delaware Sand & Gravel - Llangollen Army Creek Landfills), *New Castle County;* Harvey & Knott Drum, Inc., *Kirkwood;* New Castle Spill (once listed as Tris Spill), *New Castle County;* New Castle Steel, *New Castle County;* Tybouts Corner Landfill[2], *New Castle County;* Wildcat Landfill, *Dover;* and Old Brine Sludge Landfill, *Delaware City.* (NPL 8, PUL 1, Total 9).

FLORIDA: Alpha Chemical Corp., *Galloway;* American Creosote Works, *Pensacola;* Brown Wood Preserving, *Live Oak;* Coleman-Evans Wood Preserving Co., *Whitehouse;* Davie Landfill (once listed as Boward County Solid Waste Disposal Facility), *Davie;* Florida Steel Corp., *Indiantown;* Gold Coast Oil Corp., *Miami;* Hollingsworth Solderless Terminal Co., *Fort Lauderdale;* Kassauf Kimerling Battery (once listed as Timber Lake Battery Disposal), *Tampa;* Miami Drum Services (once listed as part of Biscayne Aquifer), *Miami;* Munisport Landfill, *North Miami;* Northwest 58th Street Landfill (once listed as part of Biscayne Aquifer), *Hialeah;* Parramore Surplus, *Mount Pleasant;* Pickettville Road Landfill, *Jacksonville;* Pioneer Sand Co., *Warrington;* Reeves Southeastern Galvanizing Corp., *Tampa;* Sapp Battery Salvage, *Cottondale;* Schuylkill Metals Corp., *Plant City;* Sherwood Medical Industries, *Deland;* 62nd Street Dump, *Tampa;* Taylor Road Landfill, *Seffner;* Tower Chemical Co., *Clermont;* Varsol Spill (once listed as part of Biscayne Aquifer), *Miami;* Whitehouse Oil Pits, *Whitehouse;* Zellwood Ground Water Contamination, *Zellwood;* Cabot/Koppers, *Gainesville;* Hipps Road Landfill, *Duval County;* Pepper Steel & Alloys, Inc., *Medley;* and Tri-City Oil Conservationist, Inc., *Temple Terrace.* (NPL 25, PUL 4, Total 29).

GEORGIA: Hercules 009 Landfill, *Brunswick;* Monsanto Corp. (Augusta Plant), *Augusta;* Olin Corp. (Areas 1, 2 & 4), *Augusta;* and Powersville, *Peach County.* (PUL 4, Total 4).

GUAM: Ordot Landfill[2], *Ordot.* (NPL 1, Total 1).

HAWAII: None.

IOWA: Aidex Corp.[2], *Council Bluffs;* Des Moines TCE (once listed as DICO), *Des Moines;* and Labounty, *Charles City.* (NPL 3, Total 3).

IDAHO: Arrcom (Drexler Enterprises), *Rathdrum;* Bunker Hill Mining & Metallurgical, *Smelterville;* Flynn Lumber Co.[5], *Caldwell;* Pacific Hide & Fur Recycling Co., *Pocatello;* and Union Pacific Railroad Co., *Pocatello.* (NPL 2, PUL 2, Total 4).

ILLINOIS: A & F Materials Reclaiming, Inc., *Greenup;* ACME Solvent Reclaiming, Inc., *Morristown;* Belvidere Municipal Landfill, *Belvidere;* Byron Salvage Yard, *Byron;* Cross Brothers Pail Recycling, *Pembrook Township;* Galesburg/Koppers, *Galesburg;* Johns-Manville Corp., *Waukegan;* LaSalle Electric Utilities, *La Salle;* Outboard Marine Corp.[2], *Waukegan;* Velsicol Chemical Corp. (Marshall Plant), *Marshall;* and Wauconda Sand & Gravel, *Wauconda.* (NPL 11, Total 11).

INDIANA: Envirochem Corp., *Zionsville;* Fisher Calo, *LaPorte;* Lake Sandy Jo (M & M Landfill), *Gary;* Lemon Lane Landfill, *Bloomington;* Main Street Well Field, *Elkhart;* Marion (Bragg) Dump, *Marion;* Midco I, *Gary;* Neal's Landfill, *Bloomington;* Ninth Avenue Dump, *Gary;* Parrot Road Dump[5], *New Haven;* Seymour Recycling Corp.[2], *Seymour;* Wayne Waste Oil, *Columbia City;* Wedzeb Enterprises, Inc., *Lebanon;* American Chemical Service, *Griffith;* Bennett Stone Quarry, *Bloomington;* Northside Sanitary Landfill, *Zionsville;* Poer Farm, *Hancock County;* and Reilly Tar & Chemical Corp., *Indianapolis.* (NPL 12, PUL 5, Total 17).

KANSAS: Arkansas City Dump[2], *Arkansas City;* Cherokee County (once listed as Tar Creek, Cherokee County), *Cherokee County;* Doepke Disposal (Holliday), *Johnson County;* and Johns' Sludge Pond, *Wichita.* (NPL 4, Total 4).

KENTUCKY: A. L. Taylor (Valley of the Drums)[2], *Brooks;* Airco[4], *Calvert;* B. F. Goodrich, *Calvert City;* Distler Brickyard, *West Point;* Distler Farm, *Jefferson County;* Lee's Lane Landfill, *Louisville;* and Newport Dump, *Newport.* (NPL 6, Total 6).

LOUISIANA: Bayou Bonfouca, *Slidell;* Bayou Sorrel[4], *Bayou Sorrel;* Cleve Reber, *Sorrento;* Old Inger Oil Refinery[2], *Darrow;* and Petro-Processors, *Scotlandville.* (NPL 3, PUL 1, Total 4).

MASSACHUSETTS: Baird & McGuire, *Holbrook;* Cannon Engineering Corp. (CEC), *Bridgewater;* Charles George Reclamation Trust Landfill, *Tyngsborough;* Groveland Wells, *Groveland;* Hocomonco Pond, *Westborough;* Industri-Plex 128 (once listed as Mark Phillip Trust), *Woburn;* New Bedford[2], *New Bedford;* Nyanza Chemical Waste Dump, *Ashland;* PSC Resources; *Palmer;* Plymouth Harbor/Cannon Engineering Corp., *Plymouth;* Re-Solve, Inc., *Dartmouth;* Silresim Chemical Corp., *Lowell;* W. R. Grace & Co., Inc. (Acton Plant), *Acton;* Wells G & H, *Woburn;* Iron Horse

Park, *Billerica;* and Sullivan's Ledge, *New Bedford.* (NPL 14, PUL 2, Total 16).

MARYLAND: Limestone Road, *Cumberland;* Middletown Road Dump, *Annapolis;* and Sand, Gravel & Stone, *Elkton.* (NPL 3, Total 3).

MAINE: F. O'Connor Site, *Augusta;* McKin Co., *Gray;* Pinette's Salvage Yard, *Washburn;* Saco Tannery Waste Pits, *Saco;* and Winthrop Landfill, *Winthrop.* (NPL 5, Total 5).

MICHIGAN: Anderson Development Co., *Adrian;* Auto Ion Chemicals, Inc., *Kalamazoo;* Berlin & Farro, *Swartz Creek;* Butterworth No. 2 Landfill, *Grand Rapids;* Cemetery Dump, *Rose Center;* Charlevoix Municipal Well, *Charlevoix;* Chem Central, *Wyoming Township;* Clare Water Supply[4], *Clare;* Cliff/Dow Dump, *Marquette;* Duell & Gardner Landfill, *Dalton Township;* Electrovoice[4], *Buchanan;* Forest Waste Products, *Otisville;* G & H Landfill, *Utica;* Grand Traverse Overall Supply Co., *Greilickville;* Gratiot County Golf Course[3], *St. Louis;* Gratiot County Landfill[2], *St. Louis;* Hedblum Industries, *Oscoda;* Ionia City Landfill, *Ionia;* K & L Avenue Landfill, *Oshtemo Township;* Kentwood Landfill, *Kentwood;* Liquid Disposal, Inc., *Utica;* Littlefield Township Dump[4], *Oden;* Mason County Landfill, *Pere Marquette Township;* McGraw Edison Corp., *Albion;* Northernaire Plating, *Cadillac;* Novaco Industries, *Temperance;* Organic Chemicals, Inc., *Grandville;* Ossineke Ground Water Contamination, *Ossineke;* Ott/Story/Cordova Chemical Co., *Dalton Township;* Packaging Corp. of America, *Filer City;* Petoskey Municipal Well Field, *Petoskey;* Rasmussen's Dump, *Green Oak Township;* Rose Township Dump, *Rose Township;* SCA Independent Landfill, *Muskegon Heights;* Shiawasee River, *Howell;* Southwest Ottawa County Landfill, *Park Township;* Sparta Landfill, *Sparta Township;* Spartan Chemical Co., *Wyoming;* Spiegelberg Landfill, *Green Oak Township;* Springfield Township Dump, *Davisburg;* Tar Lake, *Mancelona Township;* U.S. Aviex, *Howard Township;* Velsicol Chemical Corp. (St. Louis Plant), *St. Louis;* Verona Well Field, *Butler Creek;* Wash King Laundry, *Pleasant Plains Township;* Whitehall Well Field[4], *Whitehall;* Burrows Sanitation, *Hartford;* Metamora Landfill, *Metamora;* and Sturgis Municipal Wells, *Sturgis.* (NPL 41, PUL 3, Total 44).

MINNESOTA: Burlington Northern, *Brainerd/Baxter;* FMC Corp. (Fridley Plant), *Fridley;* Koppers Coke, *St. Paul;* Lehillier/Mankato, *Lehillier/Mankato;* NL Industries/Taracorp/Golden Auto, *St. Louis Park;* New Brighton/Arden Hills, *New Brighton;* Oakdale Dump, *Oakdale;* Reilly Tar & Chemical Corp.[2], *St. Louis Park;* South Andover (once listed as Andover Sites), *Andover;* Waste Disposal Engineering, *Andover;* Arrowhead Refinery Co., *Hermantown;* Boise Cascade/Onan/Medtronics, *Fridley;* General Mills/Henkel Corp., *Minneapolis;* Joslyn Manufacturing & Supply Co., *Brooklyn Center;* Macgillis & Gibbs Co./Bell Lumber & Pole Co., *New*

Brighton; Morris Arsenic Dump, *Morris;* Nutting Truck & Caster Co., *Faribault;* Perham Arsenic, *Perham;* St. Louis River, *St. Louis County;* St. Regis Paper Co., *Cass Lake;* Union Scrap, *Minneapolis;* Washington County Landfill, *Lake Elmo;* and Whittaker Corp., *Minneapolis.* (NPL 10, PUL 13, Total 23).

MISSOURI: Ellisville[2], *Ellisville;* Fulbright Landfill, *Springfield;* Minker/Stout/Romaine Creek (once listed as Arena 2: Fills 1 & 2), *Imperial;* Shenandoah Stables (once listed as Arena 1: Shenandoah Stables), *Moscow Mills;* Syntex Facility, *Verona;* Times Beach, *Times Beach;* and Quail Run Mobile Manor, *Gray Summit.* (NPL 6, PUL 1, Total 7).

MISSISSIPPI: Plastifax, Inc.[6], *Gulfport;* and Flowood[2], *Flowood.* (PUL 1, Total 1).

MONTANA: Anaconda Smelter, *Anaconda;* Libby Gound Water Contamination, *Libby;* Milltown Reservoir Sediments, *Milltown;* Silver Bow Creek, *Silver Bow/Deer Lodge Counties;* and East Helena Smelter, *East Helena.* (NPL 4, PUL 1, Total 5).

NORTH CAROLINA: Chemtronics, Inc., *Swannanoa;* Martin Marietta, Sodyedo, Inc., *Charlotte;* and PCB Spills[2], *210 Miles of Roads.* (NPL 3, Total 3).

NORTH DAKOTA: Arsenic Trioxide[2], *Southeastern.* (NPL 1, Total 1).

NEBRASKA: Phillips Chemical Co.[5], *Beatrice.*

NEW HAMPSHIRE: Auburn Road Landfill, *Londonderry;* Dover Municipal Landfill, *Dover;* Keefe Environmental Services (KES), *Epping;* Ottati & Goss/Kingston Steel Drum, *Kingston;* Somersworth Sanitary Landfill, *Somersworth;* Sylvester[2], *Nashua;* Tinkham Garage, *Londonderry;* Kearsage Metallurgical Corp., *Conway;* Savage Municipal Water Supply, *Milford;* and South Municipal Water Supply Well, *Petersborough.* (NPL 7, PUL 3, Total 10).

NEW JERSEY: A. O. Polymer, *Sparta Township;* American Cyanamid Co., *Bound Brook;* Asbestos Dump, *Millington;* Beachwood/Berkley Wells, *Berkeley Township;* Bog Creek Farm, *Howell Township;* Brick Township Landfill, *Brick Township;* Bridgeport Rental & Oil Services, *Bridgeport;* Burnt Fly Bog, *Marlboro Township;* Caldwell Trucking Co., *Fairfield;* Chemical Control, *Elizabeth;* Chemsol, Inc., *Piscataway;* Combe Fill North Landfill, *Mount Olive Township;* Combe Fill South Landfill, *Chester Township;* CPS/Madison Industries, *Old Bridge Township;* D'Imperio Property, *Hamilton Township;* Denzer & Schafer X-Ray Co., *Bayville;* Dover Municipal Well 4, *Dover Township;* Ellis Property, *Evesham Township;* Evor Phillips Leasing, *Old Bridge Township;* Fair Lawn Well Field, *Fair Lawn;* Friedman Property (once listed as Upper Freehold Site), *Upper Freehold Township;* Gems Landfill, *Gloucester Township;* Goose Farm, *Plumstead*

Township; Helen Kramer Landfill, *Mantua Township;* Hercules, Inc. (Gibbstown Plant), *Gibbstown;* Imperial Oil Co., Inc./Champion Chemicals (once listed as Imperial Oil Co., Inc.), *Morganville;* JIS Landfill, *Jamesburg/South Brunswick Township;* Jackson Township Landfill, *Jackson Township;* KINBUC Landfill, *Edison Township;* King of Prussia, *Winslow Township;* Krysowaty Farm, *Hillsborough;* Lang Property, *Pemberton Township;* Lipardi Landfill, *Pitman;* Lone Pine Landfill, *Freehold Township;* M & T Delisa Landfill, *Asbury Park;* Mannheim Avenue Dump, *Galloway Township;* Maywood Chemical Co., *Maywood/Rochelle Park;* Metaltec/Aerosystems, *Franklin Borough;* Monroe Township Landfill, *Monroe Township;* Montgomery Township Housing Development, *Montgomery Township;* Myers Property, *Franklin Township;* NL Industries, *Pedricktown;* PJP Landfill, *Jersey City;* Pepe Field, *Boonton;* Pijak Farm, *Plumstead Township;* Price Landfill[2], *Pleasantville;* Reich Farms, *Pleasant Plains;* Renora, Inc., *Edison Township;* Ringwood Mines/Landfill, *Ringwood Borough;* Rockaway Borough Well Field, *Rockaway Township;* Rockaway Township Wells, *Rockaway;* Rocky Hill Municipal Well, *Rocky Hill Borough;* Roebling Steel Co., *Florence;* Sayreville Landfill, *Sayreville;* Scientific Chemical Procesing, Inc., *Carlstadt;* Sharkey Landfill, *Parsippany/Troy Hills;* South Brunswick Landfill, *South Brunswick;* Spence Farm, *Plumstead Township;* Swope Oil & Chemical Co., *Pennsauken;* Syncon Resins, *South Kearny;* Toms River Chemical, *Toms River;* U.S. Radium Corp., *Orange;* Universal Oil Products (Chemical Division), *East Rutherford;* Vineland State School, *Vineland;* Williams Property, *Swainton;* Chemical Leaman Tank Liners, Inc., *Bridgeport;* Cooper Road, *Voorhees Township;* De Rewal Chemical Co., *Kingwood Township;* Delilah Road, *Egg Harbor Township;* Diamond Alkali Co., *Newark;* Ewan Property, *Shamong Township;* Florence Land Recontouring, Inc., Landfill, *Florence Township;* Hopkins Farm, *Plumstead Township;* Landfill & Development Co., *Mount Holly;* Nascolite Corp., *Millville;* Radiation Technology, Inc., *Rockaway Township;* Shieldalloy Corp., *Newfield Borough;* Tabernacle Drum Dump, *Tabernacle Township;* Upper Deerfield Township Sanitary Landfill, *Upper Deerfield Township;* Ventron/Velsicol, *Woodridge Borough;* Vineland Chemical Co., Inc., *Vineland;* W. R. Grace & Co., Inc. (Wayne Plant), *Wayne Township;* Wilson Farm, *Plumstead Township;* Woodland Route 532 Dump, *Woodland Township;* and Woodland Route 72 Dump, *Woodland Township.* (NPL 65, PUL 20, Total 85).

NEW MEXICO: AT & ST/Clovis, *Clovis;* Homestake Mining Co., *Milan;* South Valley[2], *Albuquerque;* and United Nuclear Corp., *Church Rock.* (NPL 4, Total 4).

NEVADA: None.

NEW YORK: American Thermostat Co., *South Cairo;* Batavia Landfill, *Batavia;* Brewster Well Field, *Putnam County;* Facet Enterprises, Inc., *Elmira;* Fulton Terminals, *Fulton;* GE Moreau, *South Glens Falls;* Hooker (Hyde Park), *Niagara Falls;* Hooker (102nd Street), *Niagara Falls;* Hooker (S-Area), *Niagara Falls;* Kentucky Avenue Well Field, *Horseheads;* Love Canal, *Niagara Falls;* Ludlow Sand & Gravel, *Clayville;* Marathon Battery Corp., *Cold Springs;* Mercury Refining, Inc., *Colonie;* Niagara County Refuse, *Wheatfield;* Old Bethpage Landfill, *Oster Bay;* Olean Well Field, *Olean;* Pollution Abatement Services (PAS)[2], *Oswego;* Port Washington Landfill, *Port Washington;* Ramapo Landfill, *Ramapo;* Sinclair Refinery, *Wellsville;* Solvent Savers, *Lincklaen,* Syosset Landfill, *Oyster Bay;* Vestal Water Supply Well 1-1, Vestal Water Supply Well 4-2 (once one site), *Vestal;* Wide Beach Development, *Brant;* York Oil Co., *Moria;* General Motors/Central Foundry Division, *Massena;* and Hudson River PCBs, *Hudson River.* (NPL 27, PUL 2, Total 29).

OHIO: Allied Chemical & Ironton Coke, *Ironton;* Arcanum Iron & Metal, *Darke County;* Big D Campground, *Kingsville;* Bowers Landfill, *Circleville;* Buckeye Reclamation, *St. Clairsville;* Chem-Dyne[2], *Hamilton;* Coshocton Landfill, *Franklin Township;* E.H. Schilling Landfill, *Hamilton Township;* Fields Brook, *Ashtabula;* Fultz Landfill, *Jackson Township;* Laskin/Poplar Oil Co. (once listed as Poplar Oil Co.), *Jefferson Township;* Nease Chemical, *Salem;* New Lyme Landfill, *New Lyme;* Old Mill (once listed as Rock Creek/Jack Webb), *Rock Creek;* Pristine, Inc., *Reading;* Skinner Landfill, *West Chester;* Summit National, *Deerfield Township;* Van Dale Junkyard[5], *Marietta;* Zanesville Well Field, *Zanesville;* Miami County Incinerator, *Troy;* Powell Road Landfill, *Dayton;* South Point Plant, *South Point;* United Scrap Lead Co., Inc., *Troy.* (NPL 18, PUL 4, Total 22).

OKLAHOMA: Hardage/Criner, *Criner;* Tar Creek (Ottawa County), *Ottawa County;* Compass Industries, *Tulsa;* Sand Springs Petrochemical Complex, *Sand Springs.* (NPL 2, PUL 2, Total 4).

OREGON: Gould, Inc., *Portland;* Teledyne Wah Chang (Albany), *Albany;* United Chrome Products, Inc., *Corvallis.* (NPL 2, PUL 1, Total 3).

PENNSYLVANIA: Blosenski Landfill, *West Caln Township;* Brodhead Creek, *Stroudsburg;* Bruin Lagoon, *Bruin Borough;* Centre County Kepone, *State College Borough;* Craig Farm Drum, *Parker;* Douglassville Disposal, *Douglassville;* Drake Chemical, *Lock Haven;* Enterprise Avenue, *Philadelphia;* Fischer & Porter Co., *Warminster;* Havertown PCP, *Haverford;* Heleva Landfill, *North Whitehall Township;* Hranica Landfill, *Buffalo Township;* Kimberton, *Kimberton Borough;* Lackawanna Refuse, *Old Forge Borough;* Lehigh Electric & Engineering Co., *Old Forge Borough;* Lindane Dump; *Harrison Township;* Lord-Shope Landfill, *Girard Township;* Malvern TCE, *Malvern;* McAdoo Associates[2], *McAdoo Borough/Kline Township;* Metal Banks, *Philadelphia;* Moyers Landfill, *Eagleville;* Old City of York Landfill,

Seven Valleys; Osborne Landfill, *Grove City;* Palmerton Zinc Pile, *Palmerton;* Presque Isle, *Erie;* Resin Disposal, *Jefferson Borough;* Stanley Kessler, *King of Prussia;* Voortman Farm, *Upper Saucon Township;* Wade (ABM) (once listed as ABM-Wade), *Chester;* Westline, *Westline;* Berks Sand Pit, *Longswamp Towship;* Dorney Road Landfill, *Upper Macungie Township;* East Mount Zion, *Springettsbury Township;* Henderson Road, *Upper Merion Township;* Mill Creek Dump, *Erie;* Industrial Lane, *Williams Township;* Taylor Borough Dump, *Taylor Borough;* Tysons Dump, *Upper Merion Township;* Walsh Landfill, *Honeybrook Township.* (NPL 30, PUL 9, Total 39).

PUERTO RICO: Barceloneta Landfill, *Florida Afuera;* Frontera Creek, *Rio Abajo;* Ge Wiring Devices, *Juana Diaz;* Juncos Landfill, *Juncos;* RCA Del Caribe, *Barceloneta;* Fibers Public Supply Wells, *Jobos;* Upjohn Facility, *Barceloneta;* Vega Alta Public Supply Wells, *Vega Alta.* (NPL 5, PUL 3, Total 8).

RHODE ISLAND: Davis Liquid Waste, *Smithfield;* Landfill & Resource Recovery, Inc. (L & RR), *North Smithfield;* Peterson-Puritan, Inc., *Lincoln/ Cumberland;* Picillo Farm², *Coventry;* Stamina Mills, Inc. (once listed as Forestdale-Stamina Mills, Inc.), *North Smithfield;* Western Sand & Gravel, *Burrillville.* (NPL 6, Total 6).

SOUTH CAROLINA: Carolawn, Inc., *Fort Lawn;* SCRDI Bluff Road², *Columbia;* SCRDI Dixiana, *Cayce;* Geiger (C&M Oil), *Rantoules;* Independent Nail Co., *Beaufort;* Kalama Specialty Chemicals, *Beaufort;* Koppers Co., Inc. (Florence Plant), *Florence;* Leonard Chemical Co., Inc., *Rock Hill;* Palmetto Wood Preserving, *Dixianna;* Wamchem, Inc., *Burton.* (NPL 3, PUL 7, Total 10).

SOUTH DAKOTA: Whitewood Creek², *Whitewood.* (NPL 1, Total 1).

TENNESSEE: Amnicola Dump, *Chattanooga;* Gallaway Pits, *Gallaway;* Lewisburg Dump, *Lewisburg;* Murray-Ohio Dump, *Lawrenceburg;* North Hollywood Dump², *Memphis;* Velsicol Chemical Corp. (Hardeman County), *Toone.* (NPL 6, Total 6).

Trust Territories: PCB Wastes², *Trust Territory of the Pacific Islands.* (NPL 1, Total 1).

TEXAS: Bio-Ecology Systems, Inc., *Grand Prairie;* Crystal Chemical Co., *Houston;* French, Ltd., *Crosby;* Harris (Farley Street), *Houston;* Highlands Acid Pit, *Highlands;* Motco, Inc.², *La Marque;* Sikes Disposal Pits, *Crosby;* Triangle Chemical Co., *Bridge City;* Geneva Industries (Fuhrmann Energy Corp.), *Houston;* Pig Road, *New Waverly;* United Creosoting Co., *Conroe.* (NPL 8, PUL 3, Total 11).

UTAH: Rose Park Slidge Pit², *Salt Lake City;* (NPL 1, Total 1).

VIRGINIA: Chisman Creek, *York County;* Matthews Electroplating[2], *Roanoke County;* Saltville Waste Disposal Ponds, *Saltville;* U.S. Titanium, *Piney River.* (NPL 4, Total 4).

VERMONT: Old Springfield Landfill, *Springfield;* Pine Street Canal[2], *Burlington.* (NPL 2, Total 2).

WASHINGTON: Colbert Landfill, *Colbert;* Commencement Bay, Near Shore/Tideflats, *Pierce County;* Commencement Bay, South Tacoma Channel, *Tacoma;* FMC Corp. (Yakima), *Yakima;* Frontier Hard Chrome, Inc., *Vancouver;* Harbor Island (Lead), *Seattle;* Kaiser Aluminum (Mead Works), *Mead;* Lakewood, *Lakewood;* Pesticide Lab (Yakima), *Yakima;* Western Processing Co., Inc., *Kent;* American Lake Gardens, *Tacoma;* Greenacres Landfill, *Spokane County;* Queen City Farms, *Maple Valley;* Rosch Property, *Roy.* (NPL 10, PUL 4, Total 14).

WISCONSIN: City Disposal Corp. Landfill, *Dunn;*Delavan Municipal Well No. 4, *Delavan;*Eau Claire Municipal Well Field, *Eau Claire City;* Janesville Ash Beds, *Janesville;* Janesville Old Landfill, *Janesville;* Kohler Co., Landfill, *Sheboygan;* Lauer I. Sanitary Landfill, *Menomonee Falls;* Lemberger Transport & Recycling, Inc., *Franklin Township;* Master Disposal Service, Inc., *Brookfield;* Mid-State Disposal, Inc. Landfill, *Cleveland Township;* Moss-American (Kerr-McGee Oil Co.), *Milwaukee;* Muskego Sanitary Landfill, *Muskego;* Northern Engraving Co., *Sparta;* Oconomowoc Electroplating Co., Inc., *Ashippin;* Omega Hills North Landfill, *Germantown;* Onalaska Municipal Landfill, *Onalaska;* Schmalz Dump, *Harrison;* Scrap Processing Co., Inc., *Medford;* Waste Research & Reclamation Co., *Eau Claire;* Wheeler Pit, *La Prairie Township.* (PUL 20, Total 20).

WEST VIRGINIA: Fike Chemical, Inc., *Nitro;* Follansbee, *Follansbee;* Leetown Pesticide, *Leetown;* West Virginia Ordance[2], *Point Pleasant.* (NPL 4, Total 4).

WYOMING: Baxter/Union Pacific Tie Treating, *Laramie.* (NPL 1, Total 1).

Subject Index